TIME SERIES ANALYSIS:
Theory and Practice 7

TIME SERIES ANALYSIS:

Theory and Practice 7

Proceedings of the (General Interest) International Conference held at Toronto, Canada, 18-21 August 1983

Edited by

O. D. ANDERSON

TSA&F, Nottingham, England

1985

NORTH-HOLLAND – AMSTERDAM · NEW YORK · OXFORD

ISBN: 0 444 87684 7

Published by:
ELSEVIER SCIENCE PUBLISHERS B.V.
P.O. Box 1991
1000 BZ Amsterdam
The Netherlands

Sole distributors for the U.S.A. and Canada:
ELSEVIER SCIENCE PUBLISHING COMPANY, INC.
52 Vanderbilt Avenue
New York, N.Y. 10017
U.S.A.

Library of Congress Cataloging in Publication Data

Main entry under title:

Time series analysis.

Proceedings of the 11th International Time Series Meeting held Aug. 18-21, 1983, Toronto.
1. Time-series analysis--Congresses. I. Anderson, O. D. (Oliver Duncan), 1940- . II. International Time Series Meeting (11th : 1983 : Toronto, Ont.)
QA280.T544 1985 519.5'5 84-25944
ISBN 0-444-87684-7 (Elsevier)

PRINTED IN THE NETHERLANDS

To
Robert
Elizabeth
Simon
and
Alexander

CONTENTS

TIME SERIES ANALYSIS: Theory and Practice 7
O.D. Anderson (editor)

INTRODUCTION

The 11th International Time Series Meeting (ITSM), and 5th North American Conference, was held August 18-21, 1983, at the Sheraton Centre in downtown Toronto, being convened immediately after the conclusion of the 1983 Joint Statistical Meetings (American Statistical Association, Biometric Society, Institute of Mathematical Statistics and Statistical Society of Canada), which were held at the same location. Thus, interested participants were able to conveniently and economically attend both events in a single trip, without changing accommodation.

Our Conference featured both invited and contributed papers, the objects being to discuss recent developments in the theory and practice of Time Series Analysis and Forecasting (TSA&F); and to bring practitioners together from diverse parent disciplines, work environments and geographical locations.

Suitable time series topics for presentation included: Statistical Methodology; Applications to Economics and in Econometrics; Rational Expectations; Government, Business and Industrial Examples; Finance and Accountancy; the Hydrosciences, such as Limnology, Hydrology, Water Quality Regulation and Control, and the Modelling of Marine Environments; Persistence and Fractional Differencing; the Geosciences, especially such areas as Oil Exploration and Seismology; Civil Engineering and allied disciplines; Point Processes; Spatial and Space-Time Processes - their theory and application - especially in Geography and related areas, such as City Planning or Energy Demand Forecasting; Biology and Ecology; Environmental Studies - Air and River Pollution; Epidemiology; Medical Applications and Biomedical Engineering; Psychology; Irregularly Spaced Data (including Outliers and Missing Observations); Robust and Nonparametric Methods; Seasonal Modelling and Adjustment, Calendar Effects; Causality; Bayesian Approaches; Distributed Lags; Box-Jenkins Univariate ARIMA, Transfer-Function, Intervention and Multivariate Modelling; State Space; Nonlinear Modelling; Identification Problems; Estimation; Diagnostic Checking; Signal Extraction; Comparative Studies; Spectral Analysis, especially for the Physical Sciences; Business Cycle and Expectations Data; Data Revisions; Computer Software and Numerical Analysis; Forecasting (including new topics such as Traffic Forecasting and Safety, and Forecasting in Agriculture); and any other areas of the subject.

This introduction lists the participants, provides the technical programme, and prints abstracts of those papers not included in these Proceedings. It also acknowledges all individuals, who helped make the Meeting a success, and gives biographical sketches for those authors that filed appropriate details.

Organising Committee (13 people)

Julia W. Ali (USA), Local Arrangements	Guest, Lexington, Kentucky
Oliver D. Anderson (UK), Convenor & General Chairman	TSA&F, Nottingham
Robert A.M. Gregson (Australia), Session Organiser	University of New England, Armidale, New South Wales
Cheng Hsiao (Canada), Visa Applications	University of Toronto, Ontario
C.W. Kenneth Keng (Canada), Session Organiser	Ontario Hydro, Toronto.
Hans Levenbach (USA), Session Organiser	American Telephone & Telegraph Co, Basking Ridge, New Jersey
Essam Mahmoud (Canada), Registration	Concordia University, Montreal, Quebec

Dan Sprevak (UK), Session Organiser	Queen's University, Belfast
Houston H. Stokes (USA), Session Organiser	University of Illinois, Chicago
Kjell Stordahl (Norway), Session Organiser	Norwegian Telecom, Oslo
W. Robert Terry (USA), Session Organiser	University of Toledo, Ohio
Cheng-jun Tian (China), Session Organiser	Shanxi University
Zhi-fang Zhang (China), Session Co-Organiser	Beijing Institute of Technology

Other Participants (a further 67 people from 12 countries)

Mukhtar M. Ali (USA)	University of Kentucky, Lexington
Christine Baufays (Belgium)	Guest, Namur
Pierre Baufays (Belgium)	Namur University
Richard H. Bee (USA)	Youngstown State University, Ohio
Tridib K. Biswas (Canada)	Ontario Hydro, Toronto
John Brode (USA)	University of Lowell, Massachusetts
Phillip A. Cartwright (USA)	University of Georgia, Athens
M.W. Luke Chan (Canada)	McMaster University, Hamilton, Ontario
S.K. Chang (Canada)	McMaster University, Hamilton, Ontario
Jean-Pierre Chanut (Canada)	University of Quebec, Rimouski
Thomas T. Cheng (USA)	Wharton School, Philadelphia
Keewhan Choi (USA)	Centers for Disease Control, Atlanta, Georgia
Pierre A. Cholette (Canada)	Statistics Canada, Ottawa, Ontario
Stephen B. Cohen (USA)	Mitre Corporation, McLean, Virginia
Melody Cole (USA)	American Telephone and Telegraph Co, Basking Ridge, New Jersey
Ross B. Corotis (USA)	John Hopkins University, Baltimore, Maryland
Eric Cozanet (France)	French Telecom, Paris
Eivind Damsleth (Norway)	Norwegian Computer Center, Oslo
Frank D.J. Dunstan (UK)	University College, Cardiff
Quang P. Duong (Canada)	Bell Canada, Montreal, Quebec
Anthony H. Elavia (USA)	Wake Forest University, Winston-Salem, North Carolina
David F. Findley (USA)	Bureau of Census, Washington DC
Henry J. Fowler (USA)	Bell Labs, Piscataway, New Jersey
John B. Guerard (USA)	Lehigh University, Bethlehem, Pennsylvania
Julie K. Guerard (USA)	Guest, Bethlehem, Pennsylvania
Jo Marie Gulledge (USA)	Guest, Baton Rouge, Louisiana
Kristin Gulledge (USA)	Guest, Baton Rouge, Louisiana
Thomas R. Gulledge (USA)	Lousiana State University, Baton Rouge
Jiwan Gupta (USA)	University of Toledo, Ohio
Zhi-gang Han (China)	Heilongjiang University, Harbin
John L. Harris (USA)	University of Illinois, Chicago
Jon Helgeland (Norway)	Norwegian Computing Center, Oslo
W. Gordon Hines (Canada)	University of Guelph, Ontario
Agnar Hoskuldsson (Denmark)	Danish Technical University, Lyngby
Gregory M. Hudak (USA)	Scientific Computing Associates, DeKalb, Illinois
Christine Jacob (France)	Ministry of Agriculture, Jouy-en-Josas
Kenneth J. Jones (USA)	Brandeis University, Waltham, Massachusetts
Myron J. Katzoff (USA)	Bureau of Census, Washington DC
Sergio G. Koreisha (USA)	University of Oregon, Eugene
F.H. (Koos) Koster (Netherlands)	Amro Bank, Amsterdam
Greg Langstaff (Canada)	University of Guelph, Ontario
Keh-Shin Lii (USA)	University of California, Riverside

Erkki Liski (Finland)	University of Tampere
Lon-Mu S. Liu (USA)	University of Illinois, Chicago
Yih-wu Liu (USA)	Youngstown State University, Ohio
Donald J. Malec (USA)	Bureau of Census, Washington DC
Ed McKenzie (UK)	University of Strathclyde, Glasgow
Guy Mélard (Belgium)	Free University, Brussels
Mike J. Miller (USA)	Department of Labor, Washington DC
Jack Y. Narayan (USA)	State University of New York, Oswego
Trygve S. Nilsen (Norway)	University of Bergen
Tarmo Pukkila (Finland)	University of Tampere
David P. Reilly (USA)	Automatic Forecasting Systems, Hatboro, Pennsylvania
Paul B. Robinson (USA)	American Telephone and Telegraph Co, Bedminster, New Jersey
Peter Rousseeuw (Belgium)	Edegem
Jack T. Schneider (South Africa)	University of Witswatersrand, Johannesburg
Richard W. Shorrock (Canada)	Bell Canada, Montreal, Quebec
Gary L. Snoop (USA)	TWA Administrative Center, Kansas City, Missouri
Timo Teräsvirta (Finland)	Research Institute of Finnish Economy, Helsinki
John E. Triantis (USA)	AT&T Long Lines, Morris Planes, New Jersey
E.G.F. (Bob) Van Winkel (Netherlands)	Technical University of Eindhoven
Leonie J.A. Van Winkel (Netherlands)	Guest, Eindhoven
A. Morris Walker (UK)	Sheffield University (Retired)
James E. Willis (USA)	Louisiana State University, Baton Rouge
Christine C. Wise (USA)	Department of Labor, Washington DC
David Wright (Canada)	University of Ottawa, Ontario
Zhi-ming Wu (China)	Shanghai Jiao Tong University.

Technical Programme (51 contributions in a single stream)

Friday, 19 August

Session 1 08.30-10.15

M.M. Ali (USA) Distributions of the Sample Autocorrelations when Observations are from a Stationary Autoregressive Moving Average Process
K.-S. Lii (USA) Simultaneous Equations Systems Identification and Estimation (with *H.S. Yang*)
S.G. Koreisha (USA) Identification of Nonzero Elements in the Polynomial Matrices of VARMA Processes (with *T.M. Pukkila*, Finland)
P.A. Cartwright (USA) Using State Dependent Models for Prediction of Time Series with Missing Observations

Session 2 10.45-12.30 "Vector ARMA Modeling" *Arranged by Professor H.H. Stokes*

K.J. Jones (USA) An Empirically derived Vector ARMA Model for Monetary and Financial Variables
L.-M. Liu (USA) Unified Econometric Model Building using Simultaneous Transfer Function Equations (with *G.B. Hudak*)
J.L. Harris (USA) Nominal and Real Sector Relationships: 1954-1980
H.H. Stokes (USA) Dynamic Adjustment of Disaggregated Unemployment Series

Session 3 14.30-16.15

T.R. Gulledge (USA) Time Series Analysis of Wheat Market Fundamentals (with *J.E. Willis*)
J.B. Guerard (USA) Mergers, Stock Prices and Industrial Production: An Empirical

Test of the Nelson Hypothesis
J.Y. Narayan (USA) Causality Testing based on Ex Ante Forecasts (with *C. Aksu*)
G. Mélard (Belgium) Illustration of the Use of a General Time Series Model

Session 4 16.45-18.30

E. Mahmoud (Canada) Empirical Results on the Accuracy of Short-Term Forecasting Techniques
D. Wright (Canada) Extensions of Smoothing Methods for Forecasting Irregularly Spaced Data
P.A. Cholette (Canada) Seasonal Moving Averages for Irregular Series with Moving Seasonality
M. Miller (USA) An Analysis of Many Time Series: How Well Do We Do? (with *J. van Erden, R. Willus & C. Wise*)

Saturday, 20 August

Session 5 08.30-10.15

Z.-M. Wu (China) Nonlinear Multi-Step Prediction of Threshold Autoregressive Models
T. Teräsvirta (Finland) Choosing between Linear and Threshold Models (with *R. Luukkonen*)
T. Pukkila (Finland) On the Frequency Domain Estimation of the Innovation Variance for a Stationary Univariate Time Series (with *H. Nyquist*, Sweden)
P.B. Robinson (USA) Traffic Engineering with Time Varying Demands by Stochastic Approximation

Session 6 10.45-12.30 "Time Series in Telecommuncations" *Arranged by K. Stordahl*

J.E. Triantis (USA) Improving Forecast Accuracy through Composite Forecasts
E. Damsleth (Norway) Estimation of the Mean Duration of Telephone Calls
M. Cole (USA) The Hybridization of Econometric and Empirical Time Series Models (with *H. Levenbach*)
K. Stordahl (Norway) Routing Algorithms based on Traffic Forecast Modelling

Session 7 14.30-16.15

R.A.M. Gregson (Australia) Invariance ahd Heterogeneity in Time Series Analysis of some Psychophysical Data
T.H. Elavia (USA) Consumption and Labor Supply Under Dynamic Optimization
F.H. Koster (Netherlands) Box-Jenkins Models and Long-Term Loans
J.E. Willis (USA) Modeling Conservation Effects on Electricity Consumption: A Case Study at Louisiana State University (with *T.R. Gulledge*)
D.P. Reilly (USA) Transfer Function Simulation, Identification and Estimation

Session 8 16.45-18.30

E. Cozanet (France) Monthly Revisions of an Annual Forecast
F.D.J. Dunstan (UK) Some Results on Fitting Time Series with Missing Observations
E.G.F. van Winkel (Netherlands) Time Series and the Standard Linear Model
J.T. Schneider (South Africa) A Time Series Technique for Locating Optimal Lags (for *P.C. Pirow*)
A. Høskuldsson (Denmark) On the Strategy of Model Building in Time Series Analysis

Sunday, 21 August

Session 9 08.30-10.15

J. Brode (USA) A Closer Look at a Proof of Time Reversibility

C. Jacob (France) Random Cycles
P. Baufays (Belgium) Multivariate Autoregressive Process with a Non-Constant Dispersion Matrix (with *J.-P. Rasson*)
P. Rousseeuw (Belgium) Robust Estimation of a Linear Trend
D.F. Findley (USA) On Polynomial Transformations to Stationarity

Session 10 10.45-12.30

E. McKenzie (UK) A Traditional Interpretation of the Forecasts of Seasonally Differenced ARIMA Processes
D. Sprevak (UK) Stochastic Modelling of Wind Speed and Direction (with *B. McWilliams*)
R.B. Corotis (USA) Prediction of Generated Wind Turbine Power with High Frequency Wind Speed Series (with *J.-J. Lou*)
A.M. Walker (UK) Some Comments on the Accuracy of Asymptotic Theory for Short Time Series

Session 11 14.30-16.15

K. Choi (USA) Definition of an Influenza Epidemic and Quantification of its Impact on US Mortality (with *S.B. Thacker*)
Z.-G. Han (China) Time Series Multi-Level AR Model and its Application
J.D. Gupta (USA) Time Series Development of Differential Equations for Characterizing Roadway Roughness (with *W.R. Terry*)
W.R. Terry (USA) A State Space Approach for Analyzing Queueing Network Problems in Unmanned Factories (with *C. Kumar*)

Session 12 16.45-18.30 *Arranged by Dr C.W.K. Keng*

C.W.K. Keng (Canada) Econometric Forecasting Modelling - A Utility Maximization Approach
M.W.L. Chan (Canada) Autoregressive Modelling of Accounting Earnings and Security Prices (with *J.L. Callen & C.C.Y. Kwan*)
T.K. Biswas (Canada) The Determination of Time of Use Rating Periods
S.-K. Chang (Canada) Stock Index, Its Future and Future Options - A Causality Analysis (with *C.W.K. Keng & C. Wu*).

Edited Abstracts (for presented papers not included in these Proceedings)

1. M.M. ALI (USA)
*Distributions of the Sample Autocorrelations when Observations are from a Stationary Autoregressive Moving Average Process**

In this paper, we examine the adequacy of the approximations to the distributions of the sample autocorrelations by their asymptotic distributions (which are normal), and also by the Edgeworth-type expansions and the four-parameter Pearson distributions. It is found that, for small samples, the asymptotic distributions provide poor approximations, while the Edgeworth expansions may be adequate in certain cases but fail for others. In comparison, the Pearson distributions seem to be better in almost all the cases considered. A major obstacle to using the Pearson distributions is obtaining the first four moments of the sample autocorrelations. Convenient expressions have been developed from which the exact moments can be computed rather easily.

2. T.K. BISWAS (Canada)
Selection of Time-of-Use Rating Periods: A Statistical Approach

The paper presents the Ontario Hydro approach to selecting rating periods for

---

* Published in Journal of Business & Economic Statistics 2 (1984), 271-278.

time-of-use rates for electricity; the objective of which is to track costs when these vary over time (diurnally and seasonally) in a recognizable and recurring manner. The approach, based on analysis of variance techniques, is designed to compare the degree of cost-tracking power for different period selections. Test statistics are developed to determine if the difference in cost-tracking power of two or more period designs is systematic or random. The approach is applied to hourly cost data that were derived by simulating the operating system of Ontario Hydro, which has used this approach to select a particular period design for introducing time-of-use rates for electricity in Ontario. The above method can be useful to other electrical utilities considering introduction of time-of-use rates.

3. K. CHOI & S.B. THACKER (USA)
Definition of an Influenza Epidemic and Quantification of its Impact on US Mortality

Epidemic influenza activity is typically defined by the qualitative assessment of morbidity, mortality, and laboratory data. We propose a quantitative definition based on the increase in mortality; and quantify the impact on mortality in the US of each epidemic since 1968, using regional and age-specific data.

4. E. COZANET (France)
Monthly Revisions of an Annual Forecast

We start with an aggregate forecast for next year, from which we obtain predictions for each month. Then, as the actual monthly values become known, divergence from these predictions allows us to revise the forecast for the whole year. The problem is to choose the predictor which, at any stage, gives the best forecast for the complete year (part of which has now been realised). The predictor should take account of the series' overall characteristics, such as trend and seasonality, as well as the fresh monthly observations. The paper concludes with an application to predicting telephone demand.

5. F.D.J. DUNSTAN (UK)
Some Results on Fitting Time Series with Missing Observations

Most of the work on analysing time series with missing observations has concentrated on the problem of estimating the parameters of a model of given order. Here the problem of estimating the order is considered. If isolated blocks of data are missing, this may not present a problem, but difficulties can arise if the data are missing randomly. The results of fitting a variety of models using different fitting methods are presented and compared.

6. T.H. ELAVIA (USA)
Consumption and Labor Supply Under Dynamic Optimization

Our object is to test the hypothesis that economic agents jointly choose current consumption and labor supply so as to maximize the present discounted value of current and future utility. This goal is met subject to the condition that policy evaluation, using the estimated decision rules, should not be subject to the Lucas (1976) critique. The results, using data for the US, reject this hypothesis (which constitutes the basis of some influential results concerning labor market behavior in Lucas and Rapping, 1969).

7. D.F. FINDLEY (USA)
On Backshift-Operator Polynomial Transformations to Stationarity for Nonstationary Time Series and their Aggregates

It is demonstrated that any two backshift operator polynomials which transform a given non-stationary time series into weakly stationary series with continuous spectral distributions must have a common divisor which has this property. It follows that the lowest-degree polynomial with this property is unique to within a constant multiple. Using this result some derivations are given, under varying assumptions, of a transformation formula used in nonstationary signal extraction. Counter-examples are presented to show that continuity assumptions, on the spectral distribution functions involved, are necessary to obtain both the uniqueness of the polynomial transformation to stationarity and this signal extraction formula.

8. J.D. GUPTA & W.R. TERRY (USA)
A Time Series Analysis of Roadway Surfaces for Computer Aided Design of Vehicle Suspension Systems

The paper describes how time series analysis might be used to develop a mathematical model for describing the surface condition of a roadway. It also describes the architecture of a computer aided design system for designing vehicle suspension systems which would use the mathematical model of roadway surface conditions as an input.

9. Z.-G. HAN (China)
Time Series Multi-Level AR Model and Its Application

We consider a series following an AR model but with time-varying coefficients that themselves follow AR models, which we call second level processes. (Evidently the idea extends to higher levels.) We obtain recursive estimation and prediction formulae which can be satisfactorily applied.

10. J.L. HARRIS (USA)
Nominal and Real Sector Relationships: 1954-1980

A vector autoregressive model is estimated for three separate US postwar periods using time series data on money, prices, interest rates, and industrial production. Changes in money and prices were found to affect the level of interest rates and output. Relationships between the nominal and real sectors were found to be significantly different in each of the postwar periods.

11. S.G. KOREISHA (USA) & T. PUKKILA (Finland)
Identification of Nonzero Elements in the Polynomial Matrices of VARMA Processes

The parameter density in the polynomial matrices of vector autoregressive moving average (VARMA) time series models is generally very low. Thus, estimation of fully parameterized high order models or of models with many variables is an ineffective, if not impractical, way of utilizing computer power. We present methods for identifying the nonzero elements in VARMA(1,1) models, and show how these procedures can be used to systematically construct models of higher order.

12. F.H. KOSTER (Netherlands)
Box-Jenkins Models and Long-Term Loans

This paper discusses medium and long-term loans granted by the commercial banks to the private sector in Holland. The security for these loans may consist of a mortgage, a guarantee, or both. The effects of fixed investment, interest rates, and wages on this kind of capital market borrowing have been examined. Box-Jenkins models were developed with an intervention introduced to model the loan-variable more realistically; and this modelling is discussed. End-of-quarter data were used for the period 1970 to 1982 (first half-year).

13. H. LEVENBACH & M. COLE (USA)
The Hybridization of Econometric and Empirical Time Series Models

In the past there have been many comparisons of the relative forecasting performance of econometric and empirical time series models. Recent advances have made it possible to enjoy the advantages of both approaches. By fitting an econometric model with a very general autocorrelation correction algorithm, and then fitting the residuals with an empirical model, it is possible to obtain efficient estimates of the econometric model's coefficients (e.g. price elasticities), and also obtain white residuals. Furthermore, the econometric and empirical models are effectively "uncoupled", so that the final econometric model is determined before the first empirical model is attempted. The software required is simple, being based on diagonally weighted least squares, and a univariate ARIMA package. This paper will outline the hybrid modeling strategy, and present an example.

14. K.-S. LII & H.S. YANG (USA)
Simultaneous Equations Systems Identification and Estimation

Methods using the Padé tables and the C-tables to identify the lags of variables in a simultaneous equations system and to estimate the coefficients in the identified model are studied. When the model is not uniquely identified, the C-tables will reduce possible competing models to just a few for further examination. Padé approximants give consistent initial values for possibly more efficient iterative procedures for estimating the parameters. Some asymptotic results on the estimation of the C-table and the Padé table are given. A few examples are presented to demonstrate the effectiveness of the method.

15. E. MAHMOUD (Canada)
Empirical Results on the Accuracy of Short-Term Forecasting Techniques

This study examined fourteen sales data series and reports the performance of short-term forecasting techniques, according to various accuracy measures. The results suggest that simple techniques perform better overall than sophisticated ones. Further, it is demonstrated that the company concerned could improve the total accuracy of its forecasts, either by combining techniques or by using a wide range of them.

16. P.C. PIROW (South Africa)
A Time Series Technique for Locating Optimal Lags

A model is developed which initially studies all pairs of variables. A technique, termed aetalocation, is developed for isolating the significant lags. Significance is determined by generating either 20 or 100 random series and testing the correlation coefficient against the maximum generated coefficient. A standard regression model is then developed which uses only the aetalocated lags. The model has given better predictions than alternatives for milk yields of dairy cows, occurrence of phthisis amongst gold miners and for inflation rates.

17. T. PUKKILA (Finland) & H. NYQUIST (Sweden)
On the Frequency Domain Estimation of the Innovation Variance for a Stationary Univariate Time Series

The innovation variance, σ^2, of a linear time series model can be estimated using periodogram ordinates. However, it is known that these can give severely biased estimates for the corresponding spectrum ordinates in small samples, as is again demonstrated. Therefore, it is believed that frequency domain estimators (fde) of σ^2 will be biased. One way of reducing the small sample bias in the periodogram ordinates is by tapering the observed time series. In this paper, some properties of an fde of σ^2 are considered both analytically and by simulation. The results indicate that, for fde's, tapering reduces bias as well as variability.

18. D.P. REILLY (USA)
Transfer Function Simulation, Identification and Estimation

Transfer Function models (Dynamic Regression) are very powerful in identifying the regression model for time series data. Univariate models have found broad acceptance but Transfer Function models are still relatively unknown. This paper approaches the problem of explaining Transfer Function modelling procedures via simulation. A clear picture of the relationships between Univariate and Transfer Function models evolves through the iterative process of Identification and Estimation. The commercial availability of automatic Univariate and Transfer Function software (AUTOBOX/AUTOBJ) has made this modelling process easier to apply.

19. P.B. ROBINSON (USA)
Traffic Engineering with Time Varying Demands by Stochastic Approximation

Telecommunicatons demand is stochastic with a time varying mean rate. This Time-of-Day (TOD) characteristic of the offered load is as important as its random pattern, for estimating the number of trunks required to provide a given grade of service. General equations for the blocking probability are developed which include the probability of retrial of blocked calls. These equations are based on a binomial approximation to the Erlang B blocking formula. All results are based on an average blocking criterion during the business day. This is preferred to a peak load criterion because it relies on mean and variance of the TOD demand rate distribution, rather than just an extreme value of that distribution. The modeling approach is directed towards the development of a certainty equivalent demand rate which is deduced from the average blocking criterion. That is, the time varying mean demand rate will be replaced by a constant demand equaling the sum of the mean demand rate and a "TOD correction factor". The certainty equivalent demand rate will have the property that Erlang B traffic at this constant rate will encounter the same number of blocked messages, on average during the business day, as will the actual traffic pattern when both are served over the same network.

20. P. ROUSSEEUW (Belgium)
Robust Estimation of a Linear Trend

A statistical procedure is called robust if it is insensitive to the occurrence of gross errors in the data. The usual least squares technique for estimation of a linear trend does not satisfy this property, because even a single outlier can totally offset the result. Therefore, the least median of squares (LMS) technique is introduced, which can resist the adverse effect of a large percentage of outliers. The latter method is illustrated on time series concerning inflation and the price of gold.

21. C.-J. TIAN (China)
*Limiting Behavior of Sample Autocovariance of a Seasonal Time Series**

We study a seasonal time series $\{x_t\}$. Under some ergodic assumptions, it is proved that the limit of the sample autocovariance function of $\{x_t\}$ exists and reveals the same periodicity, in some sense. Based on the limiting behavior, the problem of estimating the period is also discussed.

22. J. VAN ERDEN, R. WILUS, M. MILLER & C. WISE (USA)
An Analysis of Many Time Series: How Well Do We Do?

To allocate $1.5 billion in administrative funds, the Division of Actuarial Services of the US Department of Labor annually projects four key unemployment insurance program parameters (initial claims, weeks claimed, non-monetary

---

* Through a misunderstanding, this paper was not scheduled for presentation.

determinations and appeals) for each of 53 States. In this paper, we review the different ARIMA time series models and functional forms used to forecast the workload variables. We present a comparison of the inter-State differences in modeling as well as an evaluation of the success of the forecasts.

23. E.G.F. VAN WINKEL (Netherlands)
Time Series and the Standard Linear Model

In econometrics, regression equations are frequently applied to time series data, when the basic assumptions of the Standard Linear Model are violated, with the series typically showing (positive) autocorrelation. Some results of wrongly applying Ordinary Least Squares, in such cases, follow from simulations described by Granger & Newbold (1974). In this paper, we show that many of these results can be approximated analytically. Moreover, the role of the series length, n, is discussed; and we show that the inflatedness of R^2 and the underestimation of the standard errors for the regression coefficients are both simple functions of n. Simulation results agree with the analytical approximations.

24. A.M. WALKER (UK)
Some Comments on the Accuracy of Asymptotic Theory for Short Time Series

Fitting a stationary ARMA (p,q) model to a time series consisting of n equally spaced observations, by the method of generalised least squares or some approximately equivalent method, is now a standard technique, widely used in many different fields. The asymptotic distributions of the estimators of the model parameters are often used as approximations in practice. However, it is quite common for the accuracy of these approximations to be unsatisfactory even when n is moderately large. We consider some theoretical results concerning inaccuracy of these asymptotic distributions, and make some comments regarding the problem of obtaining better approximations to distributions of parameter estimators, mainly for AR models.

25. J.E. WILLIS & T.R. GULLEDGE (USA)
Modeling Conservation Effects on Electricity Consumption: A Case Study at Louisiana State University

Increasing costs for utilities have become an important factor in the budgets of individuals, businesses, and public institutions. These entities have had little control over the price of utilities they use, but some control over their expenditures has been accomplished through conservation efforts. This study concentrates on an attempt by a public institution, Louisiana State University, to reduce the rate of increase in its electrical consumption through energy conservation measures. Fluctuations in consumption due to changes in physical plant, student loads, and meteorological condition, during the months of the study period, are considered as well as the effects of identifiable system changes for conservation purposes. The resulting time series model includes transfer function and intervention components.

26. Z.-M. WU (China)
Nonlinear Multi-Step Prediction of Threshold Autoregressive Models

An approach to multi-step-ahead prediction for nonlinear threshold autoregressive (TAR) time series is introduced. Because the model is piecewise linear, the most important thing in prediction is to decide which branch of the model is appropriate for fresh data. Then the prediction values are determined by linear conditional expectation. This may be extended to long-term prediction for cyclical data. Some practical problems (such as data transformation, model diagnostic checking and actual forecasting) are also discussed.

These Proceedings

Regrettably, only about 30% of the papers offered for this Conference have eventually made the present book. Around a third were rejected before presentation; and the remainder failed to revise satisfactorily, or were otherwise withdrawn. We share the disappointment of the unpublished authors, but our aim to hold standards high must be maintained.

The Referees

We are most grateful to the following 92 specialists (from 18 countries) who helped with the refereeing, and in the reviewing process, for this volume:

A.T. Akarca (USA)
J. Aldrich (UK)
M.M. Ali (USA)
O.D. Anderson (UK)
F. Battaglia (Italy)
W.R. Bell (USA)
M. Beneke (West Germany)
T.K. Biswas (Canada)
S. Bittanti (Italy)
T.A. Boley (UK)
J.P. Burman (UK)
M.J. Campbell (UK)
M.W.L. Chan (Canada)
W.P. Cleveland (USA)
R. Coleman (UK)
R. Corotis (USA)
D.R. Cracknell (UK)
E. Damsleth (Norway)
A.G. Davenport (Canada)
J.G. De Gooijer (Netherlands)
J. Durbin (UK)
M. Farrow (UK)
J. Geweke (USA)
E.J. Godolphin (UK)
A.C. Harvey (UK)
D.F. Hendry (UK)
R.M.J. Heuts (Netherlands)
S.C. Hillmer (USA)
C. Jacob (France)
G.J. Janacek (UK)
K.J. Jones (USA)
R.H. Jones (USA)
K. Juselius (Finland)
J. Kassab (UK)
C.W.K. Keng (Canada)
P.B. Kenny (UK)
T. Kollintzas (USA)
A.A. Korabinski (UK)
H. Levenbach (USA)
L.-M. Liu (USA)
R.M. Loynes (UK)
E. Mahmoud (Canada)
R.J. Martin (UK)
G. Mélard (Belgium)
G. Million (UK)
G.E. Mizon (UK)
J.Y. Narayan (USA)
M.M. Newmann (UK)
H. Nyquist (Sweden)
L.E. Öller (Finland)
O. Österbö (Norway)
E. Page (UK)
F.C. Palm (Netherlands)
S.M. Pandit (USA)
E. Parzen (USA)
J. Pemberton (UK)
D.A. Pierce (USA)
M.B. Priestley (UK)
A.E. Raftery (Ireland)
W.D. Ray (UK)
R.J. Reed (UK)
P.B. Robinson (USA)
D.F. Scrimshaw (UK)
R. Shibata (Japan)
B.W. Silverman (UK)
J.M. Sneek (Netherlands)
T. Söderström (Sweden)
D. Sprevak (UK)
H.H. Stokes (USA)
P. Stoica (Romania)
K.D.C. Stoodley (UK)
K. Stordahl (Norway)
T. Subba Rao (UK)
G. Tauchen (USA)
S.J. Taylor (UK)
T. Teräsvirta (Finland)
W.R. Terry (USA)
R.J. Thorne (UK)
G. Thury (Austria)
D. Tjøstheim (Norway)
H. Tong (Hong Kong)
A. Turunen (Finland)
H.D. Vinod (USA)
K.F. Wallis (UK)
J.R. Walters (UK)
W. Wasserfallen (Switzerland)
D.W. Wichern (USA)
D.A.L. Wilson
D. Wright (Canada)
Z.M. Wu (China)
P.C. Young (UK) &
A. Zinober (UK).

The Authors

CELAL AKSU has a BA degree in Operations Management and an MA degree in Finance from Bogazici University (ex Robert College), Turkey; and an MBA from Syracuse University, where he is currently a PhD candidate majoring in Accounting.

OLIVER D. ANDERSON is a graduate of the UK Universities of Cambridge, London, Birmingham and Nottingham; with first class honours degrees in Mathematics and Economics, and master's degrees in Mathematics and Statistics.

He has worked in Industry, for Government, and as a Consultant Statistician; and taught in Schools, Colleges and Universities. He has lectured in over 20 countries and published some 200 items. He is an active member of a dozen professional societies (in England and abroad) concerned with Education, Mathematics, Statistics, Operations Research, Management Science, Economics and Econometrics; and, in 1979, was honoured by election to the International Statistical Institute. More recently he has become involved with Management, Marketing and Information Science.

JOHN BRODE, currently Assistant Professor of Mathematics at the University of Lowell (USA), earned his PhD in Operations Research at Harvard University. He has also taught at the ISCAE in Casablanca (Morocco) and at Wellesley College, Massachusetts. Previously, he was a Senior Research Associate at MIT, where he developed a higher-level computer language for statistical applications. His present work is on stochastic processes with no defined variance.

PHILLIP A. CARTWRIGHT holds BA and MA degrees from the Texas Christian University, and an MS and PhD from the University of Illinois at Urbana-Champaign, all in Economics. He is currently Director of the Georgia Economic Forecasting Project, University of Georgia, Athens, USA.

ROSS B. COROTIS received his SB, SM and PhD degrees from MIT, finishing in 1971. He was a professor at Northwestern University for 11 years and is currently the Hackerman Professor and Chairman of the Department of Civil Engineering at The John Hopkins University. His research interests include structural reliability and probabilistic and stochastic modeling. He has more than 60 publications and serves on national committees of ASCE, ACI and ANSI.

EIVIND DAMSLETH is a graduate in Mathematical Statistics from the University of Oslo, 1973. He has been engaged by the Norwegian Computing Center since then as a Scientist, with special responsibility for TSA&F activities.

TONY H. ELAVIA received MA degrees in Economics from the Universities of Baroda (India) and Houston (USA) in, respectively, 1978 and 1980, and a PhD from Houston in 1983. He is currently with the Economics Department at Wake Forest University, Winston-Salem, North Carolina, USA, and his research interests include macroeconomic time series forecasting and modeling.

ROBERT A.M. GREGSON is Professor of Psychology and Head of Department, University of New England, Australia. He has degrees from Nottingham and London Universities, in the UK, and was first appointed to a chair at the University of Canterbury (New Zealand) in 1967. Dr Gregson has recently published a book "Time Series in Psychology".

GREGORY HUDAK is a statistician and software developer with Scientific Computing Associates, DeKalb, Illinois. He received a BS in mathematics from Denison University and an MS in probability and statistics from Michigan State University. He is currently a doctoral candidate at the University of Wisconsin-Madison. As a program developer for the Wisconsin Multiple Time Series package, he has also been involved in research projects to analyze the air quality of New Jersey, Portland and the Los Angeles region.

CHRISTINE JACOB has a doctorate in Statistics and works on rhythms at the French Government's Biometrics Laboratory.

LON-MU LIU is an Assistant Professor in the Department of Quantitative Methods, University of Illinois at Chicago Circle. He received his Bachelor degree in Agronomy from the National Taiwan University in 1971; and then MS in Statistics (1975), MS in Computer Sciences (1977), and PhD in Statistics (1978), all from the University of Wisconsin at Madison. Before joining the University of Illinois, he served as a Senior Statistician in the Department of Biomathematics, University of California, Los Angeles.

JIANN-JONG LOU received his BS, MS and PhD, all in Civil Engineering from, respectively, the National Taiwan University (1974), Colorado State University (1979) and Northwestern University (1982). Currently, Dr Lou is a structural engineer in New York City.

RITVA LUUKKONEN is currently Research Fellow at the Institute of Occupational Health in Helsinki. She received her Master's (Statistics) degree from the University of Helsinki, and served subsequently as an assistant teacher in the Department of Statistics there. She has also participated in the development of the SURVO 76 statistical data processing system.

GUY MELARD obtained a PhD in Mathematics from the Université Libre de Bruxelles, Belgium, where he teaches statistics, methods of forecasting, quantitative methods in management, and computer science for economics. He has published about 30 papers, many of them in time series analysis.

JACK Y. NARAYAN was born in Guayana, South America. He received his BSc (Honours in Math) from Mount Allison University (New Brunswick, Canada) and his PhD in Mathematics from Lehigh University (Bethlehem, Pennsylvania). He is presently a Professor of Mathematics at the State University of New York at Oswego, and an adjunct Professor at the School of Management, Syracuse University (New York). His research interests include Time Series Analysis, Numerical Analysis and Ordinary Differential Equations.

HOUSTON H. STOKES received his BA in Economics from Cornell University in 1962 and his MA and PhD from the University of Chicago in 1966 and 1969. He has written some 40 articles and a book and is the developer of the B34S Data Analysis Program. Dr Stokes is currently Professor of Economics and Director of Graduate Studies at the University of Illinois at Chicago. His fields of specialization are applied econometrics, time series, international trade and monetary theory and policy.

KJELL STORDAHL was born in Oslo, 28 August 1945. He obtained a cand. real. (Master's) degree in Statistics at Oslo University in 1972. He then joined the Research Establishment of the Norwegian Telecommunications Administration - with sample surveys, forecasting, market analysis and demand for new services as his primary interests. Recently, however, he has mainly worked with time series analysis and queueing theory, applied to forecasting and control of telephone traffic.

TIMO TERÄSVIRTA is Research Fellow at the Research Institute of the Finnish Economy in Helsinki. He received his Master's (Statistics) and Doctor's (Econometrics) degrees from the University of Helsinki, where later he was Professor of Statistics (1976-80), with one year away as Invited Research Fellow at the Center for Operations Research and Econometrics, Université Catholique de Louvain, Belgium.

WILLIAM ROBERT TERRY is an Associate Professor in the Department of Industrial Engineering and a member of the graduate faculty in Systems Engineering at the University of Toledo, Ohio (USA). He received his BS and MS from Georgia Tech,

Atlanta; and his PhD from North Carolina State University in Raleigh. His interests are in time series analysis and control systems.

GEORGE TREVINO received a PhD (1969) from Lehigh University, Bethlehem, Pennsylvania, USA. His main field is applied mechanics, and he has taught at college level and been a senior researcher for, respectively, 9 and 5 years. Currently he heads his own research and consultancy firm in Las Cruces, specialising in turbulence and stochastic processes. He comes from San Antonio in Texas.

JOHN E. TRIANTIS is an economist in the Overseas Department of AT&T Long Lines. He received his BA in Economics from Fairleigh Dickinson University in 1974 and his MA and PhD from the University of New Hampshire in 1975 and 1978. He has written or co-authored several articles on issues ranging from forecast methodology at the company level to forecasting the demand for money.

DAVID J. WRIGHT has a BA (Mathematics) and a PhD (Control Engineering) from Cambridge University, UK. He has worked in Government and as a Private Consultant on various Forecasting projects, and has held full time university posts in both Europe and North America. He is the author of 14 papers in major international journals, and a member of 3 professional associations. He currently holds 2 government research grants.

Acknowledgements

The work of the Organising Committee was much appreciated. I would also like to thank all participants for coming, speakers for presenting their work, referees for assessing it, and authors for preparing final copy for publication. As usual, I think everyone will agree that Inez van der Heide, at North-Holland, has done an excellent job in preparing this volume for print.

Looking Forward

Although there will be no ITSMs held in 1984, we have begun to plan for 1985. Prospective authors should write to the Editor, at the address below, for details on submitting abstracts and papers.

OLIVER D. ANDERSON
TSA&F, 9 Ingham Grove, Lenton Gardens, Nottingham NG7 2LQ, England
August 1984

TIME SERIES ANALYSIS: Theory and Practice 7
O.D. Anderson (editor)

PRACTICAL EXAMPLES OF ARUMA MODELLING

Oliver D. Anderson
TSA&F, 9 Ingham Grove, Lenton Gardens, Nottingham NG7 2LQ, England

We comment on some case-studies presented at International Time Series Meetings: Jenkins (1979); Downing & Pack (1982); Raftery, Haslett & McColl (1982); O'Connor & Kapoor (1984); and Damsleth (1984). This is done in the context of ARUMA and near-ARUMA modelling.

1. INTRODUCTION

The general autoregressive integrated moving average process of order (p,d,q), which we shall abbreviate to ARIMA(p,d,q), is defined by a stochastic sequence $\{Z_i\}$, satisfying

$$(1-\phi_1 B-\ldots-\phi_p B^p)(1-B)^d Z_i = (1-\theta_1 B-\ldots-\theta_q B^q)A_i \qquad (1)$$

where $(\phi_1,\ldots,\phi_p)$ and $(\theta_1,\ldots,\theta_q)$ are two sets of real parameters, with the first subject to the stationarity condition, namely that the polynomial $1-\phi_1\zeta-\ldots-\phi_p\zeta^p$ in the complex variable ζ has no zero within or on the unit circle; and B is the backshift operator, such that B^j operating on any X_i, for instance A_i or Z_i, produces X_{i-j}. $\{A_i\}$ is a white noise sequence of independent but identically distributed normal zero-mean random variables, all with variance σ_A^2 say.

When the factor $(1-B)^d$ in (1) is replaced by any general homogeneous nonstationary operator of the form

$$U_d(b) = (1-u_1 B-\ldots-u_d B^d)$$

all of whose zeros lie precisely on the unit circle, but not necessarily all (or any) taking the value plus unity, we obtain the more general so-called ARUMA(p,d,q) class of models - see, for instance, Anderson (1980b).

Putting d = 0 in either the ARIMA or ARUMA specification retrieves the stationary ARMA(p,q) processes.

Given any series, $\{z_i: i=1,\ldots,n\}$ of length n, its sampled autocovariance, at lag k, is defined as

$$c_k^{(n)} = n^{-1}\sum_{i=1}^{n-k}(z_i-\bar{z})(z_{i+k}-\bar{z}) \qquad (k=0,1,\ldots,n-1)$$

where $\bar{z} = (z_1+\ldots+z_n)/n$ is the mean of the observed series. Associated with this is the k-th serial correlation, defined by

$$r_k^{(n)} = c_k^{(n)}/c_o^{(n)} \qquad (k=0,1,\ldots,n-1).$$

Properties of the serial correlation sequences, $\{r_1^{(n)},\ldots,r_{n-1}^{(n)}\}$, for series realisations from ARUMA models have been described, together with examples, in Anderson (1979a-1982), and in Anderson and de Gooijer (1979-1983). Attention there is focused on non-stationary ARUMA models, ie those for which $U_d(B) \neq 1$, and on nearly non-stationary ARUMA models - ones where $U_d(b)$ is replaced by $(1-\alpha u_1 B-\ldots-\alpha^d u_d B^d)$, with α "near to", but less than, unity. Thus a nearly non-stationary ARUMA(p,d,q) model is actually an ARMA(p+d,q), with a nearly non-stationary AR(d) factor - whose influence will in fact dominate the observed serial correlation structure.

The literature has long been aware of the special behaviour of the serial correlations arising from ARIMA models - for instance, see Box and Jenkins (1970). But, an appreciation of what was really going on, for these "integrated" models, only began with Wichern (1973); and this was expanded by Anderson and de Gooijer (1979-1980b). Anderson (1980a, 1980d, 1981a & 1982) has given examples of "proper" ARUMA and near-ARUMA modelling; as well as correct discrimination between ARIMA processes and other close ARMA approximations to them.

The purpose of this note is to provide further instances where the ideas help, using case-studies which have recently appeared in ITSM Proceedings.

2. APPLICATIONS

Figure 3 of Damsleth (1984), reproduced as our figure A, shows a serial correlation function which follows a "slow almost linear decay". After a usual Box-Jenkins analysis, Damsleth obtained the model

$$(1-.95B)z_i = (1-.19B)a_i. \tag{2}$$

Our approach goes through as follows. Looking at the plot, the "cross-over" from positive to negative serials occurs at "lag" K = 19.5. The length of the series is 268; so the implication is that the data is definitely stationary*, with a large AR(1) parameter.

This parameter can be immediately approximately estimated by ϕ, the solution of

$$\phi^K(1-\phi)/(1+\phi) = (n-2K)/(n-K)^2 \tag{3}$$

(see Anderson, 1979a, p.297), which yields $\phi = .86$. So the process is certainly (acceptably) stationary, as indeed Damsleth decided - but which would not be quite so evident to many practitioners (or theoreticians, for that matter).

Moreover, if (2) obtains, (3) gives the "expected" cross-over occurring at K = 37.7; and studies such as Anderson and de Gooijer (1979 & 1980a) suggest that an observed value as far away as 19.5 is not a very unlikely event, given the

---

* For an IMA(1,1) "approximation" to (2), the "expected" cross-over would be at K = 78.5 lags; with a value as low as 19.5 having a negligible probability.

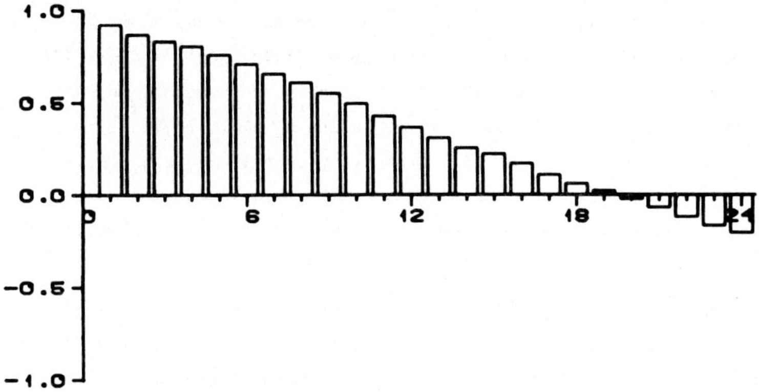

Figure A: Damsleth Serial Correlation Function (n = 268)

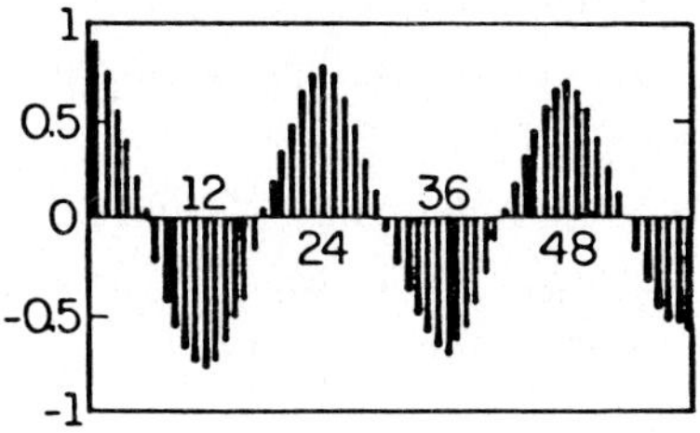

Figure B: O'Connor and Kapoor Serial Correlation Function (n = 240)

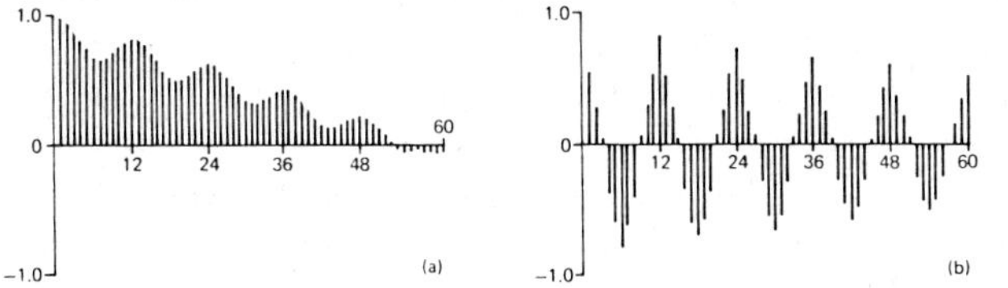

Figure C: Jenkins Serial Correlation Functions (Schematic) for (a) Original Series and (b) First Differenced Series (n = 160)

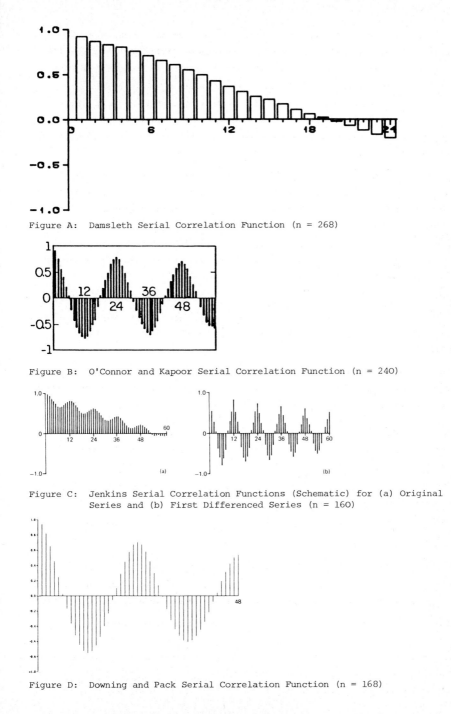

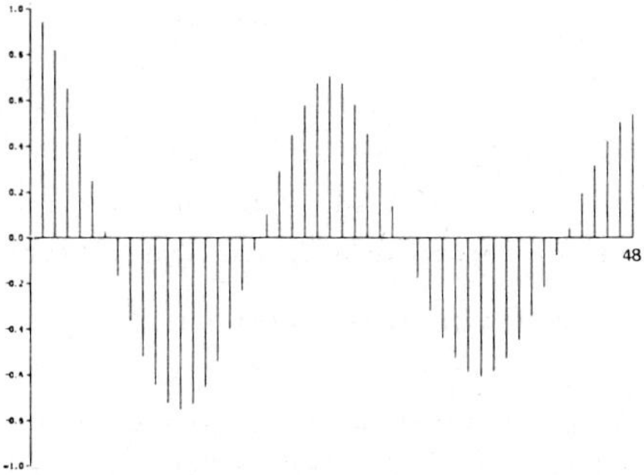

Figure D: Downing and Pack Serial Correlation Function (n = 168)

length of the series. ("Standardising" to a series length of 100, this becomes a difference of 6.79 lags; which, if displaced to higher lags, might be enough to discredit an IMA process, in favour of an ARMA alternative. But it is certainly not sufficient to reasonably disqualify model 2.)

Figure B is from O'Connor and Kapoor (1984, half of their figure 3). They produced (in passing) the Box-Jenkins model

$$(1-.74B)(1-B^{24})z_i = (1-.75B^{24})z_i. \quad (4)$$

This seems incorrect: (a) the decay is not from a value sufficiently close to unity (there is a sharp kink in the otherwise smooth sinusoidal profile, from $r_o^{(240)} = 1$ to $r_1^{(240)} \simeq .9$); and (b) we have not got a serial correlation pattern dominated by $\{r_{24}^{(240)}, r_{48}^{(240)}, \ldots\}$ following a slow linear decline.

Thus we have a situation requiring neither a nonstationary "simplifying" operator, in particular not the $(1-B^{24})$ of (4), nor a near-nonstationary factor $(1-\Phi B^{24})$, with Φ approaching unity from below.

Instead, we have the typical pattern associated with a dominant nearly nonstationary quadratic AR factor, $(1-\phi_1 B-\phi_2 B^2)$ with $\phi_1 = 2\lambda\cos\omega$ and $\phi_2 = -\lambda^2$; where, from figure B, we would take $\omega = 360/24 = 15^o$ and $\lambda = (r_{48}/r_{24})^{1/24}$. Now $r_{48}/r_{24} = .890$, according to a private communication from Dr Kapoor; so $\lambda \simeq .995$. Thus the simplifying operator should be $(1-1.92B + .99B^2)$, not $(1-B^{24})$.

Jenkins (1979, p.87) gives an example where he simplifies with the pair of factors $(1-B)$ and $(1-B^{12})$. Please see our Figure C (which is schematic, as there appear to be some errors in Jenkins' original artwork). Certainly, this time, the serial correlations immediately indicate simplification by something very close to a pair of nonstationary factors. For we get a smooth decay from something very close to unity. However, again the seasonal part should be taken care of by a quadratic factor - as is very clear from the plot of serials obtained for the first differenced series.

Note that the cross-over from positive to negative occurs between lags 53 and 54. For a series of length 160, the $(1-B)$ cross-over is expected at lag 50.9; which, if anything, would be displaced slightly to higher lag, by the peak in seasonal structure at lag 48 (which does not become negative until four lags later). I would thus be happy to accept a nonstationary $(1-B)$, the weakest type of nonstationarity, whose influence is being slightly modified by a near-nonstationary quadratic factor. This second factor, however, is certainly not nonstationary; as, if it were nonstationary, it would dominate the serial correlation pattern for the original series. It is also seen to be only near-nonstationary from the serials for the differenced data.

Downing and Pack (1982, figure 4) provide another example; where, this time, the serial correlations clearly indicate a dominant quadratic factor, plausibly nonstationary. See our figure D. The authors, however, use $(1-B^{24})$ rather than $(1-2\cos\frac{\pi}{12}B + B^2)$, which would be our choice of simplifying operator. Note that our quadratic is actually a factor of $(1-B^{24})$: it is thus a far less drastic operator.

Finally, we refer to Raftery, Haslett and McColl (1982, figure 2.5). They show a serial correlation plot, for a series of length 329, with a cross-over between lags 64 and 65. We would agree with the authors (top of p.194) that there is no evidence of non-stationarity. (The theoretical IMA(1,1) cross-over is at lag 96.36, and the difference of nearly 32 lags is certainly sufficient to reject a (1-B) possibility.) See our figure E.

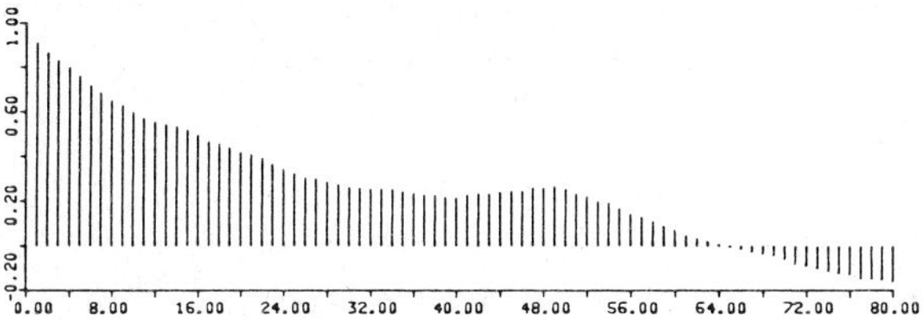

Figure E: Raftery, Haslett and McColl Serial Correlation Function (n = 329)

REFERENCES

ANDERSON, O.D. (1979a). Some sample autocovariance function results for a once integrated q^{th}-order moving average process. Statistica 39, 287-299.

ANDERSON, O.D. (1979b). Formulae for the expected values of the sampled variance and covariances from series generated by general autoregressive integrated moving average processes of order (p,d,q). Sankhyā: Indian Journal of Statistics B41, 177-195.

ANDERSON, O.D. (1979c). The autocovariance structures associated with general unit circle nonstationary factors in the autoregressive operators of otherwise stationary ARMA time series models. Cahiers du Centre d'Etudes de Recherche Opérationelle 21, 221-237.

ANDERSON, O.D. (1980a). A new approach to ARMA Modeling: Some Comments. In Analysing Time Series (Proceedings of the International Conference held on Guernsey, Channel Islands, October 1979). Ed: O.D. Anderson, North-Holland, Amsterdam & New York, 43-71.

ANDERSON, O.D. (1980b). Serial dependence properties of linear processes. Journal of Operational Research Society 31, 905-917. Correction 35 (1984), 171.

ANDERSON, O.D. (1980c). Serial dependence properties of ARUMA models. Cahiers du Centre d'Etudes de Recherche Opérationelle 22, 309-323.

ANDERSON, O.D. (1980d). An Augmented Box-Jenkins Approach with Applications to Economic Data. In Proceedings of Business and Economic Statistics Section

(Houston Meeting), American Statistical Association, 432-437.

ANDERSON, O.D. (1981a). A Time Series Case Study from Biology: Some Interplay between Theory and Practice. In Applied Time Series Analysis II, Ed: D.F. Findley, Academic Press, New York & London, 473-498.

ANDERSON, O.D. (1981b). Covariance Structure of Sampled Correlations from ARUMA Models. In Time Series Analysis (Proceedings of the International Conference held at Houston, Texas, August 1980). Ed: O.D. Anderson, North-Holland, Amsterdam & New York, 3-26.

ANDERSON, O.D. (1982). Sampled Serial Correlations from ARIMA Processes. In Applied Time Series Analysis (Proceedings of the International Conference held at Houston, Texas, August 1981). Ed: O.D. Anderson, North-Holland, Amsterdam & New York, 5-14.

ANDERSON, O.D. and DE GOOIJER, J.G. (1979). On discriminating between IMA(1,1) and ARMA(1,1) processes: some extensions to a paper by Wichern. Statistician 28, 119-133.

ANDERSON, O.D. and DE GOOIJER, J.G. (1980a). Distinguishing between IMA(1,1) and ARMA(1,1) Models: a large scale Simulation Study of two particular Box-Jenkins Time Processes. In Time Series (Proceedings of the International Conference held at Nottingham University, England, March 1979). Ed: O.D. Anderson, North-Holland, Amsterdam & New York, 15-40.

ANDERSON, O.D. and DE GOOIJER, J.G. (1980b). Distinguishing certain Stationary Time Series Models from their Nonstationary Approximations and improved Box-Jenkins Forecasting. In Analysing Time Series (Proceedings of the International Conference held on Guernsey, Channel Islands, October 1979). Ed: O.D. Anderson, North-Holland, Amsterdam & New York, 21-42.

ANDERSON, O.D. and DE GOOIJER, J.G. (1982). The Covariances between Sampled Autocovariances and between Serial Correlations for Finite Realisations from ARUMA Time Series Models. In Time Series Analysis: Theory and Practice 1 (Proceedings of the International Conference held at Valencia, Spain, June 1981). Ed: O.D. Anderson, North-Holland, Amsterdam & New York, 7-22.

ANDERSON, O.D. and DE GOOIJER, J.G. (1983). Formulae for the covariance structure of the sampled autocovariances from series generated by general autoregressive integrated moving average processes of order (p,d,q), d = 0 or 1. Sankhyā: Indian Journal of Statistics B45, 249-256.

BOX, G.E.P. and JENKINS, G.M. (1970). Time Series Analysis: Forecasting and Control. Holden-Day, San Francisco & London. (Revised Edition, 1976).

DAMSLETH, E (1984). Estimation of the Mean Duration of Telephone Calls. In Time Series Analysis: Theory and Practice 7 (Proceedings of 11th International Conference, General Interest ITSM, Toronto, Canada, 18-21 August 1983). Ed: O.D. Anderson, North-Holland, Amsterdam & New York, this volume.

DOWNING, D.J. and PACK, D.J. (1982). The Vanishing Transfer Function. In Time Series Analysis: Theory and Practice 1 (Proceedings of the International Conference held at Valencia, Spain, June 1981). Ed: O.D. Anderson, North-Holland, Amsterdam & New York, 221-246.

JENKINS, G.M. (1979). Practical Experiences with Modelling and Forecasting Time Series. In Forecasting (Proceedings of the National Conference, Cambridge University, England, July 1976). Ed: O.D. Anderson, North-Holland, Amsterdam & New York, 43-166.

O'CONNOR, M.J. and KAPOOR, S.G. (1984). Time Series Analysis of Building Electrical Load. In Time Series Analysis: Theory and Practice 5 (Proceedings of 9th International Conference held at Nottingham University, England, April 1983). Ed: O.D. Anderson, North-Holland, Amsterdam & New York, 135-144.

RAFTERY, A.E., HASLETT, H. and McCOLL, E (1982). Wind Power: A Space-Time Process? In Time Series Analysis: Theory and Practice 2 (Proceedings of the International Conference held in Dublin, Ireland, March 1982). Ed: O.D. Anderson, North-Holland, Amsterdam & New York, 191-202.

WICHERN, D.W. (1973). The behaviour of the sample autocorrelation function for an integrated moving average process. Biometrika 60, 235-239.

TIME SERIES ANALYSIS: Theory and Practice 7
O.D. Anderson (editor)

A TRADITIONAL INTERPRETATION OF THE FORECASTS OF SEASONALLY DIFFERENCED ARIMA PROCESSES

Ed. McKenzie
Department of Mathematics, University of Strathclyde, Glasgow, Scotland

The modelling of time-series as ARIMA processes is now common practice thanks to the availability of suitable computer packages. One problem for users, however, is the practical interpretation of the forecasts from such models. This is particularly unfortunate in the case of seasonally differenced models as they often contain components which have immediate intuitive appeal, viz. trend and seasonal effects. The derivation of such components from an ARIMA forecast was discussed only briefly by Box and Jenkins and neglected almost totally by subsequent textbook writers. We present here some results concerning the generation and revision of these traditional components for seasonally differenced models.

1. INTRODUCTION

The modelling of time-series as ARIMA processes is now a common practice, and there are many statistical packages available which help to achieve this. Nevertheless, a major hurdle for many users is still the practical interpretation of the identified model. Apart from the paper by Granger and Morris (1976), which dealt exclusively with non-seasonal models, and the discussions of Anderson (1975, 1977), little help is available to the user. This is particularly unfortunate in the case of seasonally differenced models because they often contain components which have immediate intuitive appeal, viz. trend and seasonal effects.

Although Box and Jenkins deal with most aspects of the ARIMA process in great detail in their book (1970), they devote comparatively less to the discussion of seasonal models, and only a little of that is reserved for interpretation. They deal only briefly with the idea of decomposing the forecast into readily understood components. Moreover, few of the long line of subsequent textbooks dealing with ARIMA models appear to discuss the matter at all. One of the reasons for this may be that the natural approach to such a decomposition is by means of the eventual forecast function (EFF), and forecast generation via the integrated form. The reason for this is that the EFF is a linear combination of deterministic functions which are usually very well understood and readily interpreted. Indeed, these functions are the usual starting point for the more ad hoc forecasting procedures such as exponential smoothing. We may recall Stern's comments (1974) that managers who require forecasts may be prepared to accept trends and seasonal effects because these correspond to familiar ideas. However, Box and Jenkins counsel against forecasting with the integrated form, because of its apparent complexity. They recommend forecasting in the simpler difference equation form.

Our purpose here is to consider interpreting seasonally differenced models in terms of the components of their EFFs. As we shall see, the T-step ahead forecast is naturally partitioned into three components. One is due to the difference operator, another to the autoregressive (AR-) operator and the third is a moving-average (MA-) term. The first two of these correspond to the EFF. The difference component is readily interpreted for most practically occurring models, whereas the AR-component is less so. We consider using those components corresponding to the difference operator. These components are usually easily understood and readily interpreted in terms familiar to most users. It is important to note that they may be derived independently of any other components of the model and may be revised and forecast without reference to these other components. Revision at time t requires only past values of the component and the latest one-step ahead forecast error.

As an illustration, consider the well-known airline model from Box and Jenkins (1970). It is given by

$$\nabla\nabla_{12}X_t = (1 - 0.4B)(1 - 0.6B^{12})a_t$$

and forecasts are generated by the usual conditional expectation arguments. The EFF for this model may be written in the form

$$\hat{X}_t(T) = a_o^t + a_1^t T + S_\tau$$

where a_o^t is the current level of the process, a_1^t the current gradient of the linear trend and S_τ is the additive seasonal factor for T months ahead, i.e. $T = \tau(\text{mod } 12)$. These two components trend and seasonal, may be revised at time t as follows:

$$a_o^t = a_o^{t-1} + a_1^{t-1} + 0.52e_t$$

$$a_1^t = a_1^{t-1} + 0.02e_t$$

$$S_k^t = S_{k+1}^{t-1} + (0.097 - 0.02k)e_t \qquad (k = 1,2,\ldots,11)$$

$$S_{12}^t = S_1^{t-1} + 0.253e_t \ .$$

Notice that whether we use the EFF to produce the forecast or not we can still obtain and use either or both of these components. Notice also that we can revise and use the trend component (a_o^t, a_1^t) independently of the seasonal component and vice versa. However, the use of either or both of the components requires that we can revise them with each new observation. Thus, in this paper we not only derive the usual trend/seasonal components from the difference equation forms but we also describe procedures for deriving the constants involved in the revision equations. In several important models we obtain the constants explicitly in terms of the model parameters. The procedures we describe are generally applicable for any ARIMA model, but we restrict our attention in the case of the difference operators to $\nabla^d\nabla_s$, and ∇_s^2. This is not done because the procedures cannot be extended, which they obviously can, but simply because the results become very complex and the work can hardly be justified in view of the rarity of occurrence of such higher order differences.

The decomposition of the EFF corresponds to a decomposition of the ARIMA model itself by means of a partial fraction expansion of the usual infinite MA-representation of the model. In recent years, such expansions have commanded attention as a means of obtaining seasonal adjustment procedures based on ARIMA models. See, for example, Hillmer and Tiao (1982), Box, Hillmer and Tiao (1978), and Burman (1980). Brewer, Hagan and Perazelli (1975) evidently consider the particular expansion described here, and Burman (1978) discusses it further. However, they are not concerned with the derivation of the revision equations for these components in general, although that may be necessary, or at least useful, for the implementation of their own procedures. Equally, we must emphasise that we are not concerned here with seasonal adjustment although the components could be used in that way. It is worth noting that the filters by which these components are generated are one-sided, whereas the usual preference in seasonal adjustment is for symmetric filters.

2. DECOMPOSITION OF THE EVENTUAL FORECAST FUNCTION

Forecasting in terms of the integrated form is discussed in detail by Box and Jenkins in Appendix A5.3 of their book (1970). The following is a very brief summary.

We suppose $\{X_t\}$ is an ARIMA process satisfying $\alpha(B)X_t = \beta(B)e_t$, where $\alpha(B) = \sum_{k=0}^{u} \alpha_k B^k$, $\beta(B) = \sum_{k=0}^{v} \beta_k B^k$ and all seasonal and difference terms are included in the α- and β-operators. For lead-times $T > v$ we have $\sum_{k=0}^{u} \alpha_k \hat{X}_t(T-k) = 0$. The solution of this difference equation is the Eventual Forecast Function (EFF) which we can write as $\hat{X}_t(T) = \sum_{k=1}^{u} b_k^t f_k(T)$. This representation of the T-step ahead forecast is valid only for $T > M = v-u$. If $M > 0$ we have for lead-times $T = 1,2,\ldots,M$

$$\hat{X}_t(T) = \sum_{k=1}^{u} b_k^t f_k(T) + \sum_{j=0}^{M-T} d_{T,j}\, e_{t-j} \tag{2.1}$$

In addition, the current value of $\underline{b}_t = (b_1^t, b_2^t, \ldots, b_u^t)'$ may be obtained from $\underline{b}_{t-1}$ via a linear equation

$$\underline{b}_t = L\,\underline{b}_{t-1} + \underline{h}\,e_t \tag{2.2}$$

where L affects the changes in the coefficients in revising the time origin from (t-1) to t. The vector $\underline{h}$ may be obtained in the form $\underline{h} = F_M^{-1}\underline{A}_M$ where the (i,j)th element of F_M is $f_j(M+i)$, $i = 1,2,\ldots, u$; $j = 1,2,\ldots, u$; $\underline{A}_M = (A_{M+1}, A_{M+2}, \ldots, A_{M+u})'$; and A_k is the coefficient of B^k in $A(B) = \beta(B)/\alpha(B)$, the usual moving-average representation of X_t. The form of L is easily derived by considering the usual forecast revision identity

$$\hat{X}_t(T) = \hat{X}_{t-1}(T+1) + A_T e_t \tag{2.3}$$

and writing the forecasts on each side in EFF-form. Indeed, if the algebraic form of A_T is known we can also derive the components of $\underline{h}$. In general, however, A_T will be obtained numerically.

We consider now the particular decomposition of interest to us here. Using equation (2.2) and the fact that

$\underline{f}(t) = L'\underline{f}(t-1)$, where $\underline{f}(t) = (f_1(t), f_2(t), \ldots, f_u(t))'$, we can derive the expression

$$\underline{f}'(1)\underline{b}_t = \sum_{j=0}^{\infty} \underline{f}'(j+1)\underline{h}B^j e_t = H(B)e_t\,. \tag{2.4}$$

Now,

$$\alpha(B)H(B) = \sum_{k=0}^{u-1}\sum_{i=0}^{k} \alpha_i \underline{f}'(k+1-i)\underline{h}B^k + \sum_{k=u}^{\infty}\sum_{i=0}^{u} \alpha_i \underline{f}'(k+1-i)\underline{h}B^k$$

and the second sum on the right hand side is zero since the components of $\underline{f}$ are the components of the EFF. Hence, $H(B) = H^*(B)/\alpha(B)$, where $H^*(B)$ is a polynomial in B of degree at most (u-1). Further, in the usual seasonal ARIMA model for a season of length s, $\alpha(B) = \phi(B)\Phi(B^s)\nabla^d\nabla_s^D$, where the orders of the operators ϕ and Φ are p and P respectively. Thus, we may obtain a partial fraction expansion of H(B) in (2.4) in the form

$$\sum_{k=1}^{u} b_k^t f_k(1) = \left\{\frac{\delta(B)}{\nabla^d} + \frac{\Delta(B)}{\nabla_s^D} + \frac{\phi^*(B)}{\phi(B)} + \frac{\Phi^*(B)}{\Phi(B^s)}\right\} e_t \tag{2.5}$$

where, in general, δ, Δ, ϕ^* and Φ^* are polynomial operators of degrees d-1, sD-1, p-1 and sP-1 respectively. Now, if we rewrite (2.1) for T=1 using (2.5) and $X_t = B\hat{X}_t(1) + e_t$ we can derive a partial fraction expansion of $\beta(B)/\alpha(B)$ of the form

$$\frac{B\delta(B)}{\nabla^d} + \frac{B\Delta(B)}{\nabla_s^D} + \frac{B\phi^*(B)}{\phi(B)} + \frac{B\Phi^*(B)}{\Phi(B^s)} + 1 + \sum_{j=1}^{M} d_{T,j-1}B^j \qquad (2.6)$$

This expansion represents a decomposition of the ARIMA process into five components: seasonal and non-seasonal difference and autoregressive components and a moving-average component. The difference components, as we shall see, may have clear and intuitively appealing interpretations, viz. trend and seasonality. We note also that each component may be generated separately from the others and need not be constructed from its own basic components. For example, the trend component, Z_t say, can be obtained directly from $\nabla^d Z_t = \delta(B)e_t$. It is not necessary to obtain the revision equations for each sub-component of that part of the EFF corresponding to ∇^d i.e. the individual coefficients of the plynomial trend.

In what follows, we assume not only that $\alpha(B)$ is of the form given above but also that $\beta(B) = \theta(B)\Theta(B^s)$ where the degrees of θ and Θ are q and Q respectively. We shall also write $A(B) = \theta(B)\Theta(B^s)/\phi(B)\Phi(B^s) = \psi(B)\Psi(B^s)$. We note that, unlike the terminology of Box and Jenkins, all the operators θ, ϕ, Θ, Φ are usually written in a "positive" form e.g. $\theta(B) = \sum_{k=0}^{q} \theta_k B^k$, where $\theta_0 = 1$.

3. THE DIFFERENCE COMPONENT

3.1. ∇_s

We consider first d = 0, D = 1 i.e. $\nabla_s X_t = A(B)e_t$. The operator ∇_s is used in an attempt to obviate the periodic nature of the seasonality. It corresponds to having a dummy variable in the EFF to mark each period of the season. Box and Jenkins denote the coefficients of these dummy variables by $\{b^t_{o,k} : k = 1,2,\ldots,s\}$. Clearly, the values $\{b^t_{o,k}\}$ act as current levels for each period of the season. Generally, however, we would prefer a single value for the current level of the process and a set of seasonal factors which are independent of this level value. To achieve this independence it is usually arranged that the seasonal factors sum to zero. We can easily derive such a structure from the values $\{b^t_{o,k}\}$ in the following way. We take as the current level estimate, a^t_o, the mean of the values $\{b^t_{o,k}\}$. The additive seasonal factors are now obtained as the deviations of the $b^t_{o,k}$ from this mean value. Thus, we define

$$a^t_o = \sum_{k=1}^{s} b^t_{o,k}/s \qquad (3.1)$$

and

$S_k^t = b_{o,k}^t - a_o^t \qquad (k = 1,2,\ldots,s)$

Clearly, the seasonal factors $\{S_k^t\}$ sum to zero.

Corresponding to ∇_s in the EFF for a lead-time $T = rs+k$ we have $a_o^t + S_k^t$. Using (2.3) we find the components defined by (3.1) are revised using

$$a_o^t = a_o^{t-1} + h_o e_t$$

$$S_k^t = S_{k+1}^{t-1} + g_k e_t \qquad (k = 1,2,\ldots,s-1) \tag{3.2}$$

and

$$S_s^t = S_1^{t-1} + g_s e_t \, .$$

Further, by considering the corresponding revision equations for $\{b_{o,k}^t\}$ and (3.1), we may easily show that $\sum_{k=1}^{s} g_k = 0$, so that $\sum_{k=1}^{s} S_k^t = 0$ for all t provided we start it so. From (3.2) we can derive the expansion

$$\frac{A(B)}{\nabla_s} = \frac{Bh_o}{\nabla} + \frac{\sum_{k=1}^{s} g_k B^k}{\nabla_s} + \text{AR-terms} + \text{MA-terms.} \tag{3.3}$$

Note that since $\sum_{k=1}^{s} g_k = 0$ the second term on the right-hand side has a common factor (1-B) in the numerator and the denominator. Solving in the usual way to obtain coefficients in the expansion we find from $B = 1$ that

$$h_o = A(1)/s \, . \tag{3.4}$$

From $B = \omega$, a complex sth root of unity other than unity itself, we obtain

$$\sum_{k=1}^{s} g_k \omega^k = A(\omega) = \psi(\omega)\Psi(1) \, . \tag{3.5}$$

Further, (3.5) holds for each of the (s-1) distinct roots, $\omega_1, \omega_2, \ldots, \omega_{s-1}$ where $\omega_k = \omega^k$ and $\omega = \exp\{2\pi i/s\}$. Solving the corresponding (s-1) equations in combination with $\sum_{k=1}^{s} g_k = 0$ yields

$$g_s = [\sum_{r=0}^{\infty} \psi_{rs} - \psi(1)/s]\,\Psi(1)$$

$$g_i = [\sum_{r=0}^{\infty} \psi_{rs+i} - \psi(1)/s]\,\Psi(1)\ , \quad (i = 1,2,\ldots,s-1)\ . \tag{3.6}$$

There are several points we may note here. Clearly $\sum_{k=1}^{s} g_k = 0$. Further, the sums on the right-hand side of (3.6) converge since under the usual assumptions $\phi(B)$ and $\theta(B)$ have their roots outside the unit circle. If we define g_o as g_i for $i = 0$ we see that $g_s = g_o$. Obviously, the sequence $\{\psi_k\}$ can be evaluated in the usual way from $\phi(B)\psi(B) = \theta(B)$ and then the sums

$z_i = \sum_{r=0}^{\infty} \psi_{rs+i}$, $(i = 0,1,\ldots,s-1)$ can be derived. The stationarity of $\theta(B)/\phi(B)$ assures a reasonable rate of convergence here. However, in the case where $p < s$, $q < s$ we can also use the relationship between ϕ, ψ and θ to show that

$$\begin{bmatrix} \phi_o & \phi_{s-1} & \phi_{s-2} & \cdots & \phi_2 & \phi_1 \\ \phi_1 & \phi_o & \phi_{s-1} & \phi_{s-2} \cdots & & \phi_2 \\ \phi_2 & \phi_1 & \phi_o & \phi_{s-1} & & \\ & \ddots & \ddots & \ddots & \ddots & \\ & & & & & \phi_{s-1} \\ \phi_{s-1} & & \phi_2 & \phi_1 & & \phi_o \end{bmatrix} \begin{bmatrix} z_o \\ z_1 \\ \vdots \\ z_{s-1} \end{bmatrix} = \begin{bmatrix} \theta_o \\ \theta_1 \\ \vdots \\ \theta_{s-1} \end{bmatrix} \tag{3.7}$$

with the obvious convention that $\theta_i = 0,\ i > q;\ \phi_i = 0,\ i > p$. If we write (3.7) as $R\underline{z} = \underline{\theta}$ and define the vector $\underline{g} = (g_s g_1 \ldots g_{s-1})'$ we have from (3.6) and (3.7)

$$R\underline{g} = [\underline{\theta} - \{\theta(1)/s\}\underline{1}]\Psi(1) \tag{3.8}$$

The (i,j)th element of R is given by $R_{ij} = \phi_{F(i,j)}$ where $F(i,j) = (i-j)$ if $j \le i$ and $(s+i-j)$ if $j > i$. Thus $\underline{g}$ may be evaluated directly without evaluating and summing subsequences of $\{\psi_k\}$.

The equation (3.7) is a most useful one so it is worth noting that when the full multiplicative seasonal ARIMA model is used we generally have $p < s$, $q < s$. However, if $\Phi = 1$ or $\Theta = 1$ or both, this may no longer be true. In any case, if $p \ge s$ and/or $q \ge s$ we can replace (3.7) by a similar form. In effect, (3.7) still holds but ϕ_k is replaced by $\sum_{r=0}^{\infty} \phi_{rs+k}$ $(k = 0,1,\ldots,s-1)$, and θ_k is replaced by $\sum_{r=0}^{\infty} \theta_{rs+k}$ $(k = 0,1,\ldots,s-1)$.

For reasons which will shortly be clear we will rewrite h_o from (3.4) and g_k from (3.6) as h_o^o and g_k^o , $(k = 1,2,\ldots,s)$.

3.2. $\nabla\nabla_s$

Perhaps the most commonly identified seasonal difference operator is $\nabla\nabla_s$. In the same way that ∇_s is used to remove periodic patterns, so ∇ may be employed in an attempt to remove trend. Corresponding to $\nabla\nabla_s$ we would expect the EFF to contain $b_{o,k}^t$ and a linear trend term. For lead-times $T = rs+k$, the EFF is $b_{o,k}^t + r\, b_1^t$ + AR-terms. We can decompose $b_{o,k}^t$ as before into $a_o^t + S_k^t$, but, in such a case, it is more intuitively appealing to express trend via a gradient defined between periods of the season rather than b_1^t which is between seasons. Thus, we may also replace b_1^t by sa_1^t . The presence of the trend changes the definiion of the other terms too. We obtain

$$a_1^t = b_1^t/s$$

$$a_o^t = \sum_{k=1}^{s} b_{o,k}^t/s - (s+1)a_1^t/2 \qquad (3.9)$$

and

$$S_k^t = b_{o,k}^t - ka_1^t - a_o^t \qquad (k = 1,2,\ldots,s) .$$

As before, using (2.3) we can derive the revision equations for these variables in the following form.

$$\begin{aligned} a_o^t &= a_o^{t-1} + a_1^{t-1} + h_o^1\, e_t \\ a_1^t &= a_1^{t-1} + h_1^1\, e_t \\ S_k^t &= S_{k+1}^{t-1} + g_k^1\, e_t \qquad (k = 1,2,\ldots,s-1) \\ S_s^t &= S_1^{t-1} + g_s^1\, e_t . \end{aligned} \qquad (3.10)$$

Also, as before, we have $\sum_{k=1}^{s} S_k^t = \sum_{k=1}^{s} g_k^1 = 0$. The partial fraction expansion given by (2.6) now becomes

$$\frac{A(B)}{\nabla\nabla_s} = \frac{Bh_1^1}{\nabla} + \frac{B\{h_o^1 + (h_1^1 - h_o^1)B\}}{\nabla^2} + \frac{\sum_{k=1}^{s} g_k^1 B^k}{\nabla_s} + \ldots$$

Solving in the usual way yields

$$h_1^1 = A(1)/s \tag{3.11}$$

$$h_o^1 = \left[\frac{s+1}{2} A(1) - A'(1)\right]/s$$

$$g_o^1 = g_s^1 = \frac{1}{s} \sum_{k=1}^{s} kg_k^o$$

and

$$g_k^1 = g_{k-1}^1 + g_k^o \qquad (k = 1,2,\ldots,s-1) .$$

3.3. $\nabla^d\nabla_s$

Difference terms such as $\nabla^d\nabla_s$ tend to imply polynomial trends of degree d. Thus models in which $d > 1$ and trend is more complex than linear are much less common in practical situations. Nevertheless, it is clear that the procedures described in the last two sections may be repeated in essentially the same way. In general, corresponding to $\nabla^d\nabla_s$ we may take $\hat{X}_t(T)$, for $T = rs+k$, as containing the terms $\sum_{i=0}^{d} a_i^t \binom{T-1+i}{i} + S_k^t$. With this model, we can show that $\hat{X}_t(1)$ contains the terms $\sum_{k=0}^{d} a_k^t + S_1^t$, and revision is made via the usual equations for S_k^t and using $a_i^t = \sum_{k=i}^{d} a_k^{t-1} + h_i^d e_t$ for $i = 0,1,\ldots,d$. Furthermore, it is clear from the above that $h_k^d = h_{k-1}^{d-1}$, $(k = 1,2,\ldots,d)$. Thus

$$h_k^d = h_o^{d-k} \qquad (k = 0,1,\ldots,d) . \tag{3.12}$$

In addition, $h_o^d = A_d(1)/s$, where $A_d(B)$ is derived recursively from $A(B)$. We may show that

$$A_d(B) = \{A(B) - A(1) \sum_{k=1}^{s} B^k/s - \sum_{k=1}^{d-1} A_k(1)B\nabla^{k-1}\nabla_s/s\}\nabla^{-d} .$$

From this we can derive $A_d(1)$ and hence obtain

$$h_o^d = \frac{(-1)^d}{s}\left\{\frac{A^{(d)}(1)}{d!} - h_o^o \sum_{i=d}^{s} \binom{i}{d} - \sum_{k=1}^{d-1} (-1)^k \binom{s+1}{d-k+1} h_o^k\right\} \qquad (3.13)$$

where $A^{(d)}(B)$ is the dth derivative of $A(B)$ with respect to B.

In particular, for quadratic trend $d = 2$ and we have

$h_2^2 = h_o^o = A(1)/s$ from (3.4), $h_1^2 = h_o^1 = \left[\frac{s+1}{2} A(1) - A'(1)\right]/s$ from (3.13), and also from (3.13) we can derive

$$h_o^2 = \left\{\frac{(s+1)(s+5)}{12} A(1) - \frac{s+1}{2} A'(1) + A''(1)/2\right\}/s .$$

To obtain the coefficients for revising the seasonal factors we define a sequence of coefficient sets $\{g_k^i : k = 1,2,\ldots,s,\ i = 0,1,\ldots,d\}$. Then $\{g_k^o\}$ is defined by (3.6) or (3.8) and $\{g_k^i\}$ is obtained from $\{g_k^{i-1}\}$ using

$$g_o^i = g_s^i = \frac{1}{s} \sum_{k=1}^{s} k g_k^{i-1}$$

and

$$g_k^i = g_{k-1}^i + g_k^{i-1} \qquad (k = 1,2,\ldots,s-1) .$$

3.4. ∇_s^D .

Not only is it rare to find values of d and D in excess of 2, but in the vast majority of applications the values are either 0 or 1. Seasonal differencing of order greater than one is far less important than first order which has already been discussed. In addition, some of the intuitive appeal of the interpretations available for first order is lost. The operator ∇_s^D suggests a polynomial trend of degree (D-1) over seasons, in fact, s different such trends, one for each period of the season. We shall develop a structure here for ∇_s^2 as an extension of that for ∇_s .

Corresponding to ∇_s^2 there are 2s seasonal coefficients $\{b_{o,k}, b_{1,k} : k = 1,2,\ldots,s\}$ appearing in the EFF. Corresponding to a lead-time of $T = rs+k$ we have $b_{o,k} + rb_{1,k}$. Thus, there are s distinct linear trends, one for each period of the season. It is by no means obvious what form of this model could be described as intuitively appealing. The usual model leading to the operator ∇_s^2 is a linear trend with multiplicative seasonal factors i.e. $\{a_o + (rs+k)a_1\}S_k$, where the factors S_k sum to s . This is a very popular forecasting model and is usually dealt with in an ad hoc way using a Holt-Winters' system. See, for example, the paper by Chatfield (1978). However, we cannot usefully equate this model with the one we have derived in $\{b_{o,k}\}$ and $\{b_{1,k}\}$ since it involves only (s+1) independent parameters and our model uses 2s such. In effect, we would require that $b_{o,k}$ and $b_{1,k}$ be functionally related. However, if we treat both $b_{o,k}$ and $b_{1,k}$ as we treated $b_{o,k}$ in Section 3.1 we obtain a fairly well-known model. We decompose $b_{o,k}$ into a current level value and an additive seasonal factor S_k such that $\sum_{k=1}^{s} S_k = 0$. Further, we decompose $b_{1,k}$ into a current gradient and a set of additive seasonal factors for it, T_k say, which also sum to zero. As before, we may arrange that gradients are defined from period to period rather than season to season. In such a case, $b^t_{o,k} + r\, b^t_{1,k}$ is replaced by $a^t_o + S^t_k + (rs+k)(a^t_1+T^t_k)$ so that we have a mixed additive-multiplicative seasonal model of the type discussed by Durbin and Murphy (1975). We may derive the new parameters in terms of $b^t_{o,k}$; $b^t_{1,k}$ and from this the revision equations. We obtain

$$a^t_o = a^{t-1}_o + a^{t-1}_1 + h_o\, e_t$$

$$a^t_1 = a^{t-1}_1 + h_1\, e_t$$

$$T^t_k = T^{t-1}_{k+1} + g_k\, e_t \qquad (k = 1,2,\ldots,s-1) \tag{3.14}$$

$$T^t_s = T^{t-1}_1 + g_s\, e_t$$

$$S^t_k = S^{t-1}_{k+1} + T^{t-1}_{k+1} + G_k\, e_t \qquad (k = 1,2,\ldots,s-1)$$

$$S^t_s = S^{t-1}_1 + T^{t-1}_1 + G_s\, e_t\ .$$

Furthermore, we may show that, as before, $\sum_{k=1}^{s} g_k = \sum_{k=1}^{s} G_k = 0$. The difference terms in the partial fraction expansion are now

$$\frac{h_1 B}{\nabla} + \frac{\{h_o+(h_1-h_o)B\}B}{\nabla^2} + \frac{\sum_{k=1}^{s} g_k B^k}{\nabla_s} + \frac{sB^s \sum_{k=1}^{s} g_k B^k + \nabla_s \left\{ \sum_{k=1}^{s} (k-1) g_k B^k + \sum_{k=1}^{s} G_k B^k \right\}}{\nabla_s^2}$$

Proceeding as before we easily obtain

$h_1 = A(1)/s^2$

$$h_o = \{sA(1) - A'(1)\}/s^2 \tag{3.15}$$

and

$g_k = g_k^o/s \qquad (k = 1,2,\ldots,s)$

where g_k^o is defined by (3.6) or (3.8).

Furthermore,

$$G_k = \{1 - \Psi'(1)/\Psi(1)\}g_k^o - \{z_k^* - \psi'(1)/s\}\Psi(1)/s \tag{3.16}$$

where

$$z_k^* = \sum_{r=0}^{\infty} (rs+k)\Psi_{rs+k} \qquad (k = 0,1,\ldots,s-1) \text{ and } G_s = G_o .$$

We note that $\{z_k^*\}$ is analogous to $\{z_k\}$ of (3.7). Indeed, if $p < s$, $q < s$, and we proceed as described leading up to (3.7) we may show that

$$R\underline{z}^* + U\underline{z} = \underline{\theta}^* \tag{3.17}$$

where $\underline{z} = (z_o, z_1,\ldots,z_{s-1})'$; $\underline{z}^* = (z_o^*,z_1^*,\ldots,z_{s-1}^*)'$;

$\underline{\theta}^* = (0, \theta_1, 2\theta_2,\ldots,(s-1)\theta_{s-1})'$; R is defined by (3.8), i.e. $R_{ij} = \phi_{F(i,j)}$; and U is defined by $U_{ij} = F(i,j)R_{ij}$. Thus, $\underline{z}^*$ may be obtained directly from (3.17) if $\underline{z}$ is known, which will be the case if $\{g_k^o\}$, which is needed in (3.16), is obtained from $\underline{z}$.

4. EXAMPLES

Very many of the models identified in the literature, and in practice, have no AR-operator i.e. $\phi(B) = 1$. In such a case the sums z_i, z_i^* are finite and we can obtain useful forms in certain important cases. In particular, if $q < s$, as is usually the case in this form of multiplicative seasonal ARIMA model, we find (3.6) given by

$$g_k^o = [\theta_k - \theta(1)/s]\Psi(1) \qquad (k = 0,1,\ldots,s-1) \tag{4.1}$$

where $g_s^o = g_o^o$. If the difference operator is $\nabla\nabla_s$ we have from (3.11)

$$g_o^1 = g_s^1 = \{2s - (s+1)\theta(1) + 2\theta'(1)\}\Psi(1)/2s \tag{4.2}$$

$$g_k^1 = g_{k-1}^1 + g_k^o \qquad (k = 1,2,\ldots,s-1).$$

In the case of ∇_s^2 , we have $z_k^* = k\theta_k$ and so

$$G_k = [1 - \Psi'(1)/\Psi(1)]g_k^o - [k\theta_k - \theta'(1)/s]\Psi(1)/s \qquad (k = 0,1,\ldots,s-1)$$

and $G_s = G_o$.

We illustrate this with the following model:

$$\nabla\nabla_{12}X_t = (1+\theta B)(1 + \Theta_1 B^{12} + \Theta_2 B^{24})e_t \ .$$

From the above, $g_k^o = -(1+\theta)\Theta(1)/s$, $\quad (k = 2,3,\ldots,s-1)$, $g_s^o = \{1 - (1+\theta)/s\}\Theta(1)$ and $g_1^o = \{\theta-(1+\theta)/s\}\Theta(1)$. Thus, we may show that $g_s^1 = (s-1)(1-\theta)\Theta(1)/2s$, and $g_k^1 = \{(1+\theta)(s-2k+1)-2\}\Theta(1)/2s$, $(k = 1,2,\ldots,s-1)$. Further, $h_1^1 = \theta(1)\Theta(1)/s$ and $h_o^1 = \{s\theta(1)(1-\Theta_1-3\Theta_2) + \Theta(1)(1-\theta)\}/2s$. These values h_o^1, h_1^1 and g_k^1, $(k = 1,2,\ldots,s)$ may be used in (3.10) to revise the value of the trend and seasonal components.

Our final model has an AR component. The model is $(1+\phi B)\nabla_{12}X_t = (1+\Theta B^{12})e_t$. The seasonal values $\{g_k^o\}$ are obtained from (3.7) which we can rewrite as $z_k + \phi z_{k-1} = 0$, $(k = 1,2,\ldots,s-1)$ and $z_o + \phi z_{s-1} = 1$. The solution is easily obtained as $z_k = (-\phi)^k/\{1-(-\phi)^s\}$, $(k = 0,1,\ldots,s-1)$, from which we can derive

$$g_k^o = \{s(-\phi)^k - \sum_{i=0}^{s-1}(-\phi)^i\}\,(1 + \Theta)/s\{1-(-\phi)^s\} \qquad (k = 0,1,\ldots,s-1)$$

and $g_s^o = g_o^o$. Further, $h_o^o = A(1)/s = (1 + \Theta)/(1+\phi)s$.

REFERENCES

ANDERSON, O.D. (1975). Time Series Analysis and Forecasting: the Box-Jenkins Approach. London: Butterworths. Chapters 14-16.

ANDERSON, O.D. (1977). The interpretation of Box-Jenkins time-series models. The Statistician 26, 127-145.

BOX, G.E.P., HILLMER, S.C. and TIAO, G.C. (1978). Analysis and modelling of seasonal time series. In Seasonal Analysis of Economic Time Series. Ed: A. Zellner, Economic Research Report, ER-1, U.S. Dept. of Commerce, Bureau of Census, Washington, DC, 309-334.

BOX, G.E.P. and JENKINS, G.M. (1970). Time Series Analysis: Forecasting and Control. San Francisco: Holden-Day.

BREWER, K.R.W., HAGAN, P.J. and PERAZELLI, P. (1975). Seasonal adjustment using Box-Jenkins models. In Proceedings of the 40th International Statistical Institute, Warsaw, 130-136.

BURMAN, J.P. (1978). Comments on "A survey and comparative analysis of various methods of seasonal adjustment" by John Kuiper. In Seasonal Analysis of Economic Time Series. See above for more details, 77-84.

BURMAN, J.P. (1980). Seasonal adjustment by signal extraction. J.R. Statist. Soc. A 143, 321-337.

CHATFIELD, C. (1978). The Holt-Winters Forecasting Procedure. Appl. Statist. 27, 264-279.

DURBIN, J. and MURPHY, M.J. (1975). Seasonal adjustment based on a mixed additive-multiplicative model. J.R. Statist. Soc. A 138, 385-410.

GRANGER, C.W.J. and MORRIS, M.J. (1976). Time series modelling and interpretation. J.R. Statist. Soc. A 139, 246-257.

HILLMER, S.C. and TIAO, G.C. (1982). An ARIMA model based approach to seasonal adjustment. J. Amer. Statist. Ass. 77, 63-70.

STERN, G.J.A. (1974). In discussion of "Experience with forecasting univariate time-series and the combination of forecasts" by P. Newbold and C.W.J. Granger, J.R. Statist. Soc A 137, 150-152.

TIME SERIES ANALYSIS: Theory and Practice 7
O.D. Anderson (editor)

SEASONAL MOVING AVERAGES FOR IRREGULAR SERIES WITH MOVING SEASONALITY

Pierre A. CHOLETTE
Statistics Canada, Ottawa, Ontario, Canada K1A 0T6

This study presents an alternative for the three by nine seasonal moving average optionally used in the X-11 (Shiskin, Young, Musgrave, 1967) and X-11-ARIMA (Dagum, 1980) seasonal adjustment methods. This alternative seasonal moving average would be particularly useful in situations of moving seasonality accompanied by irregularity. The results indicate a substantial reduction of phase-shift (bias) in the estimates of evolving seasonal factors at the end of series. The methodology of quadratic minimization used (Whittaker, 1923; Leser, 1963; and Schlicht, 1981) is then applied to the case of the X-11 three by five moving average.

INTRODUCTION

This paper provides an alternative for the three by nine seasonal moving average used in the X-11 (Shiskin, Young and Musgrave, 1967) and X-11-ARIMA (Dagum, 1980) seasonal adjustment methods. The three by nine is recommended for very irregular series, such as Building Permits. The resulting seasonal factors are stable, in the sense that they change very slowly from year to year, especially at the end of series where they "level-off".

Unfortunately, irregular series sometimes display strongly moving seasonality. In such a case, the seasonal factors estimated by the three by nine level off and sometimes lie completely under the cloud of points of the seasonal-irregular (S-I) ratios at the ends of series. In other words, the estimates show considerable phase-shift with respect to the target seasonal phenomenon and with respect to the historical estimates later available. They are biased. In our opinion, such an estimation would hardly be justifiable, especially when the seasonal component - like for Building Permits - is not only mobile but also very pronounced.

The moving average developed in Section 3 yields end estimates which are less biased and entail less phase-shift than those of the three by nine. Moreover, the average has nine terms instead of eleven, so that final "historical" seasonal factors become available in the fifth instead of the sixth year after collection of the unadjusted data.

The methodology of quadratic minimization used (Whittaker, 1923; Leser, 1963) is then applied to the case of the three by five moving average. Reductions in phase-shift are less important then for the three by nine.

Not everyone will agree on whether the averages presented constitute improvements over their X-11(-ARIMA) counterparts. Indeed, an arbitrary trade-off must always be made between the smoothing and the fitting capabilities of an average. Nevertheless, this paper shows how seasonal moving averages, which are based on an explicit and flexible model, can be derived by means of quadratic minimization.

Readers who want to know about the method before examining the results can read Section 3 in the first place. However, we believe with Ehrenberg (1982) that one should first taste the cake before enquiring about the recipe (- if at all!).

1. GRAPHICAL PRESENTATION OF THE RESULTS

a) **Case of the three by nine** - Figures 1 (a) to 1 (l) group for each month the seasonal-irregular (S-I) ratios of a Building Permit series in function of time (years). These are the S-I ratios without extremes retrieved from Table D 9 (and D 8) of the X-11-ARIMA seasonal adjustment programme. The seasonal factors estimated according to the three by nine and to the 9-term seasonal moving average presented in this paper are also displayed. The plots were given for each of the twelve months so that the reader could compare the performance of the averages under a variety of circumstances, as some months display more or less irregularity and more or less seasonal evolution than others.

The three historical estimates by the central weights of the three by nine (in years 1972, 73 and 74) reproduce moving seasonality when present in the centre of series. Such is the case for the month of January (Figure 1 (a)), February (b), April (d), May (e) June (f), July (g), August (h), September (i), October (j) and December (l). However, the average yields estimates which evolve less and less as they pertain to the ends or starts of series. This levelling of the estimates happens even in figures where the dispersion of the S-I ratios suggests that the seasonal evolution is continuing. In October, for instance, the 4 last estimates by the three by nine lie completely under the cloud of points: the 4 last S-I points are located well above the curve. The three by nine biases the evolving seasonal factors at the end of series.

On the other hand, the estimates of the proposed 9-term average also evolve at the ends and starts of series. This difference in the terminal behaviour of the two averages can be observed for the start of February (Figure 1 (b)), for the start and the end of March (c), for the end of May (e), for the starts of June (f), July (g) and August (h), for the end of September (j), for the end and the start of October (i) and for the end of December (l). The recorded evolution remains "conservative" however: In cases of rising seasonal movement, the estimates are located at the bottom of the S-I cloud of points (and not under as with the 3 by 9 which is too conservative). This conservatism protects against turning-points in the previously started movement. For instance, if in the years following the 1967-79 observation period, the seasonal factors of October display persistent decline, conservative estimates and extrapolation will not be as off-target as if the estimates had lain in the middle of the cloud of points. Conservative estimates also tend to minimize revisions to the estimates by making the revisions monotonic.

b) **Case of the three by five** - Figures 2 (a) to 2 (l) still depict the same S-I ratios of the Building Permits series. The curves, however, represent the seasonal factors estimated by the three by five of X-11 and by the 7-term seasonal moving average also presented in Section 3.

In comparing the relative performances of the averages, one must bear in mind that the series in question is very irregular and would have required longer averages. The same series was kept in order to allow comparisons between Figures 1 (a) to 2 (a), 1 (b) to 2 (b), etc.; that is between the estimates achieved by averages of different length. This comparison shows that - as desirable - the 9-term average yields smoother and more monotonic estimates than either the 7-term average or the three by five. In Figures 2 (c) and 2 (k), for instance, the central estimates reach deeper into the cloud of S-I ratios than in Figures 1 (c) and (k).

The estimates by the two averages sometimes differ at the ends or starts of months which display rapidly evolving seasonality. This situation prevails at the end of September (Figure 2 (i)), at the start of July, etc., where the 7-term average better captures the seasonal evolution. As for the 9-term, the evolving estimates by the 7-term average remain conservative with the above-described resulting advantages.

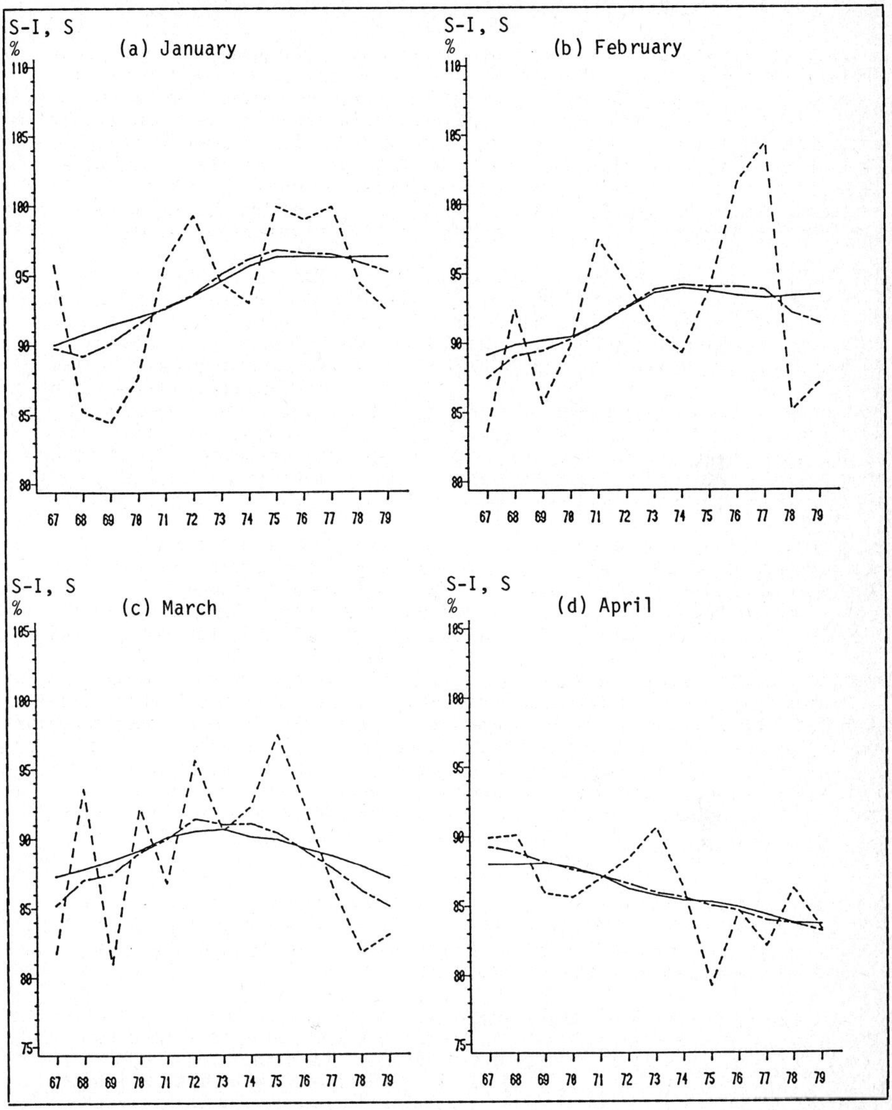

Figure 1: Seasonal-irregular S-I ratios (-----) and seasonal factors S estimated by the three by nine moving average (______) and by the 9-term seasonal moving average (-.-.-) presented in this paper

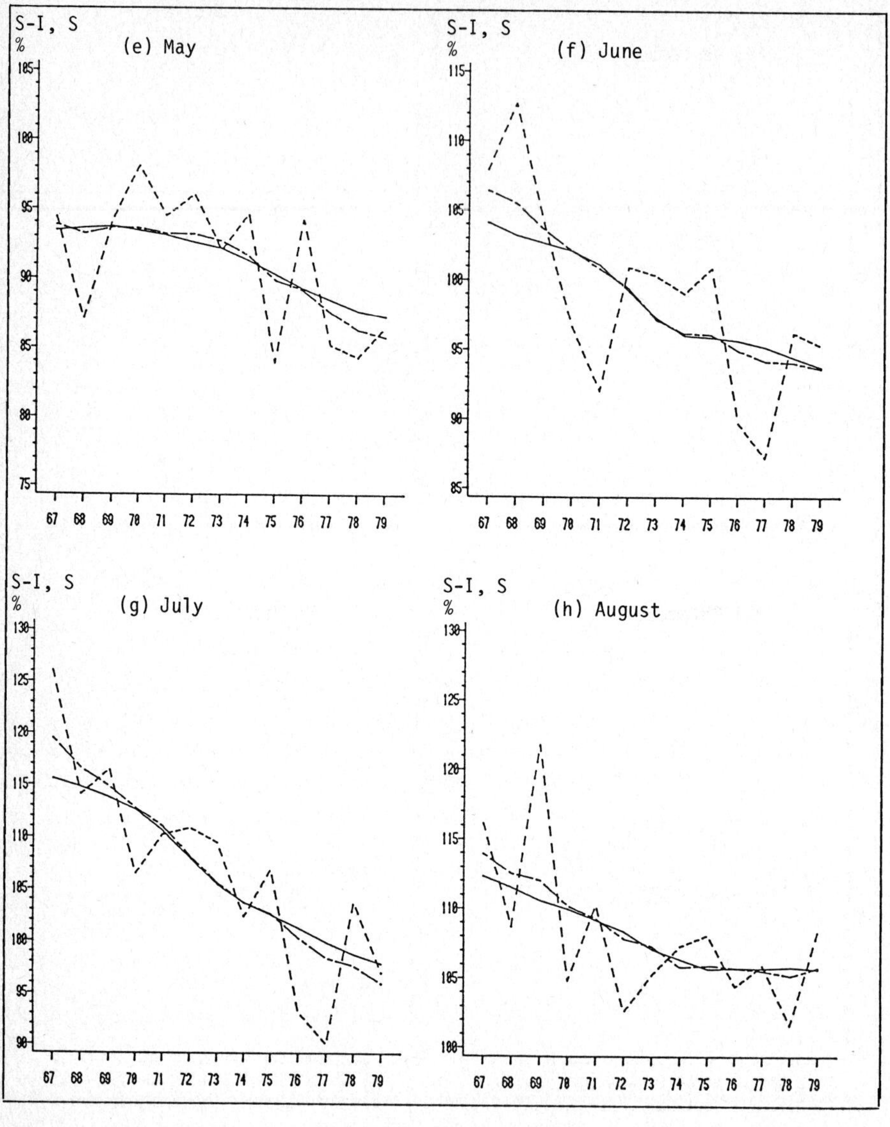

Figure 1 - continuation 1: Seasonal-irregular ratios S-I (-----) and seasonal factors S estimated by the three by nine moving average (____) and by the 9-term seasonal moving average (-.-.-) presented in this paper

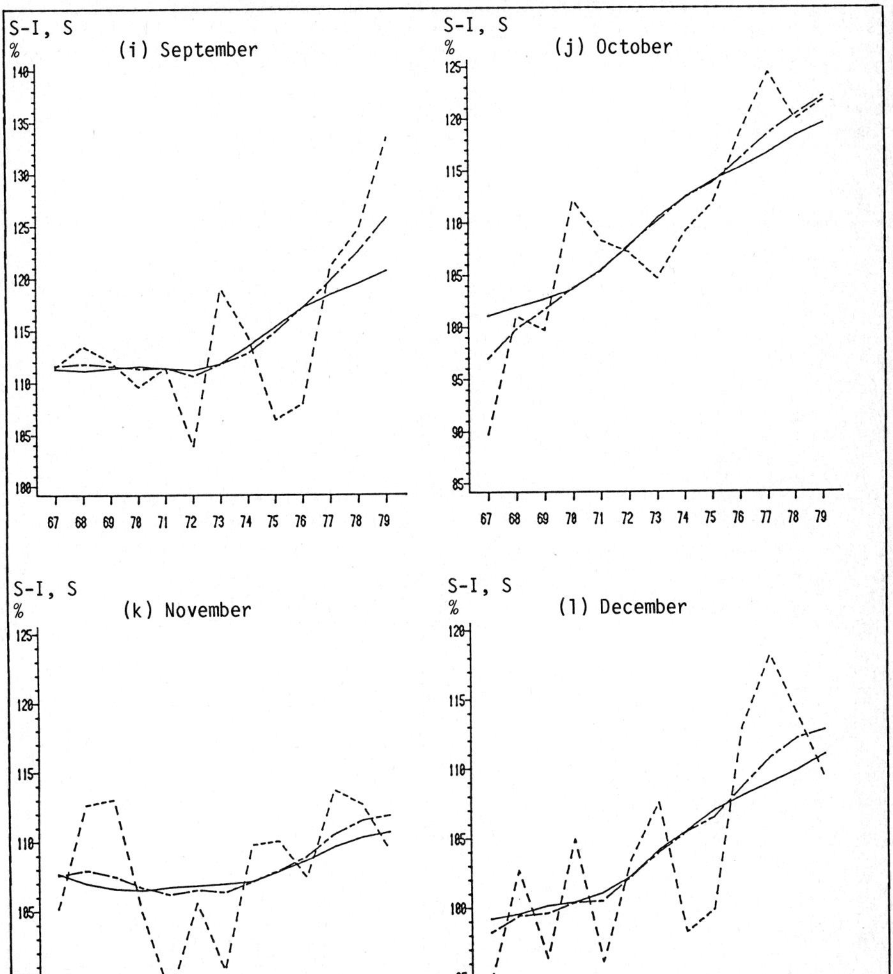

Figure 1 - continuation 2: Seasonal-irregular S-I ratios (-----) and seasonal factors S estimated by the three by nine moving average (_____) and by the 9-term seasonal moving average (-.-.-) presented in this paper

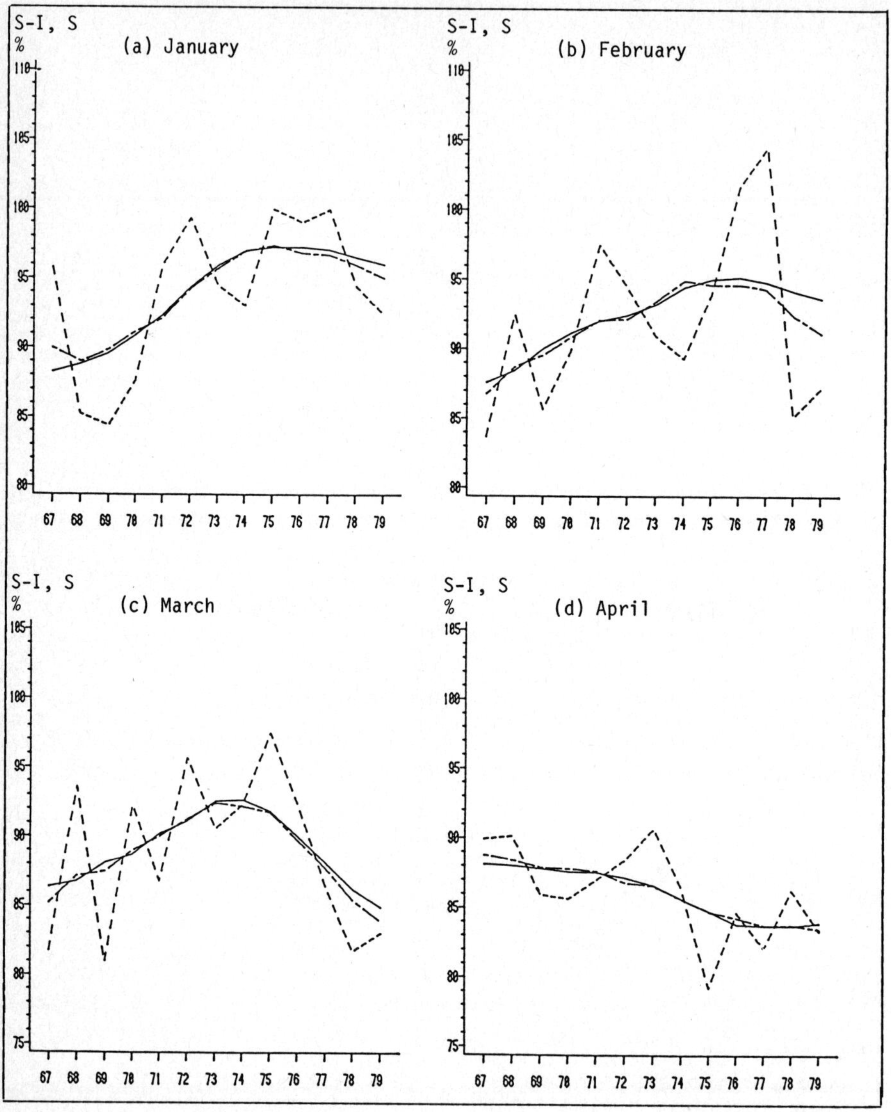

Figure 2: Seasonal-irregular S-I ratios (-----) and seasonal factors S estimated by the three by five moving average (______) and by the 7-term seasonal moving average (-.-.-) presented in this paper

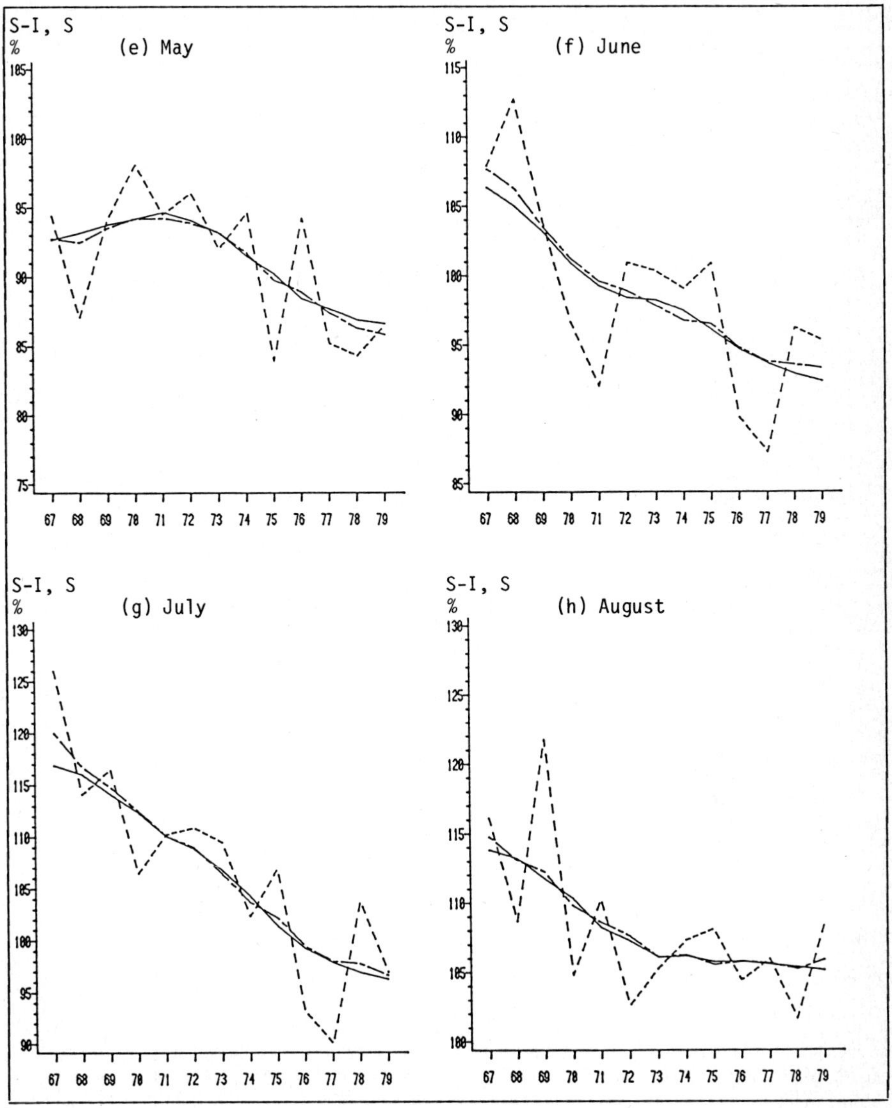

Figure 2 - continuation 1: Seasonal-irregular S-I ratios (-----) and seasonal factors S estimated by the three by five moving average (_____) and by the 7-term seasonal moving average (-.-.-) presented in this paper

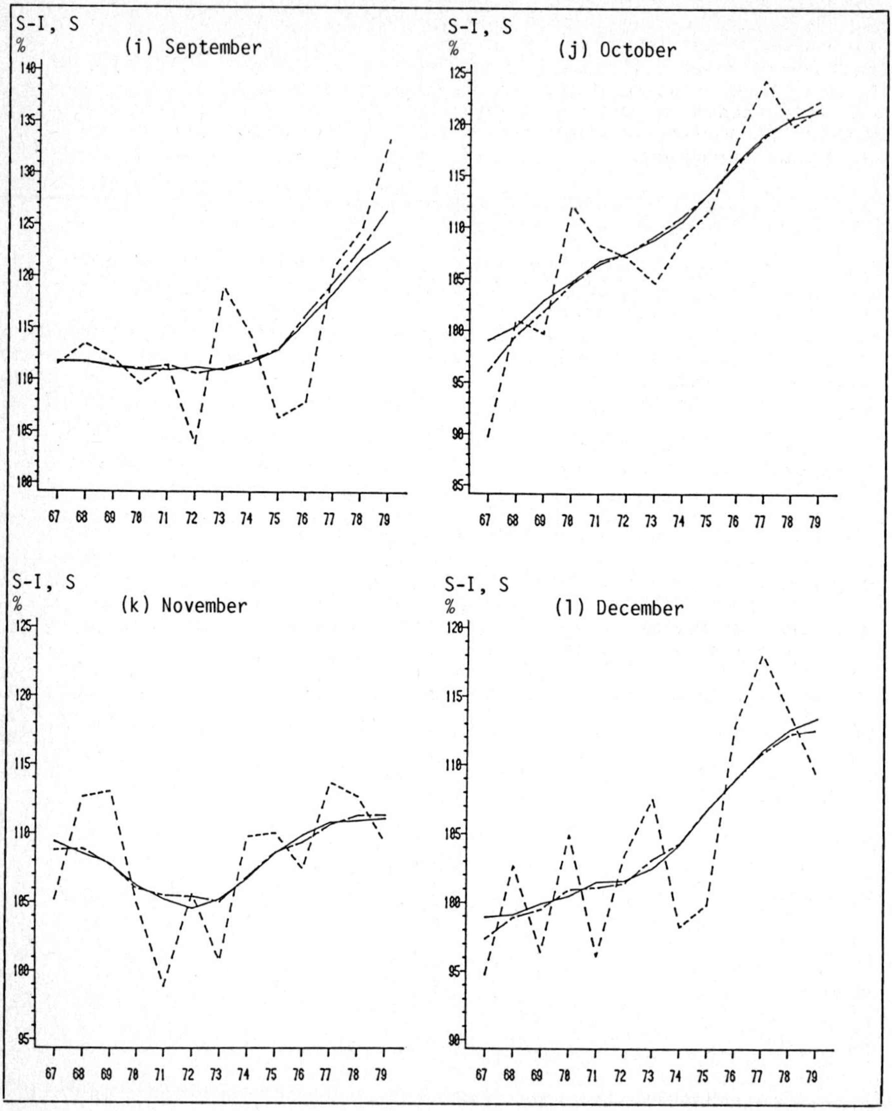

Figure 2 - continuation 2: Seasonal-irregular S-I ratios (-----) and seasonal factors S estimated by the three by five moving average (_____) and by the 7-term seasonal moving average (-.-.-) presented in this paper

However, the historical estimates from 1970 to 76, produced by the central weights of both averages, appear to be almost identical. Since most series are processed with the three by five, one could continue to use the central weights of three by five; but, substitute its non-central weights with those of the 7-term average. The resulting seasonally adjusted series would then remain the same in the centre of series - to the satisfaction of series users; but would be improved at the ends in situations of evolving seasonality (provided the series are not too irregular).

2. SPECTRAL ANALYSIS OF THE AVERAGES

Figure 3 displays the gain functions of the central and of the terminal (end) weights of the averages studied. (The gains of the other non-central weights lie between these two.) The end weights are used to estimate the seasonal factors of the last available year. Because the averages under consideration are applied to annual data (the S-I ratios), the following interpretation of the abscissa is required. Frequency 1/60 indicates a complete sine wave of 60 years; 2/60, of 30 years; 3/60, of 20 years; etc.

The three mentionned frequencies will be considered as the target seasonal frequencies which describe the evolution of the seasonal factor for a given month. A seasonal factor evolving at the rate of a full cycle per 20 years could change from being a seasonal peak month in the year, for instance, to being a trough month and then a peak month again in the 21st year.

The remaining frequencies, from 4/60 to 30/60, - especially the more rapid ones closer to 30/60 - will be associated with noise, that is with the irregular. An ideal seasonal moving average should then preserve the target frequencies and eliminate the noise frequencies.

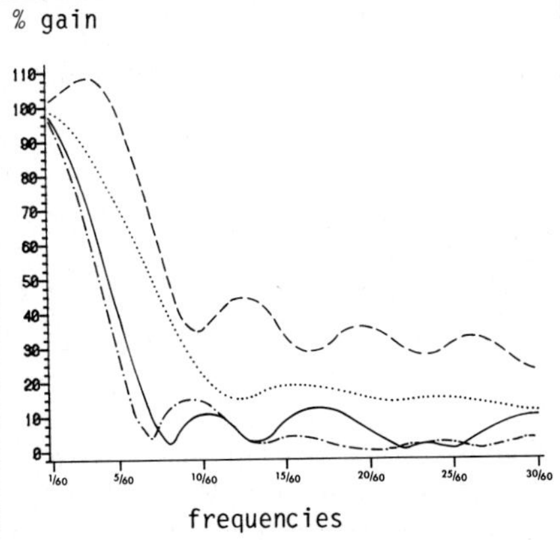

Figure 3: Gains of the central and end weights of the three by nine (-.-.- and .....) and of the 9-term (______ and -----) seasonal moving averages

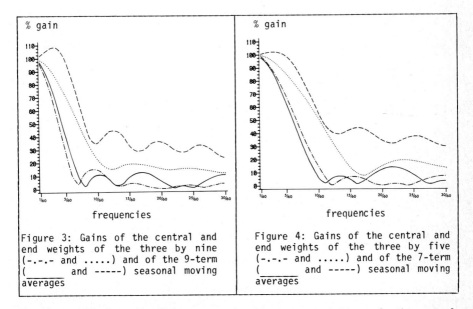

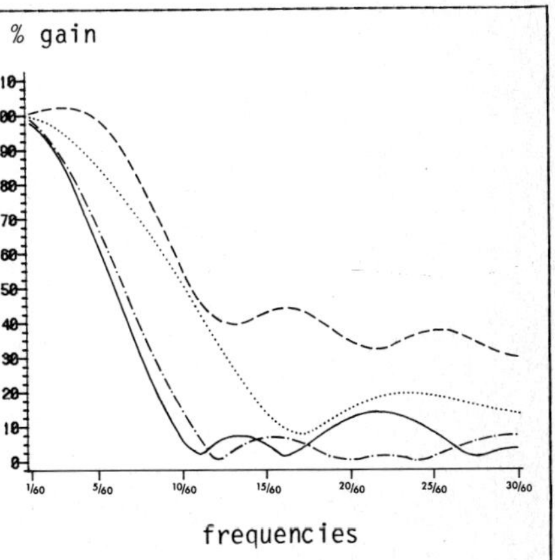

Figure 4: Gains of the central and end weights of the three by five (-.-.- and .....) and of the 7-term (______ and -----) seasonal moving averages

a) **Case of the three by nine** - According to the gains of Figure 3, the central weights of the three by nine and of the 9-term average proposed preserve close to 100% of 60-year waves; nearly, 75% of those of 30 years; and more or less the same proportions of the other waves. One concludes that the central properties of the two averages are almost identical (since the phase-shift of symmetric weights is nil).

The situation is quite different for the end weights. Both the end three by nine and 9-term averages better preserve the target seasonal frequences 1/60 to 3/60 than their respective central weights. However the end 9-term average eliminates less noise than the three by nine which has a lower gain at the noise frequencies. However, the phase-shifts recorded in Table 1 must also be considered.

At the target seasonal frequencies, the phase-shifts are much higher with the end three by nine than with the end 9-term average. Let's assume that a seasonal factor is evolving at the rate of one cycle per 60 years. Using the three by nine, the evolution of the estimated end seasonal factor will be delayed 1.8 years (with respect to the true evolution); against 0.6 years, with the 9-term average. For the first before-last estimate, the remaining delay will be 1.2 years against 0.3 with the 9-term; for the second before-last 0.7 against 0.2; and so forth. Although the target frequencies are more or less equally preserved by the averages, the phase-shifts greatly affect the results, as observed in the previous section.

Dagum (1982) measures the total revision inherent to an average by calculating the mean quadratic difference between the frequency response function of the end weights and that of the central weights at the target frequencies. This measure simultaneously takes into account the gain and the phase-shift for all the frequencies considered. This total "spectral revision" equals 13.3% for the three by nine against 7.9% for the end 9-term average. The latter consequently entails less revisions at the target seasonal frequencies.

TABLE 1: Phase-shifts recorded in number of years, at the target seasonal frequencies, by the various sets of weights of the three by nine and of the three by five and by the 9-term; and by the 7-term averages developped in this paper

	three by nine			9-term			three by five			7-term		
	frequencies											
set of weights	1/60	2/60	3/60	1/60	2/60	3/60	1/60	2/60	3/60	1/60	3/60	3/60
fifth before-last	-	-	-	-	-	-	-	-	-	-	-	-
fourth before-last	0.1	0.1	0.1	-	-	-	-	-	-	-	-	-
third before-last	0.3	0.3	0.3	0.1	0.1	0.0	-	-	-	-	-	-
second before-last	0.7	0.7	0.7	0.2	0.2	0.2	0.2	0.2	0.2	0.2	0.1	0.1
first before-last	1.2	1.2	1.2	0.3	0.4	0.5	0.6	0.6	0.6	0.3	0.3	0.3
last (end)	1.8	1.8	1.8	0.6	0.7	0.9	1.3	1.3	1.3	0.6	0.6	0.7

\- central weights

We would consequently think that the 9-term seasonal moving average presented in this paper is preferable in a situation of moving seasonality, provided the series is not too irregular. The amount of tolerable irregularity should be determined by the series subject matter experts based on experimental results.

b) **Case of the three by five** - Figure 4 displays the central and terminal (end) gains of the three by five and 7-term seasonal moving averages. Table 1 contains the respective phase-shifts recorded for the target seasonal frequencies. At these frequencies, the total spectral revision equals 7.5 % for the three by five against 3.7 % for the 7-term seasonal moving average.

In the case of the three by five, we were able to compute the total revision associated with the combined weights of the X-11 moving averages, as Dagum (1982) actually did. At the target seasonal frequencies, the revisions to the seasonally adjusted series read 31% for the 84th (last) estimate of X-11 against 15% for the X-11 with modified non-central three by five weights; 27% for the 83rd (first before-last) estimate, against 14%. Table 2 contains those results for estimates 73 to 84 and for the preliminary or current estimates 85 to 96. (We could not carry out the same calculations for the the three by nine, because the dimensions of the matrices to handle and store became too big: 132 by 132 against 84 by 84 for the three by five.)

In situations of moving seasonality, the 7-term seasonal moving average would then be preferable, provided the series does not contain too much irregularity. As a matter of fact, its end weights have a gain higher than those of the three by five at the irregular frequencies.

TABLE 2: Total revisions to the seasonally adjusted series at the target seasonal frequencies for X-11 and for X-11 with modified non-central three by five weights

	estimate												current estimate											
	73	74	75	76	77	78	79	80	81	82	83	84	85	86	87	88	89	90	91	92	93	94	95	96
X-11	22	22	22	23	23	23	25	25	25	25	27	31	53	53	53	54	54	54	57	57	56	57	59	64
modified X-11	13	14	14	14	14	14	14	14	14	14	14	15	37	38	39	39	39	38	39	38	38	38	38	37

3. CONSTRUCTING THE AVERAGES

a) **The 9-term seasonal moving average** - On one series interval considered, the observed seasonal-irregular ratios (or differences) z_t contain the seasonal s_t and the irregular e_t components to be estimated:

$$z_t = s_t + e_t, \quad t=1,\ldots,T, \tag{1}$$

where T equals 9 (for the 9-term average) and t stands for years.

Following the approach by Leser (1963), Cholette (1981) and Schlicht (1981), the desired seasonal component obeys certain criteria on the series interval. As indicated by the first right hand term of equation (2), the component minimizes the quadratic sum of first differences in consecutive years. This criterion means that the estimates are required to change as little as possible from year to year.

The seasonal component also minimizes the sum of squared second differences taken in consecutive years (second term of (2)), taken in every second year (third term of 2)) and in every third year (fourth term). These three extra criteria indicate that if the estimates do vary (despite the first criterion), the movement should be as smooth and monotonic as possible.

Finally the fitting criterion (fifth term of (2)) consists of minimizing the quadratic sum of the residuals, that is the differences between each observed S-I ratio z_t and its estimate s_t.

Taken together, the five criteria mean that the seasonal factor being sought should behave like a straight line as much as possible, but a horizontal straight line. This specification generates a monotonic (failing linearity) and "conservative" movement in the sense explained at the end of Section 2 a). The five criteria sum into the following objective function to be minimized:

$$g(s) = \sum_{t=2}^{9} (s_t - s_{t-1})^2 + \sum_{t=3}^{9} (s_t - 2s_{t-1} + s_{t-2})^2$$
$$+ \sum_{t=5}^{9} (s_t - 2s_{t-2} - s_{t-4})^2 + \sum_{t=7}^{9} (s_t - 2s_{t-3} + s_{t-6})^2 + \sum_{t=1}^{9} (s_t - z_t)^2. \qquad (2)$$

Matrix operator D_m^k of differences of order k taken on every m-th observation (k,m>0), described in Appendix, allows expressing equation (2) in matrix algebra

$$G(S) = S'D'D\,S + S'D^{2\prime}D^2\,S + S'D_2^{2\prime}D_2^2 S + S'D_3^{2\prime}D_3^2 S + (S - Z)'F'F\,(S - Z) \qquad (3)$$

$$= S'A\,S + (S - Z)'F'F\,(S - Z), \qquad (3')$$

where F stands for the identity matrix in this particular case. The normal equations corresponding to the minimization of (3) are

$$dG/dS = 2\,(A + F'F)\,S - 2\,F'F\,Z = 0. \qquad (4)$$

where $A = D'D + D^{2\prime}D^2 + D_2^{2\prime}D_2^2 + D_3^{2\prime}D_3^2$ stands for the behavioural criteria of the seasonal component and F'F for the fitting criterion. This yields solution:

$$S = (A + F'F)^{-1}\,F'F\,Z = W\,Z = \underset{T\times T}{[w_{t,i}]}\,\underset{T\times 1}{[z_i]} \qquad (5)$$

$$\Rightarrow \quad s_t = \sum_{i=1}^{T} w_{t,i}\,z_i, \quad t=1,\ldots,T.$$

Weights $w_{5,i}$ (i=1,...,9) in the middle fifth row of matrix W constitute the set of central weights of the 9-term seasonal moving average; $w_{9,i}$ in the last row, the set of end weights. These weights are found in Table 3.

We tried specification (2) on an 11- instead of a 9-year interval, with and without extra second differences taken on every fourth observation. The results did not prove any better. We kept the 9-year formulation, especially since the end weights of the three by nine only have six non-zero weights (see Dagum, 1980, Appendix B and Shiskin et al., p. 61).

The approach of quadratic minimization allows refinements from which we refrained in this paper. For instance, one could attribute less weight to the fitting of the end observations by appropriately defining diagonal matrix F of (3) (weighted least squares). One could also make the centre estimates more flexible by having

TABLE 3: Sets of weights of the 9-term seasonal moving average presented in this paper

set of weights	1 6	2 7	3 8	4 9	5
central	0.0770902 0.1198115	0.1039815 0.1160586	0.1160586 0.1039815	0.1198115 0.0770902	0.1661164
third before-last	0.0364127 0.1758711	0.0543448 0.1449283	0.0822790 0.1416822	0.1007450 0.1439253	0.1198115
sedond before-last	-0.0170832 0.1449283	0.0151792 0.2256785	0.0462253 0.1975525	0.0822790 0.1891817	0.1160586
first before last	-0.0515000 0.1416822	-0.0266116 0.1975525	0.0151792 0.3081376	0.0543448 0.2572338	0.1039815
last (end)	-0.0730059 0.1439253	-0.0515000 0.1891817	-0.0170832 0.2572338	0.0364127 0.4377453	0.0770902

TABLE 4: Sets of weights of the 7-term seasonal moving average presented in this paper

set of weights	1 6	2 7	3	4	5
central	0.0986301 0.1315068	0.1315068 0.0986301	0.1598174	0.2200913	0.1598174
second before-last	0.0248503 0.2023129	0.0716597 0.1806292	0.1225764	0.1598174	0.2381542
first before-last	-0.0258691 0.3306670	0.0090590 0.2806636	0.0716597	0.1315068	0.2023129
last (end)	-0.0583465 0.2806636	-0.0258691 0.4994424	0.0248503	0.0986301	0.1806292

them pass in the middle of the centre S-I points in Figures 1 (c) and (k) - while keeping them conservative at the ends.

b) **The 7-term seasonal moving average** - The objective function of the 7-term seasonal moving average reads:

$$g(s) = \sum_{t=2}^{7} (s_t - s_{t-1})^2 + \sum_{t=3}^{7} (s_t - 2s_{t-1} + s_{t-2})^2 + \sum_{t=5}^{7} (s_t - 2s_{t-2} + s_{t-4})^2 + \sum_{t=1}^{7} (z_t - s_t)^2. \quad (6)$$

The only difference between this equation and equation (2) relative to the 9-term average is the absence of second differences taken on third observations (4-th term of (2)). It was possible to include the term, but its presence deteriorated the results, perhaps because the second difference under consideration can be taken only once ($s_7-2s_4+s_1$) on the interval. The weights of 7-term average are found in Table 4.

4. DISCUSSION

The X-11 (Shiskin, Young and Musgrave, 1967) and X-11-ARIMA (Dagum, 1982) are the seasonal adjustment methods most widely used by statistical agencies. In the context of these methods, the following attitude prevails regarding the presence of moving seasonality in a series.

On the one hand, if the series is regular enough, the use of shorter moving averages is recommended, namely the three by three. Indeed, this average better reproduces evolving seasonality than the three by five and the three by nine, which are designed for more stable seasonal patterns. (Using the ARIMA option of the X-11-ARIMA method will further improve the end estimation of moving seasonality in regular series (Dagum, 1975)).

On the other hand however, if the series is not regular enough, the only other alternative unfortunately consists of ignoring the moving seasonal pattern and using longer averages such as the three by five and three by nine. This generates bias and phase-shift in the movement of the end seasonal estimates, as illustrated in Section 1.

We claim that it is not necessary to confine oneself in such a dilemma. In situations of moving seasonality, one can also resort to long averages which are appropriately designed, such as the 9-term and the 7-term seasonal moving averages presented in this paper. It would also be possible to develop even longer averages which would reproduce moving seasonality. An analogy is provided by many econometric textbooks. When fitting a straight line to a cloud of points, one might have to switch to estimating a constant if the data are too irregular (scattered). In the context of moving averages, however, one could also fit the straight line to more observations by making the estimation interval longer.

The proposed weights could be critized because their sums of squares are larger than those of the traditional weights: For instance, the end weights of the 9-term moving average have a quadratic sum of 0.330 against 0.193 for the three by nine (difference of 13.7%). Consequently the smoothing capacity of the former is lower. However, the smoothing capacity is only one (rough) criterion by which to assess weights. Reverting to the example of introductory econometrics, if the cloud of points does display linear evolution, one should fit a straight line (if the slope is significant) despite the higher sum of squares of the corresponding implicit weights ($X[X'X]^{-1}X'$). Indeed it can be proven that the more movement weights can reproduce, the higher their sum of squares.

APPENDIX A: Matrix Difference Operators

The matrix operator D_m of first differences taken on every m-th observation of a T-observation series is:

column m+1

$$\underset{(T-m)\times T}{D_m} = \begin{bmatrix} 1 & 0 & 0 & \dots & -1 & 0 & 0 & \dots \\ 0 & 1 & 0 & \dots & 0 & -1 & 0 & \dots \\ 0 & 0 & 1 & \dots & 0 & 0 & 1 & \dots \\ \cdot & \cdot & \cdot & & \cdot & \cdot & \cdot & \\ \cdot & \cdot & \cdot & & \cdot & \cdot & \cdot & \\ \cdot & \cdot & \cdot & & \cdot & \cdot & \cdot & \end{bmatrix}.$$

The matrix D_m^2 of second differences taken on every m-th observation reads:

$$\underset{(T-2m)\times T}{D_m^2} = \underset{(T-2m)\times(T-m)}{D_m} \quad \underset{(T-m)\times T}{D_m} = \begin{bmatrix} -1 & 0 & \cdots & \overset{\overset{\text{column } m+1}{\downarrow}}{2} & 0 & \cdots & \overset{\overset{2m+1}{\downarrow}}{-1} & 0 \\ 0 & -1 & \cdots & 0 & 2 & \cdots & 0 & -1 \\ \cdot & \cdot & & \cdot & \cdot & & \cdot & \cdot \\ \cdot & \cdot & & \cdot & \cdot & & \cdot & \cdot \\ \cdot & \cdot & & \cdot & \cdot & & \cdot & \cdot \end{bmatrix}.$$

APPENDIX B: Sets of weights of the three by nine in Shiskin et al. (1967)

set of weights	1 8	2 0	3 10	4 11	5	6	7
	0.037	0.074	0.111	0.111	0.111	0.111	0.111
central	0.111	0.111	0.074	0.037			
fourth	0.0	0.034	0.073	0.111	0.113	0.114	0.116
before-last	0.117	0.118	0.120	0.084			
third	0.0	0.0	0.034	0.075	0.113	0.117	0.123
before-last	0.128	0.132	0.137	0.141			
second	0.0	0.0	0.0	0.032	0.079	0.123	0.133
before-last	0.143	0.154	0.163	0.173			
first	0.0	0.0	0.0	0.0	0.028	0.092	0.144
before-last	0.160	0.176	0.192	0.208			
	0.0	0.0	0.0	0.0	0.0	0.051	0.112
last (end)	0.173	0.197	0.221	0.246			

APPENDIX C: Sets of weights of the three by five in Shiskin et al. (1967)

set of weights	1	2	3	4	5	6	7
central	0.067	0.133	0.200	0.200	0.200	0.133	0.067
second before-last	0.0	0.067	0.133	0.217	0.217	0.217	0.150
first before-last	0.0	0.0	0.067	0.183	0.250	0.250	0.250
last (end)	0.0	0.0	0.0	0.150	0.283	0.283	0.283

REFERENCES

AKAIKE, H. (1980). Seasonal Adjustment by a Bayesian Modelling. Journal of Time Series Analysis 1, 1-14.

CHOLETTE, P.A. (1981). A Comparison of Various Trend-Cycle Estimators. In Time Series Analysis. (Proceedings of the International Conference held at Houston, Texas, August 1980) Eds: O.D. Anderson and M.R. Perryman, North Holland, 77-87.

DAGUM, E.B. (1975). Seasonal Factor Forecasts from ARIMA Models. Proceedings of International Statistical Institute, 19th Session, Vol. 3, 206-219

DAGUM, E.B. (1980). The X-11-ARIMA Seasonal Adjustment Method. Statistics Canada, cat. 12-564E.

DAGUM, E.B. (1982a).The Effects of Asymmetric Filters on Seasonal Factor Revision. J.A.S.A., 77, 732-738.

DAGUM, E.B. (1982b). Revisions of Seasonally Adjusted Data due to Filter Changes. Invited Paper, Business and Economics Statistics Section, American Statistical Association, Ohio August 1982; Statistics Canada, Time Series Research and Analysis, Research Paper 82-07-002E.

EHRENBERG, A.S.C. (1982). Writing Technical Papers or Reports. American Statistician, 36, 326-329.

KOOPMANS, L.H. (1974). The Spectral Analysis of Time Series. Academic Press, New York.

LAROQUE, G. (1977). Analyse d'une méthode de désaisonnalisation: le programme X-11 du U.S. Bureau of the Census, version trimestrielle. Annales de l'I.N.S.E.E, 28, 105-127.

LESER, C.E.V. (1961). A Simple Method of Trend Construction. J.R.S.S. B, 23, 91-107.

LESER, C.E.V. (1963). Estimation of Quasi-Linear Trend and Seasonal Variation. J.A.S.A., 58, 1033-1043.

MACAULY, F.R. (1931). The Smoothing of Time Series. National Bureau of Economic Research, Washington.

PHLIPS. L., BLOMME, R. (1973). Analyse chronologique. Éd. Vander, Louvain.

SCHLICHT, E. (1981). A Seasonal Adjustment Principle and a Seasonal Adjustment Method Derived from this Principle. J.A.S.A., 76, 374-378.

SHISKIN, J., YOUNG, A.H, MUSGRAVE, J.C (1967). The X-11 Variant of the Census Method II Seasonal Adjustment Program. U.S. Bureau of the Census,Technical Paper No. 15.

WHITAKKER, E. (1923). On a New Method of Graduation, Proceedings of Edinburgh Mathematical Society. 41, 63-75.

TIME SERIES ANALYSIS: Theory and Practice 7
O.D. Anderson (editor)

ILLUSTRATION OF THE USE OF A GENERAL TIME SERIES MODEL

Guy Mélard
Institut de Statistique et Centre d'Economie mathématique et d'Econométrie
Université Libre de Bruxelles, Campus Plaine - CP 210, Boulevard du Triomphe
B-1050 Bruxelles, Belgium

In a recent paper, the author described a generalized univariate ARIMA model which allows time-dependent coefficients and innovation variance, deterministic trend and seasonal components, and power transformation. It subsumes models underlying various well-known forecasting schemes and permits interventions on the mean and variance. The present paper is devoted to illustrating this model with the Chatfield-Prothero data.

1. INTRODUCTION

Mélard (1982a) (referred to as M), describes a general approach to time series analysis and the corresponding software, ANSECH. The philosophy of Box and Jenkins (1976) is applied to a wider class of univariate models than the ARIMA class, called generalized ARIMA models. These result from the combination of a number of model components : autoregressive (AR) and moving average (MA) polynomials, regular and seasonal difference operators, power transformation, intervention on the variable, multiplicative and additive trends, additive seasonal, trend of the innovation variance, intervention on the innovation, and some others. In addition to the usual constant AR and MA polynomials, polynomials with time-dependent coefficients are allowed. Models underlying well-known forecasting schemes can also be represented in combination with other model components.

The parameters of a generalized ARIMA model can be estimated simultaneously. The following methods are available in the software : exact maximum likelihood (using a fast algorithm for the computation of the likelihood function of models with constant AR and MA polynomials, and another algorithm for time-dependent models), conditional maximum likelihood and an approximate unconditional method using a back-forecasting procedure.

Details of the generalized ARIMA model, and related concepts, and the methods of estimation can be found in M. In the present paper we will illustrate the use of this model by means of the Chatfield and Prothero (1973a) data, "Sales of Company X".

Section 2 serves two objects. First the best models of previous analyses are revisited. Second, our strategy is described and applied, which leads to the rejection of these models. The generalized ARIMA model should provide a solution. Its model components are briefly presented in section 3. The identification stage is the subject of section 4. An analysis featuring a multiplicative trend is performed in section 5; followed, in section 6, by another analysis which features a power transformation. In section 7, our strategy is validated ex post by the results of an extensive study of the Chatfield-Prothero data. The next three sections aim at illustrating some aspects not covered previously : a more refined submodel of the seasonal component, a model which combines a multiplicative trend with a power transformation, a model with time-dependent coefficients, a model with a deterministic additive trend, the relationship with a well-known predictor, and submodels with intervention on the innovation. In section 11 we give our conclusions. In short, time series modelling is not an easy task if we wish really to represent the data and not merely mimic the autocorrelation function.

Although a single time series is considered, the conclusion is supported by the analysis of many other time series. The additional model components shown in M and illustrated in the present paper can be helpful. Forecasts for some models are given in the appendix and compared with fresh data given by Chatfield and Prothero (1973b). The models which fit the data best are always those which provide the worst forecasts, and conversely.

2. MOTIVATIONS FOR A FURTHER ANALYSIS - STATEMENT OF A STRATEGY

Chatfield and Prothero (1973a) analyzed the monthly "Sales of Company X" (77 observations); and because of the high multiplicative seasonal variation, they worked with logarithms. Let z_t be the observed sales ($t = 1, \ldots, 77$). Then their fitted model was

$$(1 + 0.47B)\ \nabla\nabla_{12} \log z_t = (1 - 0.81B^{12})a_t$$

where B is the backshift operator, $\nabla = 1 - B$, $\nabla_{12} = 1 - B^{12}$, and the a_t are the (independent) innovations, with mean zero and variance σ^2. Using ANSECH, we obtained the following models :

<u>method with a back-forecasting procedure</u>

$$(1 + 0.45B)\ \nabla\nabla_{12} \log z_t = (1 - 0.81B^{12})\ a_t$$

<u>method based on the exact likelihood function, assuming normality of the a_t</u>

$$(1 + 0.45B)\ \nabla\nabla_{12} \log z_t = (1 - 0.73B^{12})\ a_t.$$

In their comments on Chatfield and Prothero (1973a), Box and Jenkins (1973) suggested the model

$$(1 - \phi_1 B)\ \nabla\nabla_{12}\ C_\lambda(z_t) = (1 - \Theta_1 B^{12})\ a_t$$

where $C_\lambda(z_t) \propto z_t^\lambda$, $\phi_1 = -0.5$, $\Theta_1 = 0.8$ and $\lambda = 0.25$. By using an exact maximum likelihood method, we obtained the estimates $\phi_1 = -0.46$, $\Theta_1 = 1.00$ and $\lambda = 0.23$. This model can be criticized on several counts : (a) it is borderline noninvertible since $\Theta_1 = 1$; (b) there is a highly significant residual autocorrelation at lag 11; (c) an outlier at time 74 is revealed by a residual value of -89.9, as compared to the residual standard deviation, $\hat{\sigma}$, of 34.1.

The debate about the most suitable transformation, Chatfield and Prothero (1973a, b), Box and Jenkins (1973), Tunnicliffe Wilson (1973), and about the general form of the model, Harrison (1973), Priestley (1973), is the reason for our interest in these data. Mélard (1977) has obtained a fairly good model of the form

$$(1 - \phi_1 B)\ \nabla\nabla_{12} z_t = (1 - \Theta_1 B^{12})(g_t a_t),$$

where the innovation $g_t a_t$ has a standard deviation $g_t \sigma$, and g_t is an exponential function of time. Methods of estimation, diagnostic checks and criteria for selection have been improved, however, and this model is no longer retained.

It can be supposed that better models would be obtained within the wider class of generalized ARIMA models. We need first to state our model building strategy, which is as follows. The models are generally estimated by an exact maximum likelihood method, under the assumption of normality. In any case, the problem is reduced to the minimization of a sum of squares Σ. The standard errors of the estimates are calculated by using the second partial derivatives of the log-likelihood; and, if it is necessary, by M, equation 8. The following diagnostic checks are used :

(a) The Ljung-Box test based on residual autocorrelation of orders 1 to 24; the probability of significance, calculated from the chi-squared distribution with the appropriate number of degrees of freedom, is denoted by P_{LB}; a model is rejected if $P_{LB} < 5$ %.

(b) Residual autocorrelation and partial autocorrelation coefficients : they are especially taken into account if their absolute values exceed 2.576 divided by the square root of the length of the series, which corresponds to a significance level P of 1 %.

(c) Residual outliers : they are detected if their absolute value exceeds 2.576 times the residual standard deviation $\hat{\sigma}$. Because such outliers are able to distort the fitted model, they are treated by an intervention submodel component, to be described in the next section.

(d) Four homogeneity tests, Mélard (1979), which are used in order to examine the distributional properties of the residuals with respect to time. These tests yield significance probabilities denoted by P_{FS}, P_{ME}, P_{LH} and P_{BA}. A model is rejected if any of these probabilities is lower than 10 %.

Since homogeneity tests are unfortunately rarely used in time series analysis and forecasting, we will briefly describe them and show how their use causes the Chatfield-Prothero model (1) to be rejected.

In large samples, and under the assumption of normality, about half of the residuals are expected to belong to the interval $0 \pm 0.668\hat{\sigma}$. In each of the four homogeneity tests, a 2x2 contingency table is formed by considering two criteria : 1° a residual belongs or does not belong to be 50 % probability interval $0 \pm 0.6680\hat{\sigma}$; 2° the corresponding data point has or does not have a given attribute. This attribute differs for each test, as shown in Table 1.

Table 1. Second criterion for the homogeneity tests

Name of the test	Attribute : the data point ...
FS (first-second)	... belongs to the first half of the time interval vs. to the second half
ME (middle-ends)	... belongs to the two middle quarters (vs. the two ending quarters) of the time interval
LH (low-high)	... belongs to a month with a low (vs. a high) seasonal factor
BA (below-above)	... lies below (vs. above) the median of the data values

Table 2. (a) the residuals of model (1) and (b) the corresponding data values

Time	(a)	(b)	Time	(a)	(b)	Time	(a)	(b)	Time	(a)	(b)
14	-9	118	30	-6	99	46	0	560	62	-8	392
15	-4	90	31	-55	135	47	1	612	63	-41	273
16	52	79	32	13	211	48	34	467	64	98	322
17	80	78	33	33	335	49	40	518	65	-54	189
18	-39	91	34	26	460	50	69	404	66	-5	257
19	-4	167	35	6	488	51	-7	300	67	-19	324
20	-97	169	36	-1	326	52	-24	210	68	-38	404
21	12	289	37	28	346	53	-1	196	69	18	677
22	9	347	38	84	261	54	-56	186	70	0	858
23	-7	375	39	34	224	55	-64	247	71	-11	895
24	-80	203	40	-49	141	56	-10	343	72	-2	664
25	24	223	41	21	148	57	-29	464	73	-18	628
26	-31	104	42	-53	145	58	16	680	74	-75	308
27	40	107	43	-39	223	59	7	711	75	39	324
28	36	85	44	-36	272	60	46	610	76	13	248
29	9	75	45	3	445	61	17	613	77	62	272

Table 3. Contingency table corresponding to the BA test on model (1)

data point	residual inside 50 % probability interval	residual outside 50 % probability interval
below median	12	20
above median	21	11

Let us consider the BA test on model (1). The 64 residuals and the corresponding data values are reproduced in Table 2. Given that the median lies between 273 and

289 and $\hat{\sigma}$ = 40.36, the contingency table of Table 3 is obtained. The chi-squared statistic is 5.125, with 2 degrees of freedom. Hence the approximate significance is P_{BA} = 0.07. We may conclude that there is a tendency for points below (above) the median to give rise to residuals lying outside (inside) the 50 % normal probability interval. This is a formal confirmation of a remark already made by Box and Jenkins (1973) that the logarithmic transformation is too strong for this series. The LH test leads to the approximate significance P_{LH} = 0.07. Hence the quality of the representation of the seasonal effect is also questioned.

The previously mentioned diagnostic checks can help us to weed out unsatisfactory models and to indicate what alteration in the model specification would improve the goodness of fit. At the end, there will be several models from which a single one has to be selected according to a suitable criterion. Although we include the residual variance among the model fitting statistics, our criterion will be to minimize AIC given by Ozaki (1977)

$$\mathrm{AIC} = n \left\{ \log \frac{\Sigma}{n - \ell} + \frac{2(m + 1)}{n - \ell} + \log(2\pi) + 1 \right\} \qquad (2)$$

where n is the total number of observations, ℓ is the number of these lost by differencing, m is the total number of parameters, except σ^2, and Σ is the minimized sum of squares.

The output of ANSECH corresponding to model (1) is shown in Fig. 1. The encircled numbers refer to the following explanations :

① Acknowledgment of the command phrase composed of several commands :

ESTIM	provides estimation by the maximum likelihood method
CHECK=OVAPRTHFL	controls diagnostic checks which must be applied (e.g. letter L for the Ljung-Box test)
LEAD=6	specifies the lead time of the forecasts
BOXCOX	produces the logarithm transformation with the Box-Cox normalization, i.e. multiplication by the geometric mean
DIFF	first-order regular differencing
DIFFS	first-order seasonal differencing

② Acknowledgment of the model parameters and their starting values : AR 1 corresponds to ϕ_1 and MAS 1, to Θ_1

③ Estimation method, iterations and stopping rule. The final model is given by (1)

④ Asymptotic correlation matrix of the estimates

⑤ Standard errors of the estimates, t-values and 95 % confidence limits

⑥ Computer time for the estimation procedure, on the Control Data Cyber 170-750 of the Brussels Free University, which corresponds to a cost of less than US $ 0.05

Figure 1. Computer output from ANSECH, corresponding to model (1)

```
ESTIM, CHECK=OVAPRTHFL, LEAD=6, BOXCOX, DIFF, DIFFS

  /\/\/\/\/\/\/\/\/\/\/\/\/\/\/\/\/\/\/\/\/\/\                          (1)
ANSECH 1.7, AUTHOR: G. MELARD. DATE:  83/07/29. ,  16.14.06. PROBLEM(  2): SALESX+
TITLE:  SALES OF COMPANY X.  JANUARY 1965-NOV.  1971 (CHATFIELD-PROTHERO)        SALESX+
THE CURRENT SERIES IS THE SAME AS BEFORE                            , LENGTH   83
  /\/\/\/\/\/\/\/\/\/\/\
WARNING *** MODEL FITTING IS PERFORMED WITH ONLY  77 DATA, ENDING
AT TIME       77.  6 FRESH DATA ARE RESERVED FOR EX-POST VALIDATION
 2 PARAMETERS WITH STARTING VALUES :                                     (2)
    1      AR    1 -.47000
    2      MAS   1  .81000
=== ESTIMATION BY MAXIMIZATION OF THE EXACT (LOG)LIKELIHOOD
      (FAST ALGORITHM WITH TOLERANCE  1.0E-05)
NON-LINEAR ESTIMATION:
ITER SUM OF SQ AR    1     MAS  1
 0    1166.3E+02 -.470        .810
 1    1163.4E+02 -.443        .746
 2    1163.3E+02 -.450        .737
 3    1163.3E+02 -.451        .733                                        (3)
 4    1163.3E+02 -.452        .731
 5    1163.3E+02 -.452        .730
 6    1163.3E+02 -.452        .730
ITERATION STOPS - RELATIVE CHANGE IN EACH COEFFICIENT LESS THAN   1.0000E-03
CORRELATION MATRIX
       AR    1 MAS   1                                                    (4)
AR   1    1.00
MAS  1     .33   1.00
FINAL VALUES OF THE PARAMETERS                            WITH 95% CONFIDENCE LIMITS
               NAME      VALUE           STD ERROR   T-VALUE  LOWER          UPPER
     1         AR   1   -.45235          .12276       -3.7     -.70          -.21
     2         MAS  1    .72960          .21399        3.4      .30          1.2
ESTIMATION HAS TAKEN   .6 SEC.  FOR  19 EVALUATIONS OF S.S.  (MEAN TIME=,   .033)   (5)

WARNING*** A MEAN LEVEL IS NOT INCLUDED IN THE MODEL                      (6)
THE FOLLOWING CONSTANTS ARE DEPENDENT PARAMETERS OF THE MODEL
         BOXCOX     236.57
THE FOLLOWING CONSTANTS WERE INVOLVED IN THE LEAST SQUARES ESTIMATION METHOD   (7)
         ARMA        .93126
=== SUMMARY MEASURES  <V>               EXACT AIC =                   803.641
COMPUTED SUM OF SQUARES =  116326.      ADJUSTED SUM OF SQUARES =   100884.
(BIASED) VARIANCE       =  1576.31      (UNBIASED) VARIANCE     =   1627.15     (8)
                                         STANDARD  DEVIATION    =   40.3380
=== RESIDUAL ANALYSIS WITH  64 RESIDUALS, BEGINNING AT TIME        14===
MEAN = -.816421E-02 , T-STATISTIC =    -.00          (FOR TESTING ZERO MEAN)

TABLE AGAINST TIME <RT>
       1   2   3   4   5   6   7   8   9  10  11  12
 12       -9  -4  52  80 -39  -4 -97  12   9  -7 -80
 24   24 -31  40  36   9  -6 -55  13  33  26   6  -1
 36   28  84  34 -49  21 -53 -39 -36   3   0   1  34                      (9)
 48   40  69  -7 -24  -1 -56 -64 -10 -29  16   7  46
 60   17  -8 -41  98 -54  -5 -19 -38  18  -0 -11  -2
 72  -18 -75  39  13  62
OUTLIERS <R(OR)S>                                                         (10)
  1 - 5 %     17:   79.68     20: -96.54     24: -80.28     38:  84.33     64:  98.42

TABLE OF RESIDUAL AUTOCORRELATIONS    (ASYMPT. STANDARD ERROR =   .13) <AT>
        1     2     3     4     5     6     7     8     9    10    11    12
 0    0.04  0.07 -0.17 -0.10 -0.11 -0.05 -0.27 -0.10 -0.07 -0.05  0.39  0.13   (11)
12    0.07 -0.12 -0.10  0.04 -0.05 -0.14 -0.03  0.02  0.04  0.11  0.07 -0.10
24   -0.02 -0.02
SIGNIFICANT    AUTOCORRELATIONS USING BARTLETT'S LIMITS <A(OR)S>
 .2 - 1 %       11:   .3919
  1 - 5 %        7:  -.2733

TABLE OF RESIDUAL PARTIAL AUTOCORR.    (ASYMPT. STANDARD ERROR =   .13) <PT>
        1     2     3     4     5     6     7     8     9    10    11    12
 0    0.04  0.07 -0.18 -0.10 -0.08 -0.06 -0.31 -0.15 -0.10 -0.22  0.31  0.04   (12)
12   -0.09 -0.15 -0.12  0.09 -0.16 -0.04  0.07  0.04  0.04 -0.13  0.03 -0.26
24   -0.00  0.17
SIGNIFICANT    PARTIAL AUTOCORRELATIONS <P(OR)S>
  1 - 5 %        7: -.3077            11:   .3055            24: -.2617

LJUNG-BOX  PORTMANTEAU TEST STATISTICS ON RESIDUAL AUTOCORRELATIONS <L>
 ORDER D.F.  STATISTIC  SIGNIFICANCE
  6      4     4.27         .371
 12     10    21.89         .016
 18     16    25.25         .066                                          (13)
 24     22    27.14         .206
 26     24    27.20         .295

HOMOGENEITY TESTS ON THE RESIDUALS <H>
 1ST CRITERION: NORMAL 50% INTERVAL       .  CHI SQ(2DF).  APPROXIMATE
 2ND CRITERION         .   IN OUT(BELO+ABOV).   STATISTIC . SIGNIFICANCE
(FS) FIRST HALF        .   15  17(=  9+  8) .      .625   .   .732 (FS)
     SECOND HALF       .   18  14(=  7+  7) .
(ME) MIDDLE  )QUAR-    .   17  15(=  8+  7) .      .125   .   .939 (ME)    (14)
     END     )TERS     .   16  16(=  8+  8) .
(LH) LOW   SEASONAL.       13  21(= 11+ 10) .     5.216   .   .074 (LH)
     HIGH SEASONAL.        20  10(=  5+  5) .
(BA) BELOW MEDIAN      .   12  20(= 13+  7) .     5.125   .   .077 (BA)
     ABOVE MEDIAN      .   21  11(=  3+  8) .
=== FORECASTING FROM        77  WITH FRESH DATA <F>
    DATE         OBSERVATION       FORECAST          ERROR        % ERROR
      78             260.0            283.0          -23.0            8.8
      79             304.0            427.4         -123.4           40.6
      80             390.0            551.6         -161.6           41.4
      81             614.0            859.6         -245.6           40.0   (15)
      82             783.0           1126.6         -343.6           43.9
      83             872.0           1198.3         -326.3           37.4
MEAN ERROR                    -203.9   MEAN SQUARE ERROR                     54439.5
MEAN ABSOLUTE ERROR            203.9   MEAN ABSOLUTE PERCENTAGE ERROR           35.4
```

⑦ Normalization factor of the Box-Cox procedure : G, the geometric mean of the data

⑧ Summary measures : the "computed sum of squares" (116326) is the sum of squares Σ which is minimized. By multiplying it by the square of the constant called ARMA (0.93126), we obtain the "adjusted sum of squares" (100884). Dividing the latter by the number of residuals (64) or by the number of degrees of freedom (64 - 2 = 62) yield respectively the biased and unbiased residual variance. So $\hat{\sigma}^2 = 1627$. AIC given by (2) is equal to 804.

⑨ Table of residuals against time : these are the normalized residuals, i.e. multiplied by G in this case

⑩ Residual "outliers" : since they are significant at a significance level between 1 and 5 %, they are not retained

⑪ Table of residual autocorrelation coefficients and the most significant values : the one at lag 11 is noted

⑫ Table of residual partial autocorrelation coefficients and the most significant values

⑬ Ljung-Box portmanteau test statistics : we will consider order 24 so that the significance probability of 0.21 is obtained

⑭ Homogeneity tests : the results of the BA test given previously are recalled, together with the three other tests

⑮ Forecasts compared with the fresh data and several summary statistics : comments will be given in the appendix

3. THE MODEL COMPONENTS

The generalized ARIMA model results from the combination of model components which are briefly described now :

<u>(MC1) the ARMA submodel</u>

$$\phi(B)\ z_t = \theta(B)\ a_t$$

where $\phi(B)$ and $\theta(B)$ are polynomials in B, and $\{a_t\}$ is a sequence of independent normal variables with zero mean and variance σ^2.

<u>(MC2) the regular difference operator</u> $\nabla^d = (1 - B)^d$, which, in combination with the ARMA submodel, constitutes the ARIMA model

$$\phi(B)\ \nabla^d\ z_t = \theta(B)\ a_t.$$

The other model components are shown, generally in combination with (MC1) and (MC2) :

<u>(MC3) the power transformation</u> $C_\lambda(z_t)$:

$$\phi(B)\ \nabla^d\ C_\lambda(z_t) = \theta(B)\ a_t$$

$$\text{where } C_\lambda(z_t) = \begin{cases} (z_t^\lambda - 1)/(\lambda G^{\lambda-1}) & \lambda \neq 0 \\ G \log z_t & \lambda = 0, \end{cases}$$

and G is the geometric mean of the z_t, assuming $z_t > 0$, for all t.

(MC4) the intervention on the variable y_t^I :

$\phi(B)\ \nabla^d (z_t - y_t^I) = \theta(B)\ a_t$

where y_t^I is a deterministic function of t, equal to zero except at selected times t. In M, we express y_t^I as a sum of piece-wise linear functions.

(MC5) the multiplicative trend f_t :

$\phi(B)\ \nabla^d (z_t/f_t) = \theta(B)\ a_t$

where f_t is a strictly positive deterministic function of time, normalized in such a way that $f_{d+1} f_{d+2} \ldots f_n = 1$. The normalization factor is always omitted from equations. Therefore, the symbol of proportionality $\propto$ is used in the definition of f_t and other similar functions.

(MC6) the additive trend μ_t :

$\phi(B)\ (\nabla^d\ z_t - \mu_t) = \theta(B)\ a_t$

where μ_t is a polynomial deterministic function of time which can possibly be reduced to a constant μ; otherwise, d will usually be zero.

(MC7) the additive seasonal m_t :

$\phi(B)\ (\nabla^d\ z_t - m_t) = \theta(B)\ a_t$

where m_t is a periodic deterministic function of time, with period s, such that $\Sigma_{t=1}^{s}\ m_t = 0$; m_t can be parametrized by seasonal coefficients, by coefficients in a Fourier expansion, or by a combination of both.

(MC8) the seasonal difference operator $\nabla_s^D = (1 - B^s)^D$:

$\phi(B)\ \nabla^d \nabla_s^D\ z_t = \theta(B)\ a_t.$

(MC9) the trend on the standard deviation of the innovation g_t :

$\phi(B)\ \nabla^d\ z_t = \theta(B)\ (g_t a_t)$

where g_t is a strictly positive deterministic function of time, normalized in such a way that $g_{d+1} g_{d+2} \ldots g_n = 1$.

(MC10) the evolutive ARMA submodel

$\phi_t(B)\ z_t = \theta_t(B)\ a_t$

characterized by autoregressive and/or moving average polynomials with at least one coefficient which is a deterministic function of time.

<u>(MC11) the intervention on the innovation</u>, described by y_t^M and y_t^S :

$$\phi(B)\ \nabla^d\ z_t = \theta(B)\ (y_t^S a_t + y_t^M)$$

where y_t^M and y_t^S are deterministic functions of t, expressed as sums of piece-wise linear functions, respectively equal to 0 and 1 everywhere except at some selected times t and y_t^S being such that $y_{d+1}^S y_{d+2}^S \dots y_n^S = 1$.

<u>(MC12) submodels underlying well-known forecasting schemes</u>

$$\phi(B)\ (\nabla^d \nabla_s^D\ z_t - \mu) = \theta(B)\ a_t$$

where d and D are set automatically, and the coefficients of the polynomials $\phi(B)$ and $\theta(B)$ are expressed in terms of the natural parameters of the forecasting scheme. For example, the additive Holt-Winters predictor corresponds to $d = D = 1$, $\mu = 0$, and

$$\phi(B) = 1$$

$$\theta(B) = 1 - (1 - \alpha - \alpha\beta)B + \alpha\beta(B^2 + \dots + B^{s-1}) - (1 - \alpha\beta + \alpha\gamma - \gamma)B^s + (1 - \alpha)(1 - \gamma)B^{s+1}.$$

There are some other components, not illustrated here, but given in M. All the deterministic functions are parametrized. All the parameters are usually estimated by maximum likelihood, with two exceptions : the constant μ of (MC6) and the (s - 1) parameters of m_t in (MC7) can also be estimated by least squares, based on appropriate averages of $\nabla^d\ z_t$.

<u>Figure 2.</u> Plot of the Chatfield-Prothero data

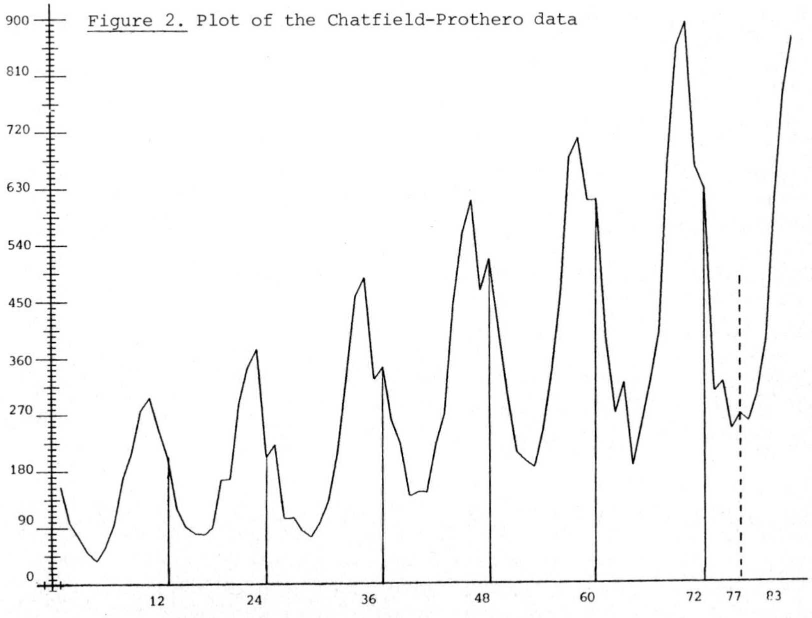

4. IDENTIFICATION

The "Sales of Company X" data, shown in Fig. 2, exhibit trend and seasonality in the mean and variance. Model components which are able to represent a trend in the mean are the regular difference operator, the seasonal difference operator combined with a constant, and the additive trend. Components which are able to represent seasonality are the seasonal difference operator and the additive seasonal. The most suitable combinations are found to be

1° (MC2) and (MC8) the regular and seasonal difference operators $\nabla\nabla_{12}$;
2° (MC2) and (MC7) the regular difference operator ∇ and the additive seasonal m_t;
3° (MC6) and (MC8) the constant μ and the seasonal difference operator ∇_{12}.

Model components which are able to represent a trend and, possibly, seasonality in the variance are

a. (MC3) the power transformation, either with $\lambda = 0$ or $\lambda \neq 0$;
b. (MC5) the multiplicative trend f_t;
c. (MC9) the trend on the standard deviation of the innovation g_t.

After some experiments, the following specifications were retained for the last three model components, as far as identification is concerned :

a. $\lambda = 0.25$
b. $f_t \propto \exp(0.0175t)$
c. $g_t \propto \exp(0.02t)$.

Several statistics are given in Table 4 : the variance $\hat{\sigma}^2$, the value of AIC, the mean, the presence of outliers corresponding to $P < 1$ %, and the autocorrelation and partial autocorrelation coefficients at lags 1, 2, 3, 11, 12, 13, and 24. The most promising combination seems to be b.2°. We will therefore start with this approach.

The very high autocorrelation at lag 11 cannot be easily explained. It is perhaps a tendency to anticipate each year by one month in order to smooth the activity, which grows too fast and with an excessive seasonal pattern. Anyway, we have checked that a few points such as outliers are not the cause of the large correlation coefficient.

5. AN ANALYSIS FEATURING A MULTIPLICATIVE TREND

If we take a multiplicative trend f_t (MC5), a regular difference operator (MC2), and an additive seasonal component m_t (MC7), and if we try to copy the ARMA models fitted in the previous studies, we find

$$(1 + 0.32B)\ \{\nabla(z_t/f_t) + 2.1 - m_t\} = (1 - 0.18B^{12})\ a_t \qquad (3)$$
$$f_t \propto \exp(0.0176t),\ \hat{\sigma}^2 = 1130,\ \text{AIC} = 776.$$

We will postpone giving details of m_t, which was estimated by least squares. This

Table 4. Selected statistics of the identification stage

Model components	$\hat{\sigma}^2$	AIC	mean	outliers (value)	autocorrelation coefficient 1	2	3	11	12	13	24	partial autocorrelation coefficient 1	2	3	11	12	13	24
$\nabla\nabla_{12}\ C_\lambda(z_t)$	3150	842	0.4	77(+150)	-.60	.36	-.22	.45	-.35	.18	-.08	-.60	.01	.00	.23	.05	-.16	-.11
$\nabla C_\lambda(z_t) - m_t$	1462	794	5.1	65(-101)	-.45	.27	-.22	.44	-.23	.11	-.19	-.45	.08	-.10	.31	.12	-.08	-.13
$\nabla_{12}\ C_\lambda(z_t)$	2620	829	57	38(+143)	.40	.54	.23	.09	-.22	-.10	-.20	.40	.45	-.10	.16	-.30	-.17	-.11
$\nabla\nabla_{12}(z_t/f_t)$	2472	832	0.1	24(-139)	-.55	.33	-.31	.44	-.35	.19	-.02	-.55	.04	-.17	.26	.02	-.16	-.06
$\nabla(z_t/f_t) - m_t$	1303	784	1.2		-.39	.20	-.26	.44	-.20	.08	-.16	-.39	.06	-.20	.40	.10	-.10	-.12
$\nabla_{12}(z_t/f_t)$	2392	822	4.7	24(-134) 38(+130)	.43	.49	.17	.12	-.17	-.06	-.24	.43	.38	-.17	.19	-.32	-.11	-.19
$(\nabla\nabla_{12}\ z_t)/g_t$	3252	846	0.6	24(-183)	-.48	.33	-.31	.39	-.22	.19	-.04	-.48	.13	-.16	.19	.05	-.09	-.01
$(\nabla z_t - m_t)/g_t$	3723	866	5.2	2(+213) 9(-183) 12(+161)	.05	.20	-.11	.26	.21	.18	.06	.05	.20	-.13	.10	.06	-.03	-.04
$(\nabla_{12}\ z_t)/g_t$	2220	820	63.0	24(-151)	.07	.11	.12	.34	-.02	.06	-.09	.07	.11	.11	.38	-.12	-.02	-.16

model is unsatisfactory for two reasons : the coefficient of B^{12} in the MA polynomial is small with respect to its standard error 0.25 and residual autocorrelation remains at lag 11, with a coefficient equal to 0.45. These defects are easily corrected, using

$$(1 + 0.36B)\ \{\nabla(z_t/f_t) + 2.1 - m_t\} = (1 + 0.50B^{11})\ a_t \qquad (4)$$
$$f_t \propto \exp(0.0174t),\ \hat{\sigma}^2 = 871,\ \text{AIC} = 759.$$

Since there remains autocorrelation at lag 24, the following model is fitted :

$$(1 + 0.42B)(1 + 0.53B^{24})\ \{\nabla(z_t/f_t) + 2.0 - m_t\} = (1 + 0.58B^{11})\ a_t \qquad (5)$$
$$f_t \propto \exp(0.0171t),\ \hat{\sigma}^2 = 647,\ \text{AIC} = 746.$$

The Ljung-Box statistic is equal to 14.98 with 21 degrees of freedom, so that $P_{LB} = 0.82$. The LH test of homogeneity gives a significance $P_{LH} = 0.002$, indicating some problems with the representation of the seasonal component. We tried some variants of this last model, without success. Hence we must search elsewhere.

6. AN ANALYSIS FEATURING A POWER TRANSFORMATION

The second choice combination which results from examination of Table 4 is composed of the power transformation (MC3) with the regular difference operator (MC2) and an additive seasonal component (MC7). If we follow the same steps as in section 5, we have first the model :

$$(1 + 0.38B)\ \{\nabla C_{0.22}(z_t) - 1.7 - m_t\} = (1 - 0.23B^{12})\ a_t \qquad (6)$$
$$\hat{\sigma}^2 = 1176,\ \text{AIC} = 779,$$

with still a highly significant autocorrelation at lag 11. The model is slightly modified and becomes

$$(1 + 0.33B)\ \{\nabla C_{0.25}(z_t) - 1.7 - m_t\} = (1 + 0.50B^{11})(1 - 0.32B^{12})\ a_t \qquad (7)$$
$$\hat{\sigma}^2 = 909,\ \text{AIC} = 764.$$

There remains autocorrelation at lag 24 and the coefficient of B^{12} is not significantly different from zero. Consequently, we obtain a somewhat better model

$$(1 + 0.43B)(1 + 0.51B^{24})\ \{\nabla C_{0.28}(z_t) - 1.7 - m_t\} = (1 + 0.58B^{11})\ a_t \qquad (8)$$
$$\hat{\sigma}^2 = 696,\ \text{AIC} = 751.$$

This model is rejected by none of the tests. The Ljung-Box statistic is equal to 18.56, with 21 degrees of freedom, which corresponds to a significance of 0.61. Model (8) is the best model obtained so far. There are at least two ways to improve the fit : to try a compromise between power transformation and multiplicative trend, and to investigate more carefully the seasonal component. These attempts are developed later.

Figure 3. Tree diagram of all the alternative models

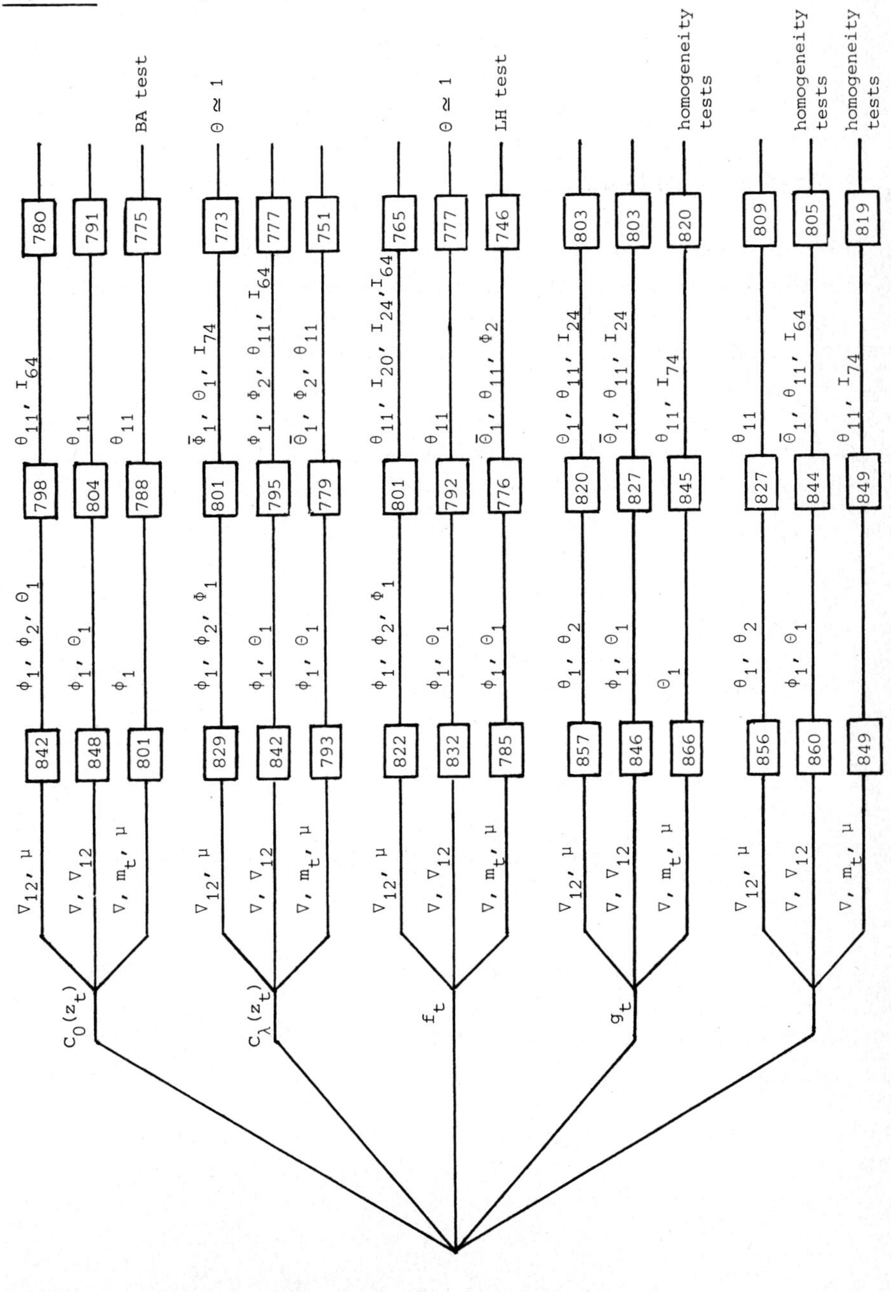

7. EX POST VALIDATION OF THE STRATEGY

Up to now, the discussion has followed a rigorous strategy, described in section 2, which has produced model (8). It seems useful to justify the strategy by showing that alternative decisions at the various stages would have led to unacceptable models, or to suboptimal models, in the sense that AIC is not minimum. As for any strategy in data modelling, we are not able to prove that our strategy is optimal. Our purpose is simply to convince the reader to try it, eventually with other model components than ours if they seem more appropriate to the data on hand, and to try other diagnostic check designed to trap the kind of deviations from the model which are feared.

To save space we present the results in the form of a tree diagram (Fig. 3) in which the forks correspond to model components and the nodes contain the value of AIC. At the end points, an indication is given about the invalidity of the final model. In accordance to our strategy, the model components are added in the following order :

1° log transformation or power transformation (MC3), or multiplicative trend (MC5), or trend on the innovation standard deviation (MC9), or nothing;

2° the model components for dealing with nonstationarity in the mean : regular (MC2) and seasonal (MC8) difference operator, additive seasonal component (MC7);

3° the simplest terms of the ARMA submodel (MC1) without paying attention to the least explainable autocorrelation coefficients, such as that at lag 11;

4° the appropriate terms of the AR and MA polynomials and, possibly, interventions (MC4) in order to cope with residual outliers.

The following additional notations are used in the diagram :

ϕ_i (θ_i) is the coefficient of B^i in the regular AR (MA) polynomial

Φ_i (Θ_i) is the coefficient of B^{12i} in the seasonal AR (MA) polynomial.

A superscript bar indicates a term which is discarded with respect to the previous model

$\phi(B)$ ($\theta(B)$) is the product of the regular and seasonal AR (MA) polynomials.

I_n denotes intervention at time n, specified by way of a function $y_t^I = I_\tau$ $(t=\tau)$ and $y_t^I = 0$ $(t \neq \tau)$.

Several models pass all the diagnostic tests. Surprisingly, a model on the raw data is among them. The strategy of section 2 is validated because the AIC reaches its minimum value, among the models that are retained, for model (8).

Models using $C_\lambda(z_t)$, f_t or no transformation at all need no further comment. In order to illustrate the introduction of g_t (MC9) and y_t^I (MC4) in a model, let us consider several examples not shown in Fig. 3. If we take $g_t = \exp\{\beta(t-\nu)\}$ with $\nu = (n+14)/2$, so that $g_{14}g_{15} \dots g_n = 1$, and a stochastic process $w_t = \nabla\nabla_{12}z_t$ such that

$$(1 - \phi_1 B)\ w_t = (1 - \theta_{11} B^{11})(g_t a_t)$$

it is possible to compute the likelihood function at each parameter point and to maximize it over the parameter space. The model obtained is

$$(1 + 0.43B)\ \nabla\nabla_{12} z_t = (1 + 0.71B^{11})(g_t a_t) \tag{9}$$

where $\beta = 0.0127$, $\hat{\sigma}^2 = 1816$, AIC = 810.

However there are outliers at times 24 and 64 and, probably as a result, the four homogeneity tests reject the model. We consider a function $y_t^I = I_{24}$ $(t = 24)$, I_{64} $(t = 64)$, 0 $(t \neq 24, 64)$ and fit a model on $(z_t - y_t^I)$, instead of z_t, with two additional parameters I_{24} and I_{64}. We obtain

$$(1 + 0.39B)\ \nabla\nabla_{12}\ (z_t - y_t^I) = (1 + 0.75B^{11})(g_t a_t) \tag{10}$$

where $\beta = 0.0137$, $I_{24} = -48$, $I_{64} = 86$, $\hat{\sigma}^2 = 1463$, AIC = 796.

The reduction of the residual variance is not surprising, but it is not our aim here. None of the four homogeneity tests rejects the model. The histograms of the residuals of models (9) and (10) are shown in Fig. 4 and confirm that normality is better achieved with the latter model. The comparatively high value of AIC for this model indicates that the approaches of the previous sections are preferable.

Figure 4. Histograms of the residuals of models (9) and (10)

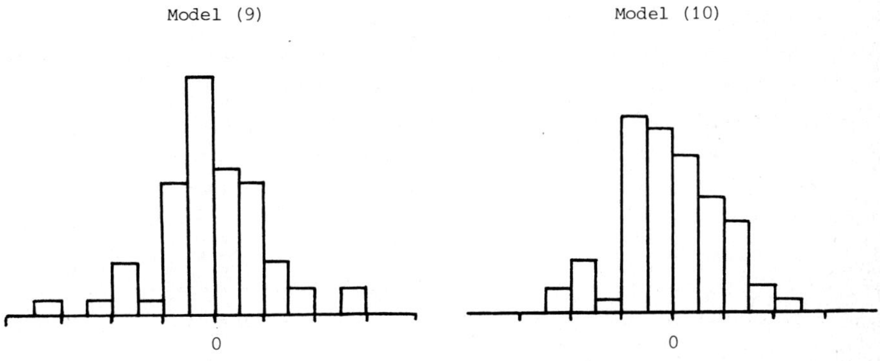

8. A MORE REFINED SUBMODEL OF THE SEASONAL COMPONENT

We have concluded earlier that the best representation of the seasonality in the "Sales of Company X" data makes use of the additive seasonal component. It requires, however, estimating 11 independent parameters. The preceding fits were obtained by the least squares method applied on $C_\lambda(z_t)$, instead of full maximum likelihood. A model of the form of (8) was fitted by the method of maximum likelihood using the seasonal coefficients of the first 11 months as parameters, with the least squares estimates as starting values in the iterative procedure.

The model obtained is

$$(1 + 0.45B)(1 + 0.58B^{24})\{\nabla C_{0.28}(z_t) - 1.7 - m_t\} = (1 + 0.60B^{11})\, a_t \qquad (11)$$

$\hat{\sigma}^2 = 634$, AIC = 747.

The seasonal coefficients contained in models (8) and (11) are given in Table 5. The fit is slightly improved despite the fact that the seasonal coefficients are not very different.

Table 5. Seasonal coefficients estimated by
(a) least squares - model (8)
(b) maximum likelihood - model (11)

Month	1	2	3	4	5	6	7	8	9	10	11	12
(a)	-5	-129	-45	-51	-28	29	75	62	103	72	13	-95
(b)	-3	-121	-45	-56	-30	21	77	64	108	68	14	-96

Table 6. Seasonal coefficient m_t^*, obtained from m_t in Table 5 (a)

Month	1	2	3	4	5	6	7	8	9	10	11	12
$\Sigma_{\tau=1}^{t} m_\tau$	-5	-134	-179	-230	-258	-229	-154	-92	11	83	96	1
m_t^*	86	-43	-88	-139	-167	-138	-63	-1	102	174	187	92

At first sight the seasonal component looks like a sinusoid perturbed at months 2, 7, 10 and 11. In order to reduce the number of parameters, we are looking for a decomposition $m_t = m_t' + m_t''$, where m_t' is a linear combination of sines and cosines and m_t'' is constant except at some specified months. The selection suggested above appeared to be bad, with an AIC of 763, although sines and cosines with periods 12 and 6 were used in m_t'. A better method is as follows : a stepwise multiple regression procedure is performed with m_t, t = 1, ..., 12, as the values of the dependent variables, and the following independent variables :

D_i (i = 1, ..., 12) $\qquad D_{it} = \begin{cases} 1 & t = i \\ 0 & t \neq i \end{cases}$

C_i (i = 1, 2, 3) $\qquad C_{it} = \cos(2\pi it/12)$

S_i (i = 1, 2, 3) $\qquad S_{it} = \sin(2\pi it/12)$.

Using line (a) of Table 5, the variables which enter into the equation are S_1, D_{12}, D_2, D_{11}, D_8, S_2, in that order, followed by D_1, D_7, D_4 and D_5. Therefore, let

$m_t' = s_1 \sin(2\pi t/12) + s_2 \sin(2\pi t/6)$

$$m''_t = \begin{cases} d_2 & t = 2 \pmod{12} \\ d_8 & t = 8 \pmod{12} \\ d_{11} & t = 11 \pmod{12} \\ d_{12} & t = 12 \pmod{12} \\ r & t \neq 2, 8, 11, 12 \pmod{12} \end{cases}$$

where $r = -(d_2 + d_8 + d_{11} + d_{12})/8$. The following model was obtained

$$(1 + 0.45B)(1 + 0.56B^{24}) \{\nabla C_{0.27}(z_t) - 1.7 - m_t\} = (1 + 0.57B^{11}) a_t \qquad (13)$$

with $s_1 = -75$ (± 3), $S_2 = 18$ ($\pm$ 3), $d_2 = -72$ (± 6), $d_8 = -16$ (± 6), $d_{11} = -9$ (± 6), $d_{12} = -94$ (± 5), $\hat{\sigma}^2 = 628$, AIC = 740.

The standard errors are shown after each parameter of m_t. Note that r is equal to 24 so that the value of d_2, d_8, d_{11}, and d_{12} should be compared to 24, not to zero. Note also that the seasonal component is extracted from the differenced transformed data. Otherwise the least squares estimates would be inefficient because of strong autocorrelation among the data. The seasonal component m_t^* relative to the transformed - not differenced - data, obtained from m_t in Table 5 (a), are shown in Table 6. An adjustment has been applied so that the sum over one year is zero.

9. A MODEL WITH BOTH MULTIPLICATIVE TREND AND POWER TRANSFORMATION

Since the model with a multiplicative trend (MC5) was our first guess, but was rejected, and a model with a power transformation (MC3), our second choice, was finally accepted, it may be supposed that a better result should come from the combination of (MC5) and (MC3). Fitting such a model to the data yields

$$(1 + 0.44B)(1 + 0.53B^{24}) [\nabla\{C_\lambda(z_t)/f_t\} + 1.6 - m_t] = (1 + 0.58B^{11}) a_t \qquad (14)$$

$\lambda = 0.71$ (± 0.13), $f_t \propto \exp(\beta t)$, $\beta = 0.011$ (± 0.003), $\hat{\sigma}^2 = 633$, AIC = 745

where the standard errors are displayed next to the corresponding parameter estimates contained in $C_\lambda(z_t)$ and f_t. The model is acceptable according to all our diagnostic checks. From the consideration of the standard errors, we can conclude that λ differs significantly from 1 and that β is significantly different from 0.

10. EXAMPLES OF SOME OTHER APPROACHES

General models with time-dependent coefficients (MC10) have not been discussed yet, although a model with a trend in the standard deviation of the innovation (MC9), such as (10), is a special case of them. Since time-dependency of the coefficients may have been the cause of the problems encountered in section 5, we will investigate that. A first attempt is the following model, with linearly-dependent coefficients,

$$(1 - \phi_{1,t}B)\ \{\nabla(z_t/f_t) - m_t - 1.1\} = (1 - \theta_{11,t}B^{11})\ a_t \qquad (15)$$

$f_t \propto \exp(0.0174t)$, $\phi_{1,t} = -0.36 - 0.0007(t - 39.5)$, $\theta_{11,t} = -0.49 - 0.0014(t-39.5)$

$\hat{\sigma}^2 = 908$, AIC = 763.

There remains autocorrelation at lag 24. Our present program is, however, unable to obtain maximum likelihood estimates for models with time-dependent coefficients, if the largest of the degrees of $\phi_t(B)$ or $\theta_t(B)$ exceeds 13. Returning to the other estimation methods, we have :

conditional method $f_t \propto \exp(0.0167t)$, $\phi_{1,t} = -0.35 - 0.0004(t - 39.5)$, $\theta_{11,t} = -0.46 - 0.0030(t - 39.5)$, $\hat{\sigma}^2 = 938$, AIC = 762;

approximate method with backforecasting $f_t \propto \exp(0.0172t)$, $\phi_{1,t} = -0.33 - 0.0015(t - 39.5)$, $\theta_{11,t} = -0.68 + 0.0034(t - 39.5)$, $\hat{\sigma}^2 = 860$, AIC = 756.

The conclusion we draw from these results is that it is unsafe to use these methods of estimation especially for the purpose of comparing high-order models.

Let us go back to model (15). Comparing the estimates of the coefficients of t in $\phi_{1,t}$ and $\theta_{11,t}$, -0.0007 and -0.0014, with their standard errors, obtained by the usual procedure, 0.0045 and 0.0071, respectively, indicates that the coefficients are not significantly variable over time.

Chatfield and Prothero (1973a) have suggested that a simple model with additive trend can represent the series fairly well. We can formalize that suggestion within the generalized ARIMA model by using the appropriate model component (MC6) $\mu_t = a + b_t$, instead of the regular difference operator. After a few trials, we have the following model :

$$(1 - 0.57B)\ \{C_{0.22}(z_t - y_t^I) - m_t - \mu_t\} = (1 + 0.52B^{11})(1 - 0.87B^{12})\ a_t \qquad (16)$$

$y_t^I = 69\ (t = 50),\ 0\ (t \neq 50)$, $\mu_t = 770 + 5(t - 40)$, $\hat{\sigma}^2 = 448$, AIC = 747.

This model passes all the tests except the LH test, for which $P_{LH} = 0.06$.

Another suggestion made by Chatfield and Prothero (1973a) was to use the Holt-Winters predictor instead of the Box-Jenkins procedure. In order to determine the optimal values of the three smoothing constants, we can exploit the ARIMA form of the model underlying the Holt-Winters method given in the description of (MC12) and fit this model by maximum likelihood. To be fair, we have included a power transformation. The following values : $\lambda = 0.25$, $\alpha = 0.50$, $\beta = 0.0008$ and $\gamma = 0.31$ were obtained, and the corresponding model is

$$\nabla\nabla_{12}\ C_{0.25}(z_t) = \{1 - 0.50B + 0.0004(B^2 + \ldots + B^{11}) - 0.85B^{12} + 0.35B^{13}\}\ a_t \qquad (17)$$

$\hat{\sigma}^2 = 1368$, AIC = 799.

There remains obvious autocorrelation at lag 11 but the probability of significance of the Ljung-Box test is 18 %. The value of AIC is, however, higher than with most of the earlier models. Consequently, we cannot retain the Holt-Winters predictor.

The last two examples will involve an intervention submodel on the innovation. There are two kinds of interventions available on the innovation : on the mean and on the variance. As an illustration of an intervention on the innovation mean, let us consider again the residuals of model (9) which contain two outliers at time 24 and 64. Instead of making an intervention directly on the variable, we can perhaps argue that the innovations at times 24 and 64 have an unknown mean, different from O. We then let $y_t^M = M_{24}$ $(t = 24)$, M_{64} $(t = 64)$, 0 $(t \neq 24, 64)$, where M_{24} and M_{64} are two parameters. Since our implementation of maximum likelihood estimation cannot be used in this case, the following estimates were obtained by an appropriate back-forecasting procedure (Mélard, 1982b) :

$$(1 + 0.45B)\ \nabla\nabla_{12}\ z_t = (1 + 0.55B^{11})\ \{g_t(a_t + y_t^M)\} \qquad (18)$$

$g_t \propto \exp(0.0130t)$, $M_{24} = -70$, $M_{64} = 56$, $\hat{\sigma}^2 = 1526$, AIC = 791.

As mentioned before, these results cannot be compared with (10) but with a model like (10) fitted by the same procedure as (18) :

$$(1 + 0.39B)\ \nabla\nabla_{12}(z_t - y_t^I) = (1 - 0.71B^{11})\ (g_t a_t)$$

$g_t \propto \exp(0.0086t)$, $I_{24} = -44$, $I_{64} = 84$, $\hat{\sigma}^2 = 1554$, AIC = 793.

We observe that the improvement of (18) is not appreciable.

Interventions on the variance of the innovation are a simple but effective way to get rid of the distortion which can be introduced because of short perturbed periods in a time series. We see that the shocks are larger during a 5-month interval of each year (months 9 to 12 and month 1). It is probably for this reason that the LH test of homogeneity rejected the Chatfield-Prothero model (1). Hence, we will try to improve that model by letting $y_t^S = k\tau$ $(t = 9, 10, 11, 12, 1 \bmod 12)$, k (elsewhere), where τ is a parameter and k is such that the condition in (MC11) is satisfied. Using again an algorithm for maximum likelihood estimation of a time-dependent model, we obtain the following model :

$$(1 + 0.43B)\ \nabla\nabla_{12}\ C_0(z_t) = (1 - 0.68B^{12})\ (y_t^S\ a_t) \qquad (19)$$

$\tau = 0.52$, $k = 1.29$, $\hat{\sigma}^2 = 2286$, AIC = 794.

There remains autocorrelation at lag 11, but the goodness of fit is improved.

11. CONCLUSIONS

In this paper we have applied the Box-Jenkins method to a wide class of models. Indeed, the use of the Box-Jenkins method is not restricted to ARIMA models. Box, Jenkins and their co-authors (e.g. Abraham and Box, 1978) have always argued for extensions of the class of models and improvements of methods, and have contributed greatly to the achievement of these objectives in a coherent framework. In this paper we have illustrated the use of a generalized ARIMA model, described in M and in the references quoted there.

Many matters are not covered : theoretical problems (e.g. the standard errors of estimates in time-dependent models, the distribution of the Ljung-Box test statistic, the validity of Ozaki's expression for AIC in order to compare our models), algorithmic problems (e.g. the reason why certain functions are normalized, the reduction of maximum likelihood estimation to least squares estimation) and practical problems (e.g. the justification of some types of parametrization, the order in which the model components are combined). Several of these problems are difficult and some of them are still unsolved. But, too often, time series modelling is restricted to looking for the AR and MA polynomials in order to get residuals which form a white noise sequence. Outliers and local perturbations, trend and seasonality in the variance, and the need for a transformation are sometimes ignored. It is particularly misleading if an interval forecast is asked for, rather than just a point forecast.

The strategy stated in section 2 and the recommendations given in section 4 can be helpful in building an overall model for a time series. The paper has shown that it can be more lengthy and time-consuming than a traditional analysis. Moreover, more complex software is needed - although not as complex as ANSECH, which is experimental and consequently contains extra submodels and methods.

The difficulties reported in section 2 with ordinary ARIMA models are not specific to the Chatfield-Prothero data. We have observed in many other instances that the homogeneity tests reject traditional ARIMA models. Fortunately, introducing a power transformation and interventions is often enough to obtain a completely satisfactory model.

The approach followed in this paper is not an isolated attempt. Besides the references already given in M we can mention Kashyap and Rao (1976), who make use of the same data set. Note also that the class of models and the estimation methods can easily be extended to include the models used by Cleveland and Tiao (1979) and Parzen and Pagano (1979)(see also Anderson, 1978). Indeed these models are special cases of evolutive ARIMA models, but with periodic coefficients.

ACKNOWLEDGEMENTS

An early version of this paper was presented at the Statistical Applications and Computing Meeting, held in parallel with the SEAS Aniversary Meeting, in Hamburg (W. Germany), September 1979.

Thanks are due to Jean Waelbroeck for his helpful comments. The author is indebted to a referee for his careful review and several useful suggestions, to another referee who provided additional references, and to the editor.

APPENDIX : THE FORECASTING PERFORMANCE OF SOME MODELS

The remarks we draw from Table A are :

1° the best fitting models are also those which produce the worst forecasts, with a mean absolute percentage error lying between 19 and 34 %;

2° the most naive method, taking as forecast the value of the same month of last year plus the average yearly increment, gives definitely the bests forecasts;

3° the only model in Table A which is as bad for forecasting as for fitting is the original Chatfield-Prothero model (1).

Table A. Forecasts obtained by several models and summary statistics

Notations : MSE = mean square error; MAPE = mean absolute percentage error

Model	$\hat{\sigma}^2$ $(x10^{-3})$	t = 78	79	80	81	82	83	MSE $(x10^{-3})$	MAPE (%)
$\nabla_{12}z_t - \mu = a_t$	3.8	257	324	404	677	858	895	1.8	6
$\nabla\nabla_{12}z_t = a_t$	4.0	340	407	487	760	941	978	14	24
Eq (1)	1.6	288	427	552	860	1127	1198	54	35
Eq (3)	1.1	287	418	545	799	1036	1101	31	29
Eq (5)	0.6	292	416	557	839	1009	1092	32	31
Eq (6)	1.2	291	408	514	710	883	923	8	19
Eq (8)	0.7	297	408	525	758	870	919	10	21
Eq (10)	1.5	316	387	549	795	963	926	17	25
Eq (13)	0.6	290	400	529	749	864	917	9	19
Eq (14)	0.6	294	417	553	827	982	1059	27	29
Eq (16)	0.4	302	454	621	816	968	1053	31	34
Eq (17)	1.4	280	387	503	735	929	974	11	19
Fresh data		260	304	390	614	783	872		

The best forecasting method is based on the model $\nabla_{12} z_t - \mu = a_t$ which was rejected at the identification stage, partly because of residual autocorrelation. Hence there is no support for the corresponding forecasts from the sample to hand. It is also possible that better forecasts can be obtained from the other models but with parameters derived empirically rather than from maximum likelihood. Again, we are unable to justify our confidence in the forecasts from these sub-optimal models. On the other hand, it may be objected that our analysis has taken account of spurious features of the time series (e.g. interventions without external justification and autocorrelation at lag 11) so that we have obtained a good fit to the data but not to the underlying process. We feel however that the criteria used for adding a parameter are more stringent than in common practice (see the significance levels given in Section 2). Perhaps still lower significance levels should have been used and an alternative information criterion to AIC, but we cannot believe that this would drastically change the respective merits of these models. The only way to resolve the matter is to conclude that the fresh data are not compatible with the data available at the fitting stage of the model building procedure.

REFERENCES

ABRAHAM, B. and BOX, G.E.P. (1978). Deterministic and forecast-adaptative time-dependent models. J. Roy. Statist. Soc. C 27, 120-130.

ANDERSON, O.D. (1978). A note on cyclic variation in the parameter of an AR(1) model. Metron 36, 73-77.

BOX, G.E.P. and JENKINS, G.M. (1973). Some comments on a paper by Chatfield and Prothero and on a review by Kendall. J. Roy. Statist. Soc. A 136, 337-345.

BOX, G.E.P. and JENKINS, G.M. (1976). Time Series Analysis, Forecasting and Control. Holden-Day, San Francisco (revised edition).

CHATFIELD, C. and PROTHERO, D.L. (1973a). Box-Jenkins seasonal forecasting : problems in a case-study. J. Roy. Statist. Soc. A 136, 295-336 (with discussion).

CHATFIELD, C. and PROTHERO, D.L. (1973b). Reply to "Some comments on a paper by Chatfield and Prothero and on a review by Kendall", by G.E.P. Box and G.M. Jenkins. J. Roy. Statist. Soc. A 136, 345-352.

CLEVELAND, W.P. and TIAO, G.C. (1979). Modeling seasonal time series. Revue d'Economie Appliquée 32, 107-129.

HARRISON, P.J. (1973). Contribution to the discussion of the paper by C. Chatfield and D.L. Prothero (1973a).

KASHYAP, R.L. and RAO, A.R. (1976). Dynamic Stochastic Models from Empirical Data. Academic Press, New York.

MELARD, G. (1977). Sur une classe de modèles ARIMA dépendant du temps. Cahiers du Centre d'Etudes de Recherche Opérationnelle 19, 285-295.

MELARD, G. (1979). Modèles ARIMA pour des séries chronologiques non homogènes. Statistique et Analyse des Données 2, 41-50.

MELARD, G. (1982a). On a deterministic sub-model for the innovation process in ARIMA models. In 1981 Proceedings of the Business and Economic Statistics Section, American Statistical Association, Washington D.C., 329-333.

MELARD, G. (1982b). Software for time series analysis. In Compstat 1982 Part I : Proceedings in Computational Statistics (Eds : H. Caussinus, P. Ettinger and R. Tomassone). Physica-Verlag, Vienna (Austria), 336-341.

MELARD, G. (1983). ANSECH - Logiciel pour l'analyse des séries chronologiques. Presses Universitaires de Bruxelles (notes de cours), Bruxelles (5th edition).

OZAKI, T. (1977). On the order determination of ARIMA models. J. Roy. Statist.Soc. C 26, 290-301.

PARZEN, E. and PAGANO, M. (1979). An approach to modeling seasonally stationary time series. J. Econometrics 9, 137-153.

PRIESTLEY, M.B. (1973). Contribution to the discussion of the paper by C. Chatfield and D.L. Prothero (1973a).

WILSON, G.T. (1973). Contribution to the discussion of the paper by C. Chatfield and D.L. Prothero (1973a).

TIME SERIES ANALYSIS: Theory and Practice 7
O.D. Anderson (editor)

ESTIMATION OF THE MEAN DURATION OF TELEPHONE CALLS

Eivind Damsleth
Norwegian Computing Center, P.O.Box 335 Blindern, Oslo 3, Norway

This paper is motivated by the belief that the mean duration of telephone calls is not constant, but varies over time. We introduce an estimator for the mean duration which does not require observation of the duration of each individual call. The properties of the estimator are discussed with regard to bias and variance. When the estimator is applied to successive periods of time, we obtain a sequence of estimated mean durations, whose autocorrelation structure is investigated. We also discuss the robustness of the estimator when the basic assumptions in the classical M/M/∞ model are not fulfilled. Finally the estimation technique is applied to real data, estimating the mean duration for each 15 minute interval over a period of almost three days. It is shown that the time series obtained can be described within an ARIMA-model framework. This leaves us with a challenging filtering problem, which we have not yet solved.

1. INTRODUCTION

In most teletraffic calculations, the distribution of the call duration is supposed to be stationary. In this paper we analyse a real data set, and show that the mean duration is not constant, but varies according to a stochastic process which can be modeled within the ARIMA framework. Our main motivation behind this study came from teletraffic simulation. We wanted a simple, but yet useful, model to describe the evolution of the expected call durations through time, a model which in turn can be included in the call-generating part of a larger simulation program. It is our belief that the calls thus generated will give a more realistic image of the process under study.

In Section 2 we discuss several estimators for the mean duration, and present some results on their mean and variance. Special emphasis is put on the robustness of the estimators when the traffic is contaminated with a few extremely long calls. In Section 3 we use the various estimators on a real data set, and it is shown that there is considerable variation in the call duration - variation which can be described within an ARIMA framework. The paper concludes with an Appendix where the more technical details are given.

2. ESTIMATORS FOR THE MEAN DURATION

Suppose the time axis be divided into intervals of length Δ, so that the interval I_t is $[(t-1)\Delta, t\Delta)$. We observe Z_t, A_t and B_t, which are the numbers, respectively, of calls in progress at the start of I_t, new calls arriving in I_t, and calls disconnected during I_t. We assume that I_t is short enough to allow the traffic to be considered as stationary during I_t. In addition we suppose, at least for the moment, the assumptions of the classical M/M/∞ model to be fulfilled, so that: calls arrive independently, and the time between arrivals is exponentially distributed with parameter (intensity) λ; call durations are exponentially distributed with parameter μ; and the system has infinite capacity.

The expected duration of a call is then given by $\eta = 1/\mu$, which is the parameter of interest.

2.1 Estimator based on observed durations

First, suppose that one was able to observe the actual durations for the B_t calls which terminated in I_t. Then we would naturaly estimate η as

$$\eta^* = \frac{1}{B_t}\sum_{j=1}^{B_t} X_j$$

where X_j is the observed duration of the j-th call terminating in I_t. Since X_j is exponentially distributed with parameter $\mu = 1/\eta$, $E(\eta^*|B_t) = \eta$, $Var(\eta^*|B_t) = \eta^2/B_t$ and $Var\ \eta^* = \eta^2 E(1/B_t) \approx \eta^2/\lambda\Delta$.

The major disadvantage of the estimator η^* is that it requires storage of the starting times for all calls in progress, which can be rather large for a big exchange. For modern, computerized exchanges this is not a severe problem, but it excludes η^* for traditional, mechanical exchanges.

2.2 Estimators based on counters

The values of Z_t, A_t and B_t can be monitored continously using only three counters. The A_t counter and B_t counter is set to 0 at the start of each interval I_t, and increased by 1 for each call which, respectively, arrives or terminates during I_t. The Z_t counter is set to 0 when the system is initiated, and increased (decreased) by 1 every time a call arrives (terminates). Alternatively, the Z_t counter can be updated using the formula $Z_{t+1} = Z_t + A_t - B_t$. Thus, we want to estimate η from Z_t, A_t and B_t.

As shown in the Appendix, we have $EZ_t = EZ_{t+1} = \lambda/\mu = \lambda\eta$, and $EA_t = EB_t = \lambda\Delta$. This suggests an estimator of the form

$$\tilde{\eta}_{c_1,c_2} = \frac{(1-c_1)Z_t + c_1 Z_{t+1}}{[(1-c_2)A_t + c_2 B_t]/\Delta} = \frac{Z_t + c_1(A_t - B_t)}{[(1-c_2)A_t + c_2 B_t]/\Delta}$$

that is the ratio of an unbiased estimator for $\lambda\eta$ and an unbiased estimator for λ. c_1 and c_2 are constants to be determined.

The simultaneous probability distribution of Z_t, A_t and B_t is derived in the Appendix, and the exact moments of η_{c_1,c_2} can be calculated for given c_1 and c_2 and for given parameter values. If $\eta\Delta$ and $\eta\lambda$ are large we can expand η_{c_1,c_2} in a Taylor series, around the expectation of each variable, to obtain

$\tilde{\eta}_{c_1,c_2} \approx \eta + (Z_t - \lambda\eta)/\lambda + [c_1/\lambda - (1-c_2)\eta/\lambda\Delta](A_t - \lambda\Delta) - (c_1/\lambda + c_2\eta/\lambda\Delta)(B_t - \lambda\Delta).$

Using the formulae for the various variances and covariances given in the Appendix, we obtain

$$\mathrm{Var}\ \tilde{\eta}_{c_1,c_2} \approx \frac{\eta}{\lambda}[1+\eta/\Delta - p(1+\eta/\Delta)^2/2 + 2p(c_1 - 1/2 + \frac{\eta}{\Delta}(c_2 - 1/2))^2]$$

where $p = 1-\exp(-\Delta/\eta)$ is the probability that a call lasts for less than one time interval.

This variance is minimized with regard to c_1 and c_2 when $c_1 - 1/2 + \frac{\eta}{\Delta}(c_2 - 1/2) = 0$. To have values for c_1 and c_2 which do not depend on the unknown parameters, it is convenient to choose $c_1 = c_2 = 1/2$, and we have the optimal estimator

$$\tilde{\eta} = \frac{Z_t + (A_t - B_t)/2}{(A_t + B_t)/2}\,\Delta = \frac{Z_t + Z_{t+1}}{A_t + B_t}\,\Delta. \qquad (1)$$

The minimum variance is given by

$$\mathrm{Var}\ \tilde{\eta} \approx \frac{\eta}{\lambda}[1 + \eta/\Delta - p(1+\eta/\Delta)^2/2]. \qquad (2)$$

Table 1 shows $\frac{\lambda}{\eta}\mathrm{Var}\ \eta_{c_1,c_2}$ for various combinations of η/λ, and various combinations of c_1 and c_2. From the table it is clear that the gain using the optimal estimator can be substantial, especially when the observation interval is short.

η/Δ	$c_1=c_2=0$ $c_1=c_2=1$	$c_1=0, c_2=1$ $c_1=1, c_2=0$	$c_1=1/2, c_2=0$ $c_1=0, c_2=1/2$	$c_1=c_2=1/2$
0.05	1.050	0.950	0.499	0.499
0.1	1.100	0.815	0.500	0.495
0.5	1.500	0.635	0.635	0.527
1.0	2.000	0.736	1.052	0.736
2.0	3.000	1.426	2.016	1.229
5.0	6.000	4.187	5.003	2.737
10.0	11.000	9.097	10.001	5.243

Table 1. $\frac{\lambda}{\eta}\mathrm{Var}\ \tilde{\eta}_{c_1,c_2}$ for various combinations of $\frac{\eta}{\Delta}$ and various combinations of c_1 and c_2.

2.3 Comparison of the two estimators

Suppose for a moment that Δ is 1, so that the length of the observation interval is the time unit. Then, using the formulae for the variances previously derived, the efficiency of $\tilde{\eta}$ versus η^* can be written

$$E = \mathrm{Var}\ \tilde{\eta}/\mathrm{Var}\ \eta^* = (1+\eta)[1-(1-e^{-1/\eta})(1+\eta)/2]/\eta. \qquad (3)$$

For η close to 0, that is when the observation interval is long compared to the mean duration, $E \approx (1-\eta^2)/(2\eta)$, $\eta << 1$. When η is large, $E \approx (\eta+1/2)/(2\eta)$, $\eta > 1$.

This implies that $E \to \infty$ when $\eta \to 0$, and that $E \to 1/2$ when $\eta \to \infty$. A closer examination shows that when $\eta > 0.54$, $E < 1$. Thus, the two estimators will be more or less equally good when the observation interval is twice the mean duration. For shorter intervals $\tilde{\eta}$ is superior, while η^* is the better when the intervals are longer. It may seem a little surprising that the efficiency can be < 1 for any parameter values. The reason, of course, is that the estimator $\tilde{\eta}$ makes use of the stationarity assumption, at least locally, while this is not so for η^*. $\tilde{\eta}$ is thus based on more information, which explains its good performance in some circumstances.

2.4 Exact moments

The above discussion requires the traffic to be rather large, for the linearization to be valid. Using the formulae in the Appendix, we can calculate the exact moments of $\tilde{\eta}$ and η^* numerically. In situations when the traffic is small, the problem arises that A_t and B_t may both be 0 with an appreciable probability. Then $\tilde{\eta}$ will be undefined if Z_t is 0, otherwise $\tilde{\eta}$ will be infinite. The same problem arises with η^*, which will be undefined when $B_t = 0$. Using formula (A5) in the Appendix, we have

$$\Pr(\tilde{\eta}\text{ undefined}) = \Pr(A_t = 0,\ B_t = 0,\ Z_t = 0) = \exp[-\lambda(\eta+\Delta)]$$
$$\Pr(\tilde{\eta}\text{ infinite}) = \Pr(A_t = 0,\ B_t = 0,\ Z_t > 0) = \exp[-\lambda(\eta+\Delta)][\exp(\lambda\eta e^{-\Delta/\eta})-1]$$
$$\Pr(\eta^*\text{ undefined}) = \Pr(B_t = 0) = \exp(-\lambda\Delta).$$

These probabilities are all very small for reasonable combinations of λ, η and Δ.

Including a second term in the Taylor expansion of (1) gives, for $\lambda\Delta$ large,

$$E\tilde{\eta} = \eta\{1 + [1-p(\eta/\Delta+1)/2]/\lambda\Delta + O(1/\lambda\Delta)^2\}.$$

Usually, the observation interval will be long compared to the mean duration, so that $\eta/\Delta < 1$. When η/Δ is small, p will be close to 1, and so will $p(\eta/\Delta+1)$. The function $p(\eta/\Delta+1)$ remains rather constant, $1 < p(\eta/\Delta+1) < 1.3$ for all $\eta > 0$. We thus have $E\tilde{\eta} \approx \eta[1 + 1/(2\lambda\Delta)]$, which shows that $\tilde{\eta}$ is asymptotically unbiased when when $\lambda\Delta \to \infty$, but that the bias can be considerable when $\lambda\Delta$ is low. It can be shown that if the estimator

$$\tilde{\tilde{\eta}} = \frac{Z_t + (A_t - B_t)/2}{(A_t + B_t + 1)/2}\,\Delta \tag{4}$$

is used instead, $E\tilde{\tilde{\eta}} = \eta + O(1/\lambda\Delta)^2$. At the same time we avoid the problem with $\tilde{\eta}$ being undefined for $A_t = B_t = 0$.

In Table 2 we give the exact expectations and variances for $\tilde{\eta}$ and $\tilde{\tilde{\eta}}$ given by (1) and (4), for some values of λ and η. For $\tilde{\eta}$ the moments are calculated in the conditional distribution given that $\max(A_t, B_t) > 0$. The moments are calculated using the exact, simultaneous distribution of A_t, B_t and Z_t given in the Appendix. The asymptotic variance given by (2) is also shown, for comparison.

η/Δ	$\lambda\Delta$	$\breve{\eta}/\Delta$ Mean	$\breve{\eta}/\Delta$ Var.	$\tilde{\tilde{\eta}}/\Delta$ Mean	$\tilde{\tilde{\eta}}/\Delta$ Var.	Asymptotic var.
0.05	1.0	.0798	.0563	.0303	.0110	.0250
	5.0	.0569	.00992	.0483	.00576	.00499
	10.0	.0526	.00316	.0491	.00265	.00250
	20.0	.0509	.00136	.0496	.00128	.00125
0.10	1.0	.157	.101	.0623	.0219	.0495
	5.0	.113	.0187	.0968	.0113	.00990
	10.0	.105	.00627	.0988	.00531	.00495
	20.0	.102	.00271	.0992	.00255	.00248
0.50	1.0	.623	.301	.324	.147	.264
	5.0	.546	.0888	.481	.0558	.0527
	10.0	.519	.0324	.491	.0274	.0264
1.00	1.0	1.14	.877	.713	.490	.736
	5.0	1.10	.285	.968	.168	.147
	10.0	1.04	.0941	.985	.0781	.0736
2.0	1.0	2.22	2.86	1.55	1.58	2.46
	5.0	2.22	1.07	1.96	.594	.492
5.0	1.0	5.64	13.5	4.12	7.86	13.7
	5.0	5.60	6.49	4.95	3.46	2.74

Table 2. Exact means and variances for the estimators $\breve{\eta}$ and $\tilde{\tilde{\eta}}$ for various combinations of η/Δ and $\lambda\Delta$.

It is clear from Table 2 that, except for the rather unrealistic situation $\lambda\Delta = 1$, the estimator $\tilde{\tilde{\eta}}$ outperforms $\breve{\eta}$. The bias is much smaller, and decays faster with increasing $\lambda\Delta$, and the variance for $\tilde{\tilde{\eta}}$ is less than for $\breve{\eta}$. The variance for $\tilde{\tilde{\eta}}$ is also converging faster to its asymptotic value.

2.5 Autocovariance for the estimator

Suppose that the estimator $\breve{\eta}$, or preferably $\tilde{\tilde{\eta}}$, is used to estimate the mean duration η in successive intervals. This will give a time-series of estimates for η, $\{\breve{\eta}_t: t=1,2,\ldots\}$. If the traffic is stationary during the period of observation, this series will be stationary, with an autocorrelation structure. Using, as before, the Taylor expansion and the results from the Appendix, we get

$$\mathrm{Cov}(\breve{\eta}_t,\breve{\eta}_{t-k}) \approx \frac{\eta}{\lambda}[1 - p(1+\eta/\Delta)/2]^2(1-p)^{k-1}, \quad k \geq 1$$

where again $p = 1 - \exp(-\Delta/\eta)$ is the probability that a call shall have a duration less than Δ. The covariances consist of one atypical value at lag 1, and follow an exponentially decaying pattern after that. This theoretical pattern of the autocorrelation function is of the same form as that for an ARMA(1,1)-process. The autocorrelations for the $\{\tilde{\eta}_t\}$ series are given by

$$\varrho_k = \begin{cases} 1 & k = 0 \\ \dfrac{[1 - p(1+\eta/\Delta)/2]^2 (1-p)^{k-1}}{[1 + \eta/\Delta - p(1+\eta/\Delta)^2]} & k \geq 1 \end{cases}$$

It is notable that the ϱ_k are independent of the arrival parameter λ.

In Table 3 the values of ϱ_1 and ϱ_2 are given for some values of η/Δ. The associated parameters in the ARMA-representation $(1-\varphi B)(\eta_t - \eta) = (1-\theta B)a_t$ are given too.

η/Δ	ϱ_1	ϱ_2	φ	θ
0.01	0.490	0	0	-0.938
0.05	0.452	0	0	-0.633
0.10	0.409	0	0	-0.519
0.50	0.234	0.032	0.135	-0.105
1.00	0.184	0.068	0.368	0.191
2.00	0.137	0.083	0.607	0.482
5.00	0.076	0.062	0.819	0.875
10.0	0.043	0.039	0.905	0.969

Table 3. Autocorrelations at lag 1 and 2 for the $\{\tilde{\eta}_t\}$ series, with the associated parameters for an ARMA(1,1)-representation. 0 implies a value < 0.001.

2.6 Robustness

So far, the deductions have been made under the assumptions of pure Poisson traffic. Now, suppose that the traffic is contaminated with a small proportion of other traffic, with much longer durations. In this section we study what effects this will have on the estimators.

For convenience, we assume that the contaminating traffic is also Poisson, and introduce the parameters λ_i and η_i, respectively the arrival intensity and expected duration, for process i (i=1,2), and $\mu_i = 1/\eta_i$; where process 1 is the ordinary traffic and process 2 is the contamination. Usually $\lambda_2 \ll \lambda_1$ and $\eta_2 \gg \eta_1$. We also assume that the two processes are independent.

In the following we consider the estimator $\tilde{\eta}$, for simplicity. Since the arguments are made asymptotically, the results for $\tilde{\tilde{\eta}}$ would be the same.

Let us, as before, observe an interval of length Δ, and let Z, A and B be the number of, respectively, calls in progress at the start of the interval, new calls arriving in the interval and calls disconnected during the interval.

Then $Z = Z_1 + Z_2$, $A = A_1 + A_2$ and $B = B_1 + B_2$, where Z_1, Z_2, A_1, A_2, B_1 and B_2 are the (unobservable) corresponding values for processes 1 and 2, respectively, and

$$\tilde{\eta} = \frac{Z + (A-B)/2}{(A+B)/2}\Delta = \frac{Z_1+Z_2 + (A_1+A_2 - B_1-B_2)/2}{(A_1+A_2 + B_1+B_2)/2}\Delta.$$

Expanding in a Taylor series as before, we obtain

$$\begin{aligned}\tilde{\eta} \approx \eta &+ (Z_1-\lambda_1\eta_1)/\lambda + (A_1-\lambda_1\Delta)(1-\eta/\Delta)/(2\lambda) - (B_1-\lambda_1\Delta)(1+\eta/\Delta)/(2\lambda)\\ &+ (Z_2-\lambda_2\eta_2)/\lambda + (A_2-\lambda_2\Delta)(1-\eta/\Delta)/(2\lambda) - (B_2-\lambda_2\Delta)(1+\eta/\Delta)/(2\lambda)\end{aligned}$$

where $\eta = (\lambda_1\eta_1+\lambda_2\eta_2)/(\lambda_1+\lambda_2)$ and $\lambda = \lambda_1+\lambda_2$ are the mean duration and arrival intensity for the combined process. Thus, $\tilde{\eta}$ is still asymptotically unbiased for the mixed process, and some computation gives

$$\text{Var }\tilde{\eta} \approx \frac{\eta}{\lambda}[1 + \eta/\Delta - p(1+\eta/\Delta)^2/2] \tag{5}$$

where, as before, $p = (\lambda_1\eta_1p_1 + \lambda_2\eta_2p_2)/(\lambda_1\eta_1 + \lambda_2\eta_2)$ is the probability that a call lasts less than the interval length. p_1 and p_2 are the corresponding values for the individual processes, given by $p_i = 1 - \exp(-\Delta/\eta_i)$, $i = 1,2$. Comparison of (5) with (2) shows that the formula for the variance remains the same for the contaminated traffic, when the parameters are defined for the mixed process.

For the estimator η^*, a similar argument gives

$$\text{Var }\eta^* \approx \frac{\eta^2}{\Delta\lambda}(1 + 2\frac{(\beta-\alpha)^2}{\alpha(1-\alpha)})$$

where $\alpha = \lambda_2/\lambda$ and $\beta = \lambda_2\eta_2/\lambda\eta$ are the proportions that the contaminating traffic makes up of the arrivals and traffic volume, respectively.

The contamination thus leads to an increase in the variance of η^*, and the increase can be considerable. If, for example, $\alpha = 0.001$ and $\beta = 0.033$, so that 0.1 % of the calls comes from a distribution with approximately 34 times as long durations as the ordinary traffic, then Var η^* will be approximately 3 times larger, compared with uncontaminated traffic with the same parameters. Obviously, this will change the relative efficiency in $\tilde{\tilde{\eta}}$'s favour.

Now, suppose that $\lambda_2 \to 0$ and $\eta_2 \to 0$ in such a way that $\lambda_2\eta_2 \to C$. This implies that the contamination permanently occupies C lines. This will be relevant in exchanges where some lines are permanently assigned to subscribers, for use within an inter-office network or for data-transmission between a terminal and a data center. The above asymptotic considerations will not be valid in this situation. These (infinitely long) calls will never end, and will thus not enter the estimator η^*. η^* will then estimate η_1, the mean duration of the ordinary traffic, and will not be influenced by the permanently occupied lines. This will not be so for $\tilde{\eta}$. We have

$$\tilde{\eta} = \frac{Z_t + (A_t-B_t)/2}{(A_t + B_t)/2}\Delta$$

or the similar definition of $\tilde{\tilde{\eta}}$. Since the permanently occupied lines are always busy, A_t and B_t will refer only to arrivals and disconnections of ordinary traffic. Z_t, on the other hand, will be the sum of the number of ordinary calls in progress and the C permanently occupied lines. Therefore

$$\tilde{\eta} = \frac{C + Z_t + (A_t - B_t)/2}{(A_t + B_t)/2}\,\Delta$$

where Z_t is the number of lines carrying ordinary traffic. Another Taylor expansion gives

$$E\,\tilde{\eta} \approx \eta_1/(1-\beta), \quad \mathrm{Var}\,\tilde{\eta} \approx \frac{\eta_1}{\lambda_1}\left[1 + \frac{\eta_1}{\Delta(1-\beta)^2} - \frac{p_1}{2}\left(1+\frac{\eta_1}{\Delta(1-\beta)}\right)^2\right]$$

where $\beta = C/(C+\lambda_1\eta_1)$ is the proportion of the permanently occupied lines of the total traffic, and the parameters λ_1 and η_1 refer to the ordinary traffic. In this situation, η_1 will be the parameter of interest, and $\tilde{\eta}$ may overestimate η_1 seriously if β is not close to zero.

For a specific exchange, C will normally be known. The problem above can then be avoided by reducing Z_t in the formula for $\tilde{\eta}$ (or $\tilde{\tilde{\eta}}$) by C. Then Z_t, as well as A_t and B_t, will refer to ordinary traffic, and the estimator will have the properties outlined above.

3. APPLICATION ON REAL DATA

We have applied the various techniques to an actual set of data, consisting of measurements from an ARM 20 exchange in Copenhagen, Denmark, for the period from 15.37 on Friday October 1 to 10.53 on Monday October 4, 1976. The original database contains sufficient information to compute $\tilde{\eta}$ as well as η^*.

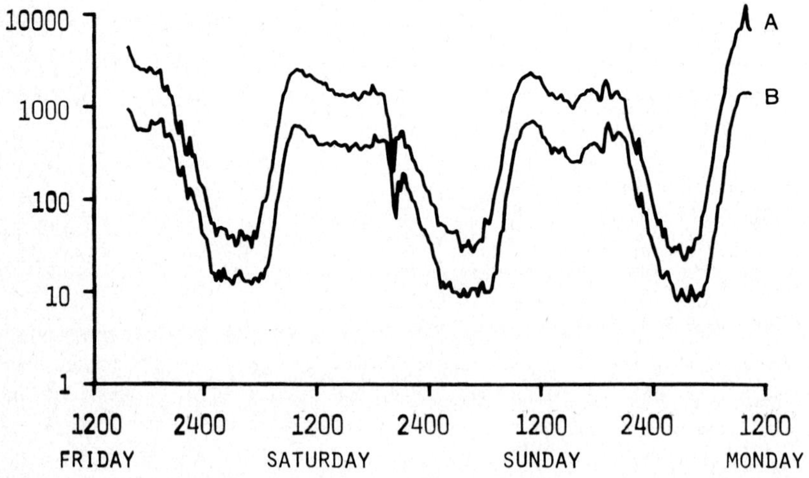

Figure 1. Arrivals (A) and traffic (B) during the observation period.

Figure 1 shows the number of observations in each 15 minute period (Δ = 900), and the number of calls in progress, observed every 15 minutes. Note the logarithmic scale. There is a distinct 24-hour pattern, with low traffic during the night, and there is a strong increase in the traffic towards the end of the period, which includes the "busy hour" on Monday morning. On Monday morning, between 9.55 and 10.15, the data includes an overload experiment, where artificial calls were introduced into the exchange. The purpose of the experiment was to study the effects on the exchange, not on the network. So, the generated calls were terminated immediately, having almost no duration. This brought the mean duration down for the period.

The traffic through the exchange is to some extent contaminated, and the hypothesis of pure Poisson traffic is dubious. A total of 300,500 calls were registered during the observation period. Out of these 214, or 0.071 %, had a duration of more than 1 hour, and 29, 0.010 %, lasted more than 2 hours. Assuming exponential durations with mean 3 minutes, the probabilities of these events are approximately 2.10^{-7} % and 4.10^{-16} % respectively. Table 4 shows the number of terminated calls according to their duration, with special emphasis on the long calls.

Duration in minutes	<60	[60-90)	[90-120)	[120-1000)	⩾1000
Number	300,321	158	27	19	10
% of total	99.93	0.053	0.009	0.006	0.003

Table 4. Distribution of the duration of terminated calls, with special emphasis on the very long ones.

Although the number of calls with duration more than 1 hour is small, these will on the average occupy 12 lines. Compared to the mean traffic of 275 lines occupied, this amounts to appr. 4.4 %, and considerably more during low-traffic periods. With reference to the previous robustness discussions, this contamination may have a notable effect on the estimation.

In Figure 2 we show the estimated mean duration for each 15 minute period, estimated by four more or less different methods. Figure 2(a) gives the estimate η^*, that is the mean of all calls terminated during each interval. The most pronounced feature of the figure is the extremely large value for the interval [2100-2115) on Saturday night, 940.3 seconds. This is because three of the 471 calls disconnected during this interval had a duration of almost 30 hours, that is more than 100,000 seconds. This of course leads to a very large mean value.

Figure 2(c) shows the pattern of the estimator $\overset{\approx}{\eta}$, as defined earlier. Comparison

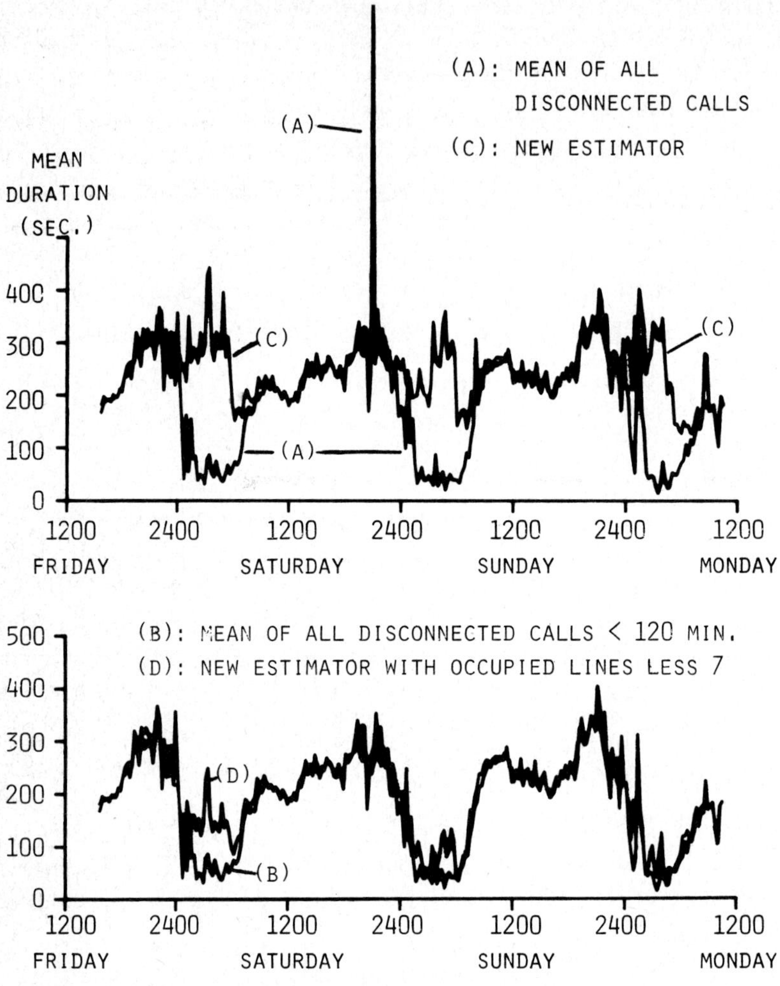

Figure 2. Estimated mean duration for each 15 minute period (see text).

of (a) and (c) shows that the major difference is that the estimated durations in (c) do not drop as low during the night as do the ones in (a). Remembering that the traffic is to some extent contaminated by very long calls, this is in accordance with the theoretical results. During the night the traffic is low, and most of the calls are short. The estimator η^*, which is based on the mean of the calls actually disconnected, will estimate the mean duration of the ordinary (short) traffic. $\tilde{\tilde{\eta}}$, on the other hand, will always estimate the mean of the total, mixed traffic. When the traffic is low during the night, the proportion of the very long calls will increase, and the mean duration of the mixed traffic may, in fact, increase even if the mean of the ordinary traffic decreases. This behaviour leads to a cross-correlation between the two estimated series of 0.26 only.

Figure 2(b) is very similar to Figure 2(a). The estimator is still η^*, but the calls lasting more than 2 hours have been excluded. This does not result in large changes; the large value on Saturday night disappears and a few other values are slightly reduced.

There are 7 lines constantly occupied during the whole observation period. Following the argument in Section 2.5 we re-calculated the $\tilde{\tilde{\eta}}$-estimates with Z_t reduced by 7 for the whole period. The resulting series is shown in Figure 2(d). This series now estimates the ordinary traffic, excluding the contamination. So does the estimates in Figure 2(b), and a visual comparison of the two series shows a rather close agreement. This agreement can also be seen from the cross-correlation, which is 0.91!

Table 5 gives the grand mean, variance around the mean and cross-covariance matrix for the four series of estimates.

	Mean	Variance	Cross-correlations			
			a	b	c	d
a - Mean of all disconnected calls	190.9	10962	1.0	0.90	0.26	0.84
b - Mean of all calls lasting < 120 min.	186.9	8718		1.0	0.28	0.91
c - The estimator $\tilde{\tilde{\eta}}$	248.5	3286			1.0	0.47
d - The estimator $\tilde{\tilde{\eta}}$ with Z_t less 7	198.4	5500				1.0

Table 5. Means, variances and cross-correlations between the four series of estimates (see text).

3.1 Time series analysis of the duration process

Now, suppose that the true arrival intensity and mean duration are not constant, but vary over time, so that the intensity and mean duration in interval I_t are λt and η_t, respectively. It is fairly obvious that the (unobservable) η_t series will be highly autocorrelated, and a look at Figure 2 sheds considerable doubts as to whether it is stationary. Figure 3 shows the estimated autocorrelation function for the series (d), based on the estimator $\tilde{\tilde{\eta}}$ with Z_t reduced by 7. There is a slow almost linear decay, confirming the near non-stationarity. Usual Box-Jenkins analysis results in the model

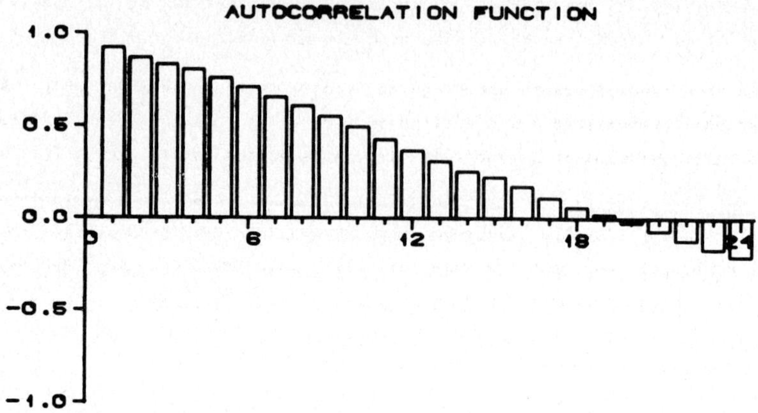

Figure 3. Autocorrelation function for the series with estimated mean durations, using method (d).

$$\underset{(0.02)}{(1-0.95B)}(\tilde{\tilde{\eta}}_t - \underset{(27)}{198}) = \underset{(0.07)}{(1-0.19B)}a_t, \quad \sigma_a = 28.97.$$

We are thus in the following situation: The unobservable series η_t has an unknown correlation structure. The observed estimated series, $\tilde{\tilde{\eta}}_t$, is almost unbiased for η_t, and has a known correlation structure, conditional on λ_t and η_t. This structure is given in Section 2, and involves η_t as well as λ_t. The unconditional correlation structure for the estimated series, though not known, can be described by the ARMA-model above.

Question: How to filter the $\tilde{\tilde{\eta}}_t$ series to obtain an optimal, smooth estimate of η_t?

APPENDIX: SIMULTANEOUS DISTRIBUTION

A.1 Assumptions

We use the same notation that was introduced in Section 2; and, as before, suppose the assumptions of the classical M/M/∞ model to be fulfilled.

A.2 Conditional distribution

We first derive the conditional distribution of B_t given $Z_t = z$ and $A_t = a$. To simplify notation, in the following we omit the subscripts t. Define the events

C_k: Exactly k of the z lines occupied at the start of I are disconnected during I, $k=0,1,\ldots,z$.

D_k: Exactly k of the calls arriving during I are also disconnected during I, $k=0,1,\ldots,a$.

Then

$$\Pr(B=b|Z,A) = \sum_{k=\max(0,b-a)}^{\min(z,b)} \Pr(C_k|Z,A)\Pr(D_{b-k}|Z,A) \tag{A1}$$

since all the C (and D) events are disjoint, and C_i is independent of D_j for all i and j, because C_i refers to calls having arrived before the start of I, whilst D_j refers to calls arriving in I, and calls arrive independently.

When the duration of a call is exponential with a given parameter, the remaining duration, conditional on the call being in progress at a certain point in time, will be exponential too, with the same parameter. Thus Pr(a call, in progress at the start of I, is disconnected during I) is

$$p = \int_0^{\Delta} \mu e^{-\mu t} dt = 1 - e^{-\mu\Delta}.$$

Since the calls are independent, C_k will be binomial with

$$\Pr(C_k|Z) = \binom{z}{a} p^k (1-p)^{z-k} \quad (k=0,1,\ldots,z). \tag{A2}$$

Let s be a random time point in I. Then Pr(a call arriving at s is disconnected during I) is

$$\int_0^{\Delta-s} \mu e^{-\mu t} dt = 1 - e^{-\mu(\Delta-s)}.$$

So Pr(a call arriving in I is disconnected during I) is

$$q = \int_0^{\Delta} (1-e^{-\mu(\Delta-s)})/\Delta \, ds = 1 - (1-e^{-\mu\Delta})/(\mu\Delta).$$

Again, calls are independent, so D_k will also be binomial with

$$\Pr(D_k|A) = \binom{a}{k} q^k (1-q)^{a-k} \quad (k=0,1,\ldots,a). \tag{A3}$$

When (A2) and (A3) are inserted in (A1), we obtain the conditional distribution

$$\Pr(B=b|Z,A) = \sum_{k=\max(0,b-a)}^{\min(z,b)} \binom{z}{k} p^k (1-p)^{z-k} \binom{a}{b-k} q^{b-k} (1-q)^{a-b+k} \quad (b=0,1,\ldots,z+a) \tag{A4}$$

A.3 Unconditional distribution

Z and A are independent. Z is Poisson with parameter λ/μ, and A is Poisson with parameter $\lambda\Delta$. This, combined with (A4), gives the required probability density

$$\Pr(Z=z,\ A=a,\ B=b) = \frac{(\lambda/\mu)^z (\lambda\Delta)^a}{z!\quad a!} e^{-\lambda(\Delta+1/\mu)}.$$

$$\sum_{k=\max(0,b-a)}^{\min(z,b)} \binom{z}{k} p^k (1-p)^{z-k} \binom{a}{b-k} q^{b-k} (1-q)^{a-b+k} \quad (0\leq a,\ 0\leq z,\ 0\leq b\leq z+a). \tag{A5}$$

A.4 Moments

As shown above, we can write B = C + D, where the conditional distributions of C and D, given Z and A, are independent binomial with parameters (z,p) and (a,q), respectively. Then $E(B|Z,A) = zp + aq$ and $\mathrm{Var}(B|Z,A) = zp(1-p) + aq(1-q)$.

This, and the usual formulae for operations on conditional moments, gives $E(B|Z) = \lambda\Delta + p(z-\lambda/\mu)$, $Var(B|Z) = zp(1-p) + \lambda\Delta q$ and $Cov(A,B|Z) = \lambda\Delta q$. Further, $EB = \lambda\Delta$, $Var\ B = \lambda\Delta$, $Cov(A,B) = \lambda\Delta q = \lambda\Delta - p\lambda/\mu$ and $Cov(A,Z) = p\lambda/\mu$. Finally, $EA = E(A|Z) = \lambda\Delta$, $Var\ A = Var(A|Z) = \lambda\Delta$, $Cov(A,Z) = 0$, $EZ = \lambda/\mu$ and $Var\ Z = \lambda/\mu$.

A.5 Autocovariance

Let us reintroduce the subscript t for Z, A and B. Then $Z_t = Z_{t-1} + A_{t-1} - B_{t-1}$, and we find

$$Cov(Z_t,Z_{t-1}) = Var\ Z_t + Cov(A_t,Z_t) - Cov(B_t,Z_t) = (\lambda/\mu)(1-p) = (\lambda/\mu)\exp(-\mu\Delta)$$

$$Cov(Z_t,Z_{t-k}) = (\lambda/\mu)(1-p)^k = (\lambda/\mu)\exp(-\mu\Delta k).$$

For the disconnections we get $Cov(B_t,B_{t-k}) = 0$, and the arrival process is uncorrelated, by definition.

For the cross-covariances, we find in a similar way

$$Cov(Z_t,A_{t-k}) = \begin{cases} 0 & k \leqslant 0 \\ (\lambda/\mu)p(1-p)^{k-1} & k \geqslant 1 \end{cases}$$

$$Cov(B_t,Z_{t-k}) = \begin{cases} 0 & k < 0 \\ (\lambda/\mu)p(1-p)^{k} & k \geqslant 0 \end{cases}$$

$$Cov(B_t,A_{t-k}) = \begin{cases} 0 & k < 0 \\ \lambda\Delta - (\lambda/\mu)p & k = 0 \\ (\lambda/\mu)p^2(1-p)^{k-1} & k \geqslant 1. \end{cases}$$

TIME SERIES ANALYSIS: Theory and Practice 7
O.D. Anderson (editor)

ROUTING ALGORITHMS BASED ON TRAFFIC FORECAST MODELLING

Kjell Stordahl
Norwegian Telecommunications Administration Research Establishment, P O Box 83, N-2007 Kjeller, Norway

Installation of digital exchanges in the telephone network has opened up new possibilities for implementing advanced automatic control actions in the system. The objective of routing algorithms is to optimize the traffic flow in the network. In this paper, forecasts of the traffic on the routes in the network are based on the interarrival process and the holding time process of calls. Different forecasting models for the processes are analyzed and used as an input to the routing algorithm procedures. The evaluation of the models is carried out by simulation of a telephone network. A similar approach has also been used by Kårstad and Stordahl (1983).

1 INTRODUCTION

In February 1979 Bell Canada introduced centralized dynamic routing in a local area consisting of nine SPC switching machines in the Toronto metropolitan network. Results from the trial and simulation experiments are given in Szybicki (1979a, b), Cameron (1980) and Regnier (1983).

Figure 1.1 shows a network with centralized dynamic routing.

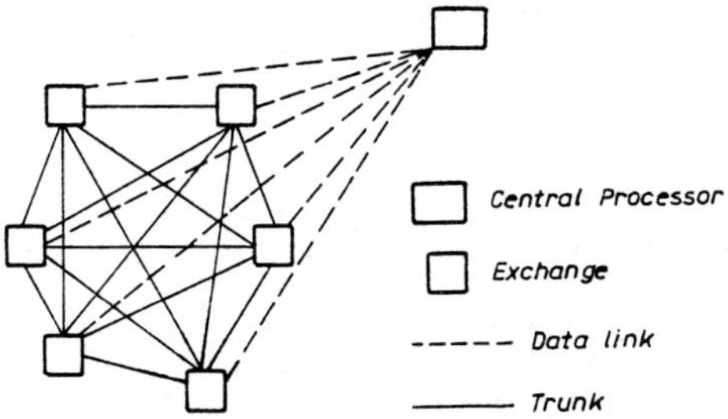

Figure 1.1 Network with centralized traffic routing

Between the exchanges there is a fully meshed network. Therefore, there will be a huge number of routing possibilities. The central processor collects data from exchanges on the actual trunk states in the different trunk groups, and at equidistant time intervals it calculates optimal alternative paths and sends back the information to the exchanges. Because of the time required for transmission and processing there will be a time delay between the times when the trunk states have been recorded at each exchange and when a new routing scheme is received at the exchanges. The fact that it is generally an advantage to get an indication of future traffic, makes it very important to base the calculation of alternative paths on forecasts of the traffic on each trunk group.

2 TRAFFIC FORECAST MODELLING

2.1 The traffic process

Telephone traffic follows a certain pattern during day and night. In a control model it is of course desirable to have forecasts of the traffic along different routes in the network.

In Stordahl (1980) it is shown that the traffic process can be described as an autoregressive process or a random walk depending on the length of the observation period Δ.

This is also confirmed in Torp (1982) where data from an ARM-exchange are used to model the traffic process on trunk groups. The following models have been found when the observation period Δ = 10 sec.

$$U_t = U_{t-1} + a_t \tag{2.1}$$

$$U_t - \xi = \phi_1(U_{t-1} - \xi) + a_t \tag{2.2}$$

where U_t is number of occupied trunks at time t, ξ is mean number of occupied trunks and a_t is white noise at time t.

An alternative model of the traffic process on trunk groups is expressed by the mean number of occupied trunks V_t during the observation period t. Then

$$V_t = V_{t-1} + a_t \tag{2.3}$$

$$V_t - \xi = \phi_1(V_{t-1} - \xi) + a_t. \tag{2.4}$$

It is, however, possible to improve the forecasting model by separating the traffic process into an interarrival process and a holding time process and then make separate forecasts of these processes.

2.2 The holding time process

Measurement in the Norwegian telephone network has shown great variations in the holding time during day and night. The holding time is shortest during the night, longest in the afternoon.

In Szybicki and Bean (1979a) it is shown that the holding time process $\{H_t\}$ can be described as an autoregressive process of second order

$$H_t-h = \phi_1(H_{t-1}-h)+\phi_2(H_{t-2}-h)+a_t \tag{2.5}$$

if the observation period is 15 minutes, where h is the expected holding time.

If, however, the observation period is short, i.e. about 10-30 seconds, the holding time $\{h_t\}$ can be expressed by an IMA(1,1)-model.

$$h_t = h_{t-1} + a_t - \theta a_{t-1}. \tag{2.6}$$

The model fits very well when there is congestion in the network which causes a number of calls with very short holding times.

For every observation period we have to estimate the holding time. This can be done by calculating the average of all calls which are completed during the period. This means:

$$h_t^{(1)} = \frac{1}{n_t} \sum_{i=1}^{n_t} X_i \tag{2.7}$$

where X_i is the length of call no i, and n_t is the number of completed calls in the observation period.

Using this formula it is necessary to record the start and end times for all calls. Because of the limited memory in the control routing processor and the limited capacity on the data links, the sampling of all holding times is not feasible. An alternative estimator for the mean holding time is then given by

$$h_t^{(2)} = \frac{2A_{t-1}+\Lambda_t-\Gamma_t}{(2-C_1)\Lambda_t+C_1(\Gamma_t+1)}\Delta \tag{2.8}$$

where A_{t-1} is number of occupied trunks at time t-1, Λ_t is number seizures during interval no t, Γ_t is number of releases during interval no t, and C_1 is a constant.

In the simulation model the values $C_1 = 0$ and $C_1 = 1$ have been used. The estimator is analyzed in more detail by Damsleth (1983).

It is important to underline that the variance of the estimators (2.7) and (2.8) increases when the observation period, Δ, decreases.

2.3 Number of releases

The forecast of the traffic on a trunk group can be defined as traffic at time t plus forecast of the number of arrivals minus forecast of number of releases during the forecast period.

The number of releases, Γ_t, during the period t may be given by the IMA(1,1)-model

$$\Gamma_t = \Gamma_{t-1} + a_t - \theta a_{t-1}. \tag{2.9}$$

It is, however, possible to make a better forecasting model of the number of releases taking into account the holding time. In a telephone network the holding time distribution is nearly exponential. Hence, a forecast of the number of releases during a forecasting period of length τ is expressed by $A_t \cdot \tau/\hat{h}_t(1)$, where $\hat{h}_t(1)$ is the one step a head forecast of the holding time.

2.4 Number of seizures

Like the holding time process, the interarrival process also has a characteristic day and night variation. In addition events in the community may give peak traffic in some areas. In Torp (1983) it is shown that the stochastic variation of the number of seizures during an interval Δ of about 15 seconds can be described as an IMA(1,1)-model.

$$\Lambda_t = \Lambda_{t-1} + a_t - \theta a_{t-1}. \tag{2.10}$$

In some cases it is necessary to use a different length for the forecasting interval τ and the observation interval Δ. The interarrival intensity $\lambda = \Lambda/\Delta$ is then used as a basis for the forecasting.

2.5 Different forecasting models

Suppose the actual time is T and the one step ahead forecast of the traffic is $A(T+\tau)$. Since the observation period of the interarrival process, Δ_1, and the holding time process, Δ_2, may be of different size, we denote the period which includes the interval $[T, T+\tau]$ by t_1+1 and t_2+1 respectively.

The following forecasting models were evaluated before the simulation based on an overload experiment of an ARM exchange was run.

1) $$\hat{A}(T+\tau) = \hat{U}_t(1) \qquad (2.11)$$

where $\hat{U}_t(1)$ is the one step ahead forecast based on (2.1) and (2.2) respectively.

2) $$\hat{A}(T+\tau) = \hat{V}_t(1) \qquad (2.12)$$

where $\hat{V}_t(1)$ is the one step ahead forecast based on (2.3) and (2.4) respectively.

3) $$\hat{A}(T+\tau)=A_t+\frac{\tau}{\Delta_1}\hat{\Lambda}_{t_1}(1)-\frac{\tau}{\Delta_2}\hat{\Gamma}_{t_2}(1) \qquad (2.13)$$

where $\hat{\Lambda}_{t_1}(1)$ and $\hat{\Gamma}_{t_2}(1)$ are the one step ahead forecasts of the interarrivals and releases based on (2.10) and (2.9) respectively.

4) $$\hat{A}(T+\tau) = A_t + \frac{\tau}{\Delta_1}\hat{\Lambda}_{t_1}(1)-\frac{\tau}{h}A_t \qquad (2.14)$$

where h is a non-adaptive estimator of the expected holding time.

5) $$\hat{A}(T+\tau) = A_t + \frac{\tau}{\Delta_1}\hat{\Lambda}_{t_1}(1)-\frac{\tau}{\hat{H}_{t_2}(1)}A_t \qquad (2.15)$$

where $\hat{H}_{t_2}(1)$ is the one step a head forecast based on (2.5).

6) $$\hat{A}(T+\tau) = A_t + \frac{\tau}{\Delta_1}\hat{\Lambda}_{t_1}(1)-\frac{\tau}{\hat{h}_{t_2}(1)}A_t \qquad (2.16)$$

where $\hat{h}_{t_2}(1)$ is the one step ahead forecast based on (2.6).

The forecasting models were compared by use of the root mean square estimator. All forecasting models gave fairly good predictions when the traffic load was normal. In an overload situation, especially the forecasting models (2.14) and (2.15) turned out to give poor forecasts of the traffic. In such a situation a number of short holding times are generated. It is therefore important to use a forecasting model which adaptively reestimates the holding times.

When the autoregressive and moving average parameters are estimated, only simple calculations are necessary for finding the one step ahead forecasts. Since the forecasting models are expressed by autoregressive models with one or two parameters or the IMA(1,1)-model, it is not necessary to store much traffic data in the central processor. The one step ahead forecast for the interarrival process is given by:

$$\hat{\Lambda}_{t_1}(1) = (1-\theta)\Lambda_{t_1} + \theta\hat{\Lambda}_{t_1-1}(1). \qquad (2.17)$$

Here, $\hat{\Lambda}_{t_1}(1)$ is the last forecast and Λ_t the last observed value. In a similar way the one step ahead forecast for the holding time process is expressed by:

$$\hat{h}_{t_2}(1) = (1-\theta)h_{t_2} + \theta\hat{h}_{t_2-1}(1). \qquad (2.18)$$

Equation (2.17) and (2.18) may also be formulated by Kalman filter theory. (Knutsson 1979).

3 THE CONTROL STRATEGIES

3.1 Different control strategies

Some protection against instabilities should be introduced in a network. That is, we want the probability of connecting a call on alternative routes to drop off more rapidly as the carried load builds up. This can be attained by reserving some trunks in each trunk group for direct traffic only. An effective way to improve the traffic flow is to combine trunk reservation and adaptive traffic routing techniques.

3.2 The routing algorithms

These routing algorithms are developed by Szybicki and Bean (1979b). In a fully meshed network a call is first offered the direct route. If this route is blocked, the call is routed to an alternative path.

Let $P_{ij}(\tau)$ be a forecast for the number of idle trunks in the trunk group between exchanges i and j, τ time units ahead. If the direct route is occupied, the call may be routed via the exchange k which maximizes min $(P_{il}(\tau), P_{lj}(\tau))$ for all tandem exchanges l. This means:

$$R_l(\tau) = \min\,(\hat{P}_{il}(\tau), \hat{P}_{lj}(\tau)) \qquad (3.1)$$

$$R_k(\tau) = \max_l\,(R_l(\tau), 0) \qquad (3.2)$$

In a network with heavy traffic load it may be better to use a routing algorithm which selects the route via exchange l with probability $p_l(\tau)$:

$$p_l(\tau) = \frac{R_l(\tau)}{\sum_j R_j(\tau)} \qquad (3.3)$$

This routing algorithm will never systematically place overflow calls on the trunk group with highest number of free trunks.

Taking into account the trunk reservations, the forecast τ time units ahead for the number of free trunks between exchanges i and j for an overflow call is expressed by:

$$\hat{P}_{ij}(\tau) = N_{ij} - \hat{A}_{ij}(T+\tau) - X_{ij} \tag{3.4}$$

where N_{ij} is the number of trunks bewtween i and j, X_{ij} is the number of reserved trunks for direct routing i to j, and $\hat{A}_{ij}(T+\tau)$ is the one step a head forecast for the traffic between i and j.

The different forecasting models for the traffic given by equations (2.11) to (2.16) may be substituted into equation (3.4).

4 THE SIMULATION STUDY

4.1 The simulation model

The programme describes a model of a network of exchanges connected by trunk groups. Each exchange has a queue where a call may wait for a free outgoing trunk.

We have simulated a fully meshed network with six exchanges and 20 trunks between each ordered pair of exchanges. The traffic is generated in such a way that the total offered traffic between each pair of exchanges follow the traffic pattern profile in Figure 4.1. Figure 4.2 shows the holding time variation. Two types of traffic are investigated: (1) The total offered traffic between each pair of exchanges is the same; and (2) The total offered traffic towards one of the exchanges is 10 percent higher than the traffic towards the other exchanges (focused overload).

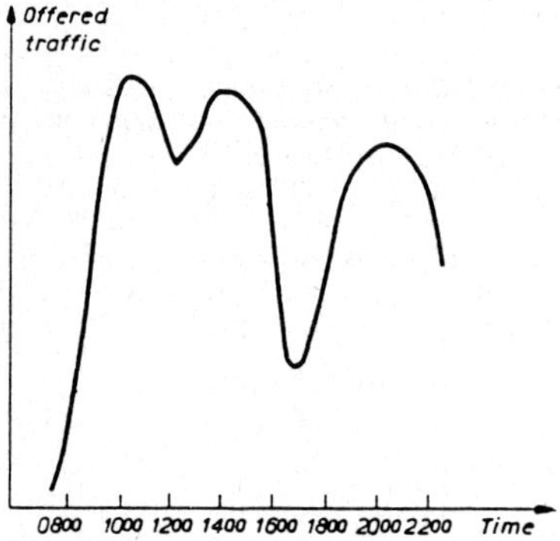

Figure 4.1 Expected offered traffic as a function of time (day and night variation of offered traffic)

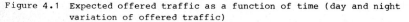

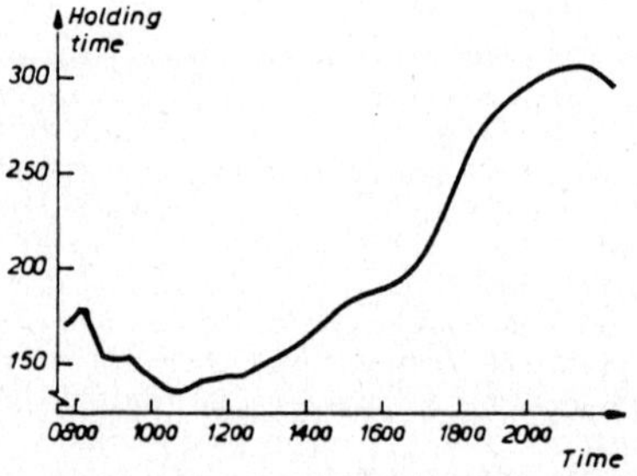

Figure 4.2 Expected holding time as a function of time (day and night variation)

4.2 The length of the simulation and the number of repeated simulations

In each programme-run approximately 90,000 calls are generated. The holding times and interarrival times are exponentially distributed. The parameters in the exponential distributions change in such a way that the mean holding time and the mean offered day time traffic satisfy the curves in Figure 4.1 and Figure 4.2.

These changes in the parameters make it difficult to calculate the mean percentage of lost calls based on one programme-run only. Therefore it has been necessary to carry out 10 independent runs to calculate the standard deviation and the mean percentage of lost calls.

5 THE SIMULATION RESULTS

In this section the effects of the following control strategies and different combinations of them will be considered: (1) Fixed alternative routing, (2) trunk reservation, (3) routing based on different forecast models.

It is important to underline that the results are based on the assumptions given in Section 4.

5.1 Trunk reservation

To describe the traffic load we have used a parameter α called the traffic factor. The variation of the traffic load is shown in Figure 4.1. Different α values represent different levels of the curve in Figure 4.1. Let a point on the curve which is represented by $\alpha = 1.00$ be given by (t,χ), then the corresponding point on the curve represented by $\alpha = 1.08$ is given by $(t,1.08\chi)$, for $\alpha = 1.16$ the point has the coordinates $(t,1.16\chi)$. In two simulations, with different α-values, the traffic factor α represents the variation in the offered traffic without regard to the simulation time.

As a reference we have investigated fixed alternative routing. We have gone through repeated simulations, with the traffic factor α at 1.00, 1.08 and 1.16. Table 5.1 lists the results. The standard deviation of the percentage of lost calls is estimated.

Table 5.1 The total percentage of lost calls as a function of the traffic factor α

Traffic factor α	1.00	1.08	1.16
Percentage of lost calls	4.9	12.5	18.4
Standard deviation	0.19	0.26	0.24

The probability that the estimate of the percentage of lost calls shall diverge by more than twice the standard deviation from the true value is approximately 0.05, if the model is correct. With those small standard deviations, it is obvious that the length of the simulations and the number of simulations are sufficient.

Trunk reservation greatly improves the network capacity. Table 5.2 shows how the percentage of lost calls changes as a function of the number of reserved trunks. The traffic factor α is equal 1.08.

The Table shows that there is an optimal number of trunks which should be reserved. This number will of course depend on the traffic and the number of trunks in a trunk group. As already mentioned, all the trunk groups in the network have 20 trunks.

Table 5.2 The total percentage of lost calls as a function of the number of reserved trunks, α = 1.08, fixed alternative routing.

The number of resesved trunks	0	1	2	3
Percentage of lost calls	12.5	4.2	2.9	3.5
Standard deviation	0.26	0.07	0.04	0.05

5.2 Routing based on forecasts

Now we introduce routing based on forecasts. First we study two different forecasting methods. The methods are based on equation (2.15) and (2.16) respectively. The forecasting model F1 is defined by:

The time between the observations of the interarrival process is 15 seconds.
θ = 0.9 in the forecast model (2.17) for the interarrival process.
The time between the observations of the holding time process is 15 minutes.
$\phi_1 = 0.6$, $\phi_2 = 0.3$, h = 199 in the forecasting model for the holding time process.
The look ahead time τ in equation (2.15) is τ = 10 seconds.

The forecasting model F2 is defined by:
The time between the observations of the interarrival process is 15 seconds.
θ = 0.9 in the forecasting model (2.17) for the interarrival process.
The time between observations of the holding time process is 30 seconds.
θ = 0.8 in the forecasting model (2.18) for the holding time process.

The forecast time τ in equation (2.16) is τ = 10 seconds.

Table 5.3 The total percentage of lost calls as a function of the traffic factor α. The routing is based on method F1 or F2. No reserved trunks.

Forecast method	The traffic factor α	1.00	1.08	1.16
F1	Percentage of lost calls	2.81	5.69	8.93
	Standard deviation	0.08	0.07	0.09
F2	Percentage of lost calls	1.13	2.94	5.40
	Standard deviation	0.05	0.06	0.06

Table 5.3 shows that the forecasting method F2 is considerably better than F1. As pointed out in Section 2, it is important to reestimate the holding time frequently.

5.3 Focused overload

As mentioned in Section 4.1 the different control strategies are investigated during the time when focused overload occurs in the network. In the simulation study we have generated very high traffic towards one of the six exchanges.

The total sum of the generated traffic is the same as before. In addition to forecast methods F1 and F2, forecast methods F3, F4 and F5, defined below, are considered.

Forecasting model F3 is defined by:

The time between the observation of the interarrival process is 15 seconds.
$\theta = 0.9$ in the forecast model (2.17) for the interarrival process.
The time between the observations of the holding time process is 5 minutes.
$\phi_1 = 0.7$, $\phi_2 = 0.2$, $h = 199$ in the forecasting model (2.5) for the holding time process.
The forecast time τ in equation (2.15) is $\tau = 10$ seconds.

Forecasting model F4 is defined by F3 with the exception of θ which is 0. This means that the forecast for the interarrival process is based only on what has happened during the last 15 seconds.

Forecasting model F5 is defined by F3 with the exception of ϕ_1 and ϕ_2, which are $\phi_1 = 1$ and $\phi_2 = 0$. This means that the forecast for the holding time process is based only on what has happened during the last 5 minutes.

Table 5.4 shows the effect of the different forecast methods.

Table 5.4 The total percentage of lost calls as a function of the traffic factor α. Very high traffic towards one of the exchanges (focused overload). No reserved trunks.

Forecast method	The traffic factor α	1.00	1.08	1.16
F2	Percentage of lost calls	1.26	3.18	5.39
	Standard deviation	0.05	0.07	0.05
F3	Percentage of lost calls	1.80	4.79	8.13
	Standard deviation	0.05	0.07	0.14
F4	Percentage of lost calls	1.68	4.77	8.26
	Standard deviation	0.04	0.10	0.12
F5	Percentage of lost calls	1.97	5.28	8.95
	Standard deviation	0.07	0.09	0.13

Comparing the results in Table 5.3 and 5.4 we see that the forecast method F3 is better than F1, in spite of the fact that F3, as opposed to F1, is studied during focused overload conditions. The only difference between F1 and F3 is that the forecast period for the holding time process is 15 minutes for F1 and 5 minutes for F3. We may therefore conclude that the forecast period for the holding time should be relatively short.

In forecast model F5 the forecast for the holding time process is based only on what has happened during the last 5 minutes. From Table 5.4 we see that this is not sufficient, the forecast method F4 is better. In F4 the forecast of the interarrival process is based on what has happened during the last 15 seconds only, but the holding time forecasts are as in F3.

6 CONCLUSIONS

A number of different forecasting models have been evaluated by simulation experiments. The performance of some of the models is described in Section 5. The simulation experiments show that the forecasting model F2 is best.

Introduction of routing of the traffic based on forecasts improves the amount of carried traffic in the network.

In addition, trunk reservation is an effective way of controlling the traffic flow. This protection mechanism also increases carried traffic when the traffic load in the network is high.

7 ACKNOWLEDGEMENT

The author is indepted to T Kårstad for work with the simulation programme.

REFERENCES

CAMERON W.H., GALLOY P., GRAHAM W.J. (1980). "Report on the Toronto advanced routing concept trial". (Telecommunication Network Planning Conference, Paris 1980).

DAMSLETH E. (1983). "Estimation of the mean duration of telephone calls". (ITSM 11, Toronto 1983, In these Proceedings).

KNUTSSON Å. (1979). "Modelling for automatic network management: An adaptive time-recursive parameter estimator and traffic predictor". (ITC 9, Torremolinos 1979).

KÅRSTAD T., STORDAHL K. (1983). "Centralized routing based on forecasts of the telephone traffic". (ITC 10, Montreal 1983).

REGNIER J., BLONDEAU P., CAMERON W.H.(1983). "Grade of service of a call-routing system". (ITC 10, Montreal 1983).

STORDAHL K. (1980). "Analysis of telecommunication data by use of ARIMA-models". In Forecasting Public Utilities (Proceeding of the International Conference held at Nottingham University, England, March 1980). Ed: O.D. Anderson, North Holland, Amsterdam & New York, 81-99.

SZYBICKI E., LAVIGNE M.E. (1979a). "The introduction of an advanced routing system into local digital networks and its impact on the networks economy, reliability and grade of service". (ISS, Paris 1979).

SZYBICKI E., BEAN A.E. (1979b). "Advanced traffic routing in local telephone networks; performance of proposed call routing algorithms". (ITC 9, Torremolinos 1979).

TORP T. (1982). "Analysis of time series from an ARM 20 exchange". (Norwegian Computing Center No 721).

TIME SERIES ANALYSIS: Theory and Practice 7
O.D. Anderson (editor)

IMPROVING FORECAST ACCURACY THROUGH COMPOSITE FORECASTS

John E. Triantis
Overseas Department, AT&T Long Lines, 201 Littleton Road, Morris Plains, N.J. 07950, USA

Forecasting in a public utility environment usually necessitates use of econometric models that measure responses to changes in exogenous variables, especially rates. On the other hand, alternative types of time series models can produce superior forecasts. In order to reflect exogenous changes in the forecast and benefit from the forecast of another model, we derive a composite prediction from the two alternative forecasts. In this study, we estimate an OLS model with ARIMA on its residuals and a state space model. We generate and investigate the individual forecasts and then combine these to obtain a more accurate prediction.

1. INTRODUCTION

Demand modeling and forecasting of international message telecommunication service (IMTS) has taken on a new dimension in the research effort as numerous market forces have become the dominating influence on IMTS demand. On the other hand, changes in economic conditions abroad, foreign network limitations, the complexities of coordinating installation of international facilities, advances in cable and satellite technology, expanded use of direct dial, enhanced transmission quality, and improved completion rates affect the extent to which demand for IMTS is stimulated or satisfied.

In a regulated environment, demand is usually modeled with econometric models which identify the structure that generates the demand data (see Taylor 1978). That is, econometric models identify the major factors influencing the decision to place an overseas call and estimate their impact on demand. However, due to their abstract nature, econometric models cannot identify and explicitly incorporate all the factors affecting demand. Furthermore, many of the elements that affect demand may not be quantified and/or forecast with any degree of confidence. Thus, one is lead to investigating models that use not only the information content of the explanatory variables but, also, the past behavior of the demand series and the forecast error structure.

Forecasters have been searching for a statistical tool which combines the ideas of econometric models, univariate ARIMA models, transfer functions, and multiple time series models while avoiding the shortcomings of the individual models (see Mahmoud 1983, Makridakis and Hibon 1979). In response to such need, forecasters have adopted the state space method which was developed for industrial control applications. With this technique, one can identify and estimate a dynamic model for multiple time series taking into account the effects of all leading, lagging, or feedback relationships within and between the component series. State space models handle (or avoid) the well known problems in econometrics, are superior to ARIMA models and transfer functions - in terms of conceptual framework - and can handle multiple time series forecasting at least as well as multivariate time series models.

The purpose of this paper is threefold: we first aim to construct econometric and state space models for IMTS to a relatively large country. The second objective is to analyze the predictive performance of the two models and the third goal is to explore potential benefits from combining forecasts from the two alternative models. In Section 2 we introduce the basic demand model which is estimated using OLS. The basic model is expanded in Section 3 by using ARIMA on the residuals of the OLS model and estimation results are shown. The state space model is briefly discussed in Section 4 along with empirical results and Section 5 contains the composite forecast methodology and the outcome of combining the forecasts. Section 6 concludes the paper with a brief summary and questions raised in this investigation that need further studying.

2. THE BASIC MODEL

For purposes of exposition, total demand for IMTS to the country under study is viewed as arising from a business and residence market that responds to the economic factors of rate levels, fluctuations in economic activity, and bilateral trade between the US and the particular country. The model specification for the traffic of messages billed in the US incorporates a traditional functional form that explains demand by the multiplicative influence of a well tested (in telephone demand) set of key exogenous variables each raised to the power of their respective elasticity coefficients.

Assuming no lagged adjustments, the exogenous factors hypothesized to explain IMTS demand for messages (OMSG) to the country being studied are: Real telephone rates (RRI), real US personal income (YP72), real bilateral trade (RTR), and seasonal factors (S). The basic model can be represented in a log-linear form as follows:

$$\begin{aligned} Ln(OMSG)_t &= b_0 + b_1 Ln(RRI)_t + b_2 Ln(YP72)_t + b_3 Ln(RTR)_t \\ &\quad + b_4(S1) + b_5(S2) + b_6(S3) + U_t \end{aligned}$$

where $b_1<0$, $b_2>0$, $b_3>0$ are expected and the seasonal dummies are defined as: S1 = 1 0 0 -1, S2 = 0 1 0 -1, S3 = 0 0 1 -1 for each of the years in the sample.

Demand theory dictates that changes in the real price affect inversely the quantity demanded, while economic activity - measured through real personal income - has a positive impact on demand for the reason that changes in economic activity affect the consumer's willingness to place an overseas call. Real trade affects positively IMTS demand and it underlies movements in capital flows, exchange rates, and to some extent interest rates.

Using quarterly data from 1970:1 to 1980:4 we estimate the demand function using OLS and the results are the following:

$$\begin{aligned} Ln(OMSG) = .6564 &\underset{(2.295)}{-.3184Ln(RRI)} + \underset{(4.248)}{.8844Ln(YP72)} + \underset{(7.412)}{.4358Ln(RTR)} \\ &\underset{(4.569)}{-.0357S1} \underset{(3.597)}{-.0331S2} + \underset{(12.470)}{.1171S3} \end{aligned}$$

$\bar{R}^2 = .922$, D-W = 1.1195, SER = .0297.

SER is the standard error of the regression and numbers in parentheses are t-statistics.

3. LEAST SQUARES WITH ARIMA

The presence of serial correlation in the residuals of the econometric model is indicative of some type of specification error which is, usually, translated into a large forecast error because the behavior of the residuals of the regression is largely left unexplained. Thus, it seems that a regression model used in conjunction with an ARIMA model on the residuals would constitute an improvement over the OLS model in terms of predictive performance.

Estimating an OLS model in tandem with a time series model for the residuals is, essentially, a generalization of the Cochrane-Orcutt procedure for estimating autoregressive processes in least squares regression. It is an iterative process which involves the following steps:

a. Estimate the original regression model to get fitted values and residuals, $\hat{Y}$ and $\hat{U}$ respectively.
b. Identify and estimate an ARIMA model of the residuals $\hat{U}$. This gives fitted values $\hat{W}$.
c. Calculate a whitened dependent variable, Z, by $Z = Y-\hat{W}$.
d. Re-estimate the regression using Z as the dependent variable and obtain $\hat{Z}$, its fitted value.
e. Obtain the new set of residuals $E = Y-\hat{Z}$.
f. Iterate the process from step (b) until the values of the residuals become stable in the sense that: $\Sigma(E-W)^2 / \Sigma(E^2) < \lambda$ where W are the residuals from the previous iteration and λ is some pre-specified limit.

The estimation results of the OLS with ARIMA model are shown below with t-statistics in parentheses.

$$\begin{aligned}\text{Ln(OMSG)} = 1.649 &- \underset{(3.451)}{.4218\text{Ln(RPI)}} + \underset{(4.629)}{.8491\text{Ln(YP72)}} + \underset{(7.816)}{.4049\text{Ln(RTR)}} \\ &- \underset{(5.156)}{.0355\text{S1}} - \underset{(3.777)}{.0307\text{S2}} + \underset{(13.73)}{.1137\text{S3}} + E\end{aligned}$$

$\bar{R}^2 = .994$, D-W = 2.044, SER = .0262.

SER is the standard error of the regression and E is estimated by the moving average model

$$E_t = -.002 + (1 + .526B)\, a_t; \quad \chi^2(23) = 16.219, \quad SE(a_t) = .152 .$$

The estimation results obtained from the OLS with ARIMA model are superior to the OLS results in terms of reduced standard errors of the regression coefficients and the residual and improved $\bar{R}^2$ and Durbin-Watson statistic. Increased forecast accuracy is also achieved by the OLS with ARIMA model as shown in Table 1.

4. THE STATE SPACE MODEL

State space models provide a unifying framework by bringing together concepts from Canonical Correlation Analysis, Control Theory, Kalmal Filtering, and Statistical Decision Theory (see Akaike 1976 and Mehra 1979). A key concept used in these models is the state vector which, for a multidimensional stochastic linear system, is defined as the basis vector of the linear space spanned by the components of the predictors of present and future values of the system based on its present and past values. More intuitively, the state vector of a system is made up of all the information from the past and present and future behavior.

From this definition of the state vector, it is apparent that a state space model encompasses more information than that contained in the set of explanatory variables of an OLS regression model. And, since this additional information content is projected into the future, one may reasonably expect greater forecast accuracy. A further advantage of state space models is that they are much less restrictive than OLS econometric models whose rigid structure is, usually, imposed on the data with little regard to the effects of several factors which are not explicitly included in the set of explanatory variables.

The form of the state space model developed is: $V_{T+1} = F.V_T + G.U_{T+1}$ and $Y_T = H.V_T$, where V_T is the state vector at time T, Y_T is the vector of observations of the time series at time T, U_{T+1} is the residual one step ahead forecast error for the observations at time T+1, F is the transition matrix obtained by canonical correlation analysis, G is the innovation input matrix, and H is the output matrix whose elements are the regression coefficients of Y_T on V_T.

It should be noted that the size of the transition, of the innovation input, and of the output matrices depends on the dimensions of the observation vector Y_T and the state vector V_T.

G and H are 4 x 4 identity matrices, while F is estimated as:

$$\begin{bmatrix} .9328 & .0450 & -.0087 & .0098 \\ -.0310 & -.0351 & -.4782 & .0448 \\ -.0065 & -.0313 & .4510 & .4198 \\ .3196 & -.2834 & -.1505 & .2995 \end{bmatrix} .$$

The residual mean vector of Ln(OMSG) is zero, and the normalized correlations for the residuals at different lags are within 2 standard deviations of zero with the exception of those at lags 3, 4, and 15. The goodness of fit for 16 degrees of freedom is 374.308 for the original data and 192.676 for the differenced data. The R^2 for the Ln(OMSG) variable is .9936 for the original data and .2589 for the differenced data.

5. THE COMPOSITE FORECAST

The studies by Cooper and Nelson (1975), Granger and Newbold (1977), Uri (1979), and Shih (1980) were motivated by the fact that although the criteria for selecting a forecasting model are extensively discussed (see, for example, Chambers, Mullik, and Smith 1971, Makridakis and Wheelwright 1979, and Fildes and Howell 1979) there is no theoretical basis for assessing the relative accuracy of various forecasting models. Because the OLS with ARIMA and the state space models account for various influences differently and may perform equally well under different circumstances, one is led to the conclusion that a combined forecast from the two techniques can improve forecast accuracy. Thus, we expect the composite forecast to be an improvement over the OLS with ARIMA forecast while, at the same time, we are able to use the elasticity estimates from the regression-ARIMA model. In Levenbach's (1980) words, "a forecaster's judgement is bound to improve with the benefit of as many approaches as can be gathered together".

The predictive performance of the individual forecasts can be improved if the separate models' forecasts contain different information. That is, if the individual forecasts differ substantially, the composite forecast is likely to predict more accurately than each of its components because the different assumptions, variables, or relationships between variables are blended together and reflected in the composite forecast. Similarity of forecasts is judged by the correlations between forecasting errors of each of the models fitted to the historical data. If the correlations between the errors of the two models are high, this implies that there is little to be gained by combining the two forecasts because they basically reflect the same information. On the other hand, if the errors are weakly correlated, this implies that the forecasts have different information contents and one would benefit from combining the forecasts.

The correlation coefficient of .24 between the error series of the OLS with ARIMA and the state space models shows that the predictions of each model contain information omitted from the other model's predictions. Improved forecast accuracy is normally achieved by simply combining the information content of the two models in a linear composite where the weights are determined by the regression:

$$OMSG_t = b_1(LSBJ)_t + b_2(SS)_t + U_t$$

and where OMSG is the actual US billed message variable, LSBJ is the series of predicted values under OLS with ARIMA estimation, SS is the series of predicted values from the state space model, and for unbiased estimators, $b_1 + b_2 = 1$.

The actual weights obtained from the above regression are $b_1 = .497555$ and $b_2 = .502407$, they are statistically significant, and their sum is not statistically different from 1. The normalized weights to be used in deriving the composite forecast are $W_1 = .497574$ and $W_2 = .502426$ for the OLS with ARIMA and the state space forecasts, respectively.

To assess the relative accuracy of the individual and composite forecasts, we generate 7-quarter ex-post forecasts from the OLS with ARIMA and the state space models and then compare the different measures of forecast accuracy. To eliminate the error associated with incorrect values for exogenous variables, their actual values are used for the ex-post forecast period. The accuracy of the three forecasts is judged by the forecast root mean square error (RMSE), Theil's U coefficient, and the average percent deviation (APD).

In Table 1 we present the percent deviations of each model's prediction for the 1981:1 to 1982:3 period along with each forecast's RMSE, U, and APD.

Table 1. Ex-Post Forecast Percent Deviations

	OLS	OLS with ARIMA	State space	Composite
1981:1	- 3.19	-3.34	- .25	-2.01
2	- 3.36	-2.87	-5.11	-3.83
3	- 4.04	-3.52	-5.63	-4.42
4	- 5.15	-4.24	-1.98	-3.26
1982:1	-10.04	-8.81	.94	-4.61
2	- 9.03	-7.76	-1.21	-4.94
3	- 2.67	-1.38	5.51	1.59
RMSE	1,483.22	1,277.94	941.37	926.36
U	.060140	.051816	.038169	.037561
APD	- 5.35	-4.56	-1.10	-3.07

The ex-post forecast results show that the OLS forecast is less accurate than the forecast obtained from the OLS with ARIMA model in terms of all the forecast accuracy criteria. However, the OLS with ARIMA forecast is dominated by the state space forecast in terms of RMSE, Theil's U, and average percent deviation. While the composite forecast is superior to the OLS with ARIMA forecast, the dominance over the state space forecast is only with respect to the RMSE and Theil's U; its average percent deviation is larger than that of the state space model.

6. SUMMARY AND CONCLUSION

In this study, we estimate a basic OLS model for IMTS demand. This is then expanded to include a moving average model for the residuals. The OLS with ARIMA model represents an improvement over the OLS model not only in terms of sample statistics but, also, in terms of ex-post forecast performance. Although the OLS with ARIMA model yields more reliable elasticity estimates, its ex-post forecast accuracy is less than that of the state space forecast.

For the purpose of taking advantage of the information content included in the two different models and to satisfy regulatory requirements, we combine the two forecasts into a linear composite which is superior to the OLS with ARIMA forecast. And, since the composite forecast is not completely dominated by the state space forecast, we conclude that using a composite forecast for IMTS demand is superior to using the individual forecasts.

The nature of the state space technique ensures that identifiable models only are used with a parsimonious representation where past behavior, feedback effects and correlated residuals are taken into account automatically. Although the state vector can be written in a form similar to regression equations, the coefficients cannot simply be interpreted as elasticities and one may object to using state space results that do not enhance the confidence placed in the elasticity estimates. Nonetheless, using a state space model in conjunction with an OLS model aids in gaining a better understanding of the structure that generates the data which can lead to improved forecast accuracy.

Although the weights used in deriving the composite forecast are uniquely determined by the regression of the actuals for the dependent variable on the fitted values generated by the two different models, the combination itself may not be optimal in the sense that there may be other models which yield better ex-post forecast results. For instance, the combination of forecasts from some univariate ARIMA and state space models may give more accurate predictions. Furthermore, another state space model may be developed such that different weights can be placed on each individual forecast.

ACKNOWLEDGEMENTS

The author acknowledges the support of John J. Cotter in this project. The models, ideas, and conclusions expressed in this paper represent the personal views of the author and do not necessarily reflect the views, policies or practices of AT&T Long Lines.

REFERENCES

AKAIKE, H. (1976). Canonical correlations analysis of time series and the use of an information criterion. In Advances and Case Studies in System Identification. Ed: R. Mehra and D. G. Lainiotis, Academic Press, New York, 363-387.

COOPER, J.P. and NELSON, C.R. (1975). The ex-ante prediction performance of the St. Louis and FRB-MIT-PENN econometric models and some results on composite predictors. Journal of Money, Credit, and Banking 7, 1-31.

FILDES, R. and HOWELL, S. (1979). On selecting a forecasting model. In Forecasting (TIMS Studies in the Management Sciences 12). Ed: S. Makridakis and S.C. Wheelwright, North-Holland, New York, 299-312.

GRANGER, C.W.J. and NEWBOLD, P. (1977). Forecasting Economic Time Series, Academic Press, New York.

LEVENBACH, H. (1980). A comparative study of time series models for forecasting telephone demand. In Forecasting Public Utilities (Proceedings of the International Conference held at Nottingham University, England, March 1980). Ed: O.D. Anderson, North-Holland, Amsterdam & New York, 153-164.

MAHMOUD, E. (1983). Empirical results on the accuracy of short-term forecasting techniques, paper presented at the 11th International Time Series Meeting (ITSM) in Toronto, Canada, August 1983.

MAKRIDAKIS, S. and HIBON, M. (1979). Accuracy of forecasting, an empirical investigation. J. Roy. Statist. Soc. 142, 97-145.

MAKRIDAKIS, S. and WHEELWRIGHT, S.C. (1979). Forecasting: Framework and Overview. In Forecasting (TIMS Studies in the Management Sciences 12). Ed: S. Makridakis and S.C. Wheelwright, North-Holland, New York, 1-15.

MEHRA, R.K. (1979). Kalman filters and their applications to forecasting. In Forecasting (TIMS Studies in the Management Sciences 12). Ed: S. Makridakis and S.C. Wheelwright, North-Holland, New York, 75-94.

SHIH, E.T.H. (1980). Comparison of forecasting techniques. In Proceedings of Business and Economic Statistics Section (American Statistical Association Meetings held in Houston, Texas, August 1980), 213-218.

TAYLOR, L.D. (1978). The demand for telephone service: a survey and critique of the literature. Bell system monograph.

URI, N.D. (1975). A mixed time series/econometric approach to forecasting peak load. Journal of Econometrics 9, 155-171.

TIME SERIES ANALYSIS: Theory and Practice 7
O.D. Anderson (editor)

INVARIANCE AND HETEROGENEITY IN TIME SERIES ANALYSIS OF SOME PSYCHOPHYSICAL DATA

Robert A. M. Gregson
Department of Psychology, University of New England, Armidale, N.S.W. 2351
Australia

Psychophysical experiments are characteristically bivariate sequences, on each trial of which there are a vector of physical input variables and a corresponding set of numerically-encoded responses.
In studies relating chemical concentrations in odourous mixtures to perceived constituent odour intensities there is interaction between components, mutually reducing intensities, and there are individual differences between observers in the sequential ARMA structure of their responses. Time series analyses may be used in conjunction with planned experiments, systematically varying components in mixtures, to examine the invariance of transfer functions within an individual and the stable differences between observers.

INTRODUCTION

The problems reviewed here were thrown into prominence by exploratory analyses of experiments investigating what happens when a mixture of two odourous substances is presented to a human observer. The observer attempts to describe, using any numerical response chosen from a closed range, the perceived intensity of each of the two odours which are detected. This experiment therefore constitutes a time series with two inputs and two outputs on each trial; there are over fifty trials in one experimental session. Consideration is restricted to two substances and two intensity judgments, because both experimentally and theoretically the more complex cases have received little formal attention, though they arise commonly in the work of the expert perfumier, and indeed in everyday life though not with the controlled regularity of the laboratory procedures. Experimental studies showing that components in a mixture typically inhibit the perceived intensity of one another, so that the intensity of the whole is judged to be less than the sum of the constituent perceived intensities judged singly in the context of the mixtures, or judged in isolation, have been devised since Passy (1895).

If one simply plots the regression of perceived rated intensities upon component concentrations without any regard to sequential processes then one obtains a typical psychophysical function which is approximately a straight line for log(psychological intensity as a numerical rating with a defined zero) regressed upon log(concentration in ml/litre). This is the psychophysical scaling procedure used with what is called stimulus magnitude estimation, of which our experiments are a special case (Marks, 1974). Such analyses fail to capture information on the dynamics of the process and are very noisy regressions in the case of olfaction. This is not surprising since the human observer shows transient changes in sensitivity such as adaptation and recovery (which may induce nonstationarity), responds like a filter to recent stimulus sequences, shows autoregression in the choice of responses, shows cross-coupling between different sensory dimensions, and can exhibit feedback effects yet may not effectively employ externally provided feedback in terms of how accurately or consistently he or she is performing in a psychophysical task (Gregson, 1980, 1982, 1983a, 1983b, 1983c).

The general situation faced here has of course been attacked from various perspectives; if one has no control over the multivariate input-output time series then the methods reviewed by Geweke (1982) and Parzen (1982) are pertinent. However the experimental psychologist can manipulate the input variables, and replicate closely any realization of the time series. The internal causal links are partly known and modelling can thus be constrained within a restricted range of alternatives. We are still left with the inevitable compromises between statistical goodness of fit and psychological interpretability of the model structure and parameters. As the hybrid strategies available to the psychologist have been powerfully extended by time series computing, many long-standing problems could be reexamined and redefined. The general indeterminacy of any multivariate black box has been analysed by Fossard (1977); the best strategy is still to accept as a working model the simplest structure that makes good sense, given (in as much detail as possible) what we already know.

In a psychophysical experiment there is not actually a one-one correspondence between stimulus values and response values; in the examples used here it would not be proper to expect the same odourous substance at a fixed chemical concentration always to produce the same judged intensity. We do not function like weighing scales or ammeters, we can show marked hysteresis and may have local discontinuities and multiple representations in the mappings from stimuli onto responses. Either these second-order problems are averaged out, or they are taken seriously as some intrinsic characteristics of the system which need explanation. The problems have been around since the 1860s and have produced what are from a statistician's viewpoint piecemeal approaches toward time series or even control theory modelling (Gregson, 1983a for a review). We are still faced with questions about the generality of any process model, as we want to know if all observers behave consistently and if all observers behave in the same way. The answer to the first is 'sometimes' and to the second is 'no'.

The Box-Jenkins p,d,q notation for AR, differencing and MA parameters respectively will be used, but the results reported here are only part of the complete cycle of estimation, identification, revision which is sometimes called the Box-Jenkins philosophy. The problems of the order of p,d,q have all been separately anticipated in experimental psychology but not answered at once in a time series analysis framework. Let us take the differencing question first; it has been argued (Krantz, 1972; Laming, 1982) that human response processes are not capable of making absolute responses but only relative ones, or that the process linking responses to internal storage operates upon two successive representations of the same stimulus (Green, Luce and Duncan, 1977) and that the scale of responses is adjusted continually relative to some central reference value (Parducci, 1963).

Notwithstanding that all such effects analysed outside time may be the unidentified consequences not of, or not just of, $d \neq 0$, in psychophysical transfer functions, but also of MA effects of low order, it is expedient to set up some competing models in which $d = 0$ and $d = 1$ for fixed p and q, and look at residuals. The dangers of over-differencing have been emphasised by O.D. Anderson (1977) and there are no good psychological reasons for looking beyond $d = 1$.

Even if we establish that $d = 1$ representations are optimal in some circumstances, generalizations are risky; it is likely that the time interval between trials, δt, which is used in a psychophysical experiment, will alias some components of high frequency but little interest, but δt may also itself be a variable in the process and thus alter the estimates of other parameters, such as p,d,q. In this context δt is fixed at 30 seconds for all the data which facilitates comparisons within the data set but inhibits generalization of the results. There is a distinction between biological and engineering systems which has perhaps not received sufficient attention; when we intervene in a psychological system the act of data collection, however miniscule, is itself an input to the system which induces synchronous process components, or disturbs the natural latencies of feedbacks. It is well established that strong periodic inputs to manual tracking tasks will induce behaviour not shown in response to aperiodic inputs. Little is known of the dynamic structure of sequential cognitive processes so the corresponding questions are open, though such phenomena as spontaneous alternation of two responses, contrast and assimilation to previous responses, and perseveration with the same response, are found as weak local sequence properties. A number of recent cases where attempts to model sequential structure in psychological experiments results in the postulation of a multistate model (Petzold, 1981; Wagner and Baird, 1981; Ward, 1982) suggests that local phases within states that are input-dependent might be expected in long experiments, hence the attractiveness of SETAR (Tong and Lim, 1980) models for the psychologist. One can turn this conjecture around and observe that psychophysical experiments are vigilance tasks in which some signals are missed due to stochastic central processes, which are themselves input independent. All this hints that we must expect large residuals to a best fit of any linear time series model, but there is virtue in trying.

A moving average mapping from stimulus values u_{j-k}, u_{j-k+1}, ..., u_{j-1}, u_j onto response y_j with unknown state variables $x_{j-k}, \ldots, x_j$ intervening is quite appropriate for psychophysical data; responses are strongly to the local stimulus u_j presented on the current trial and the effect of previous stimuli rapidly falls off, which is what Wiener conjectured systems with evolutionary survival capacity would be like. In olfaction there has been a tendency in the literature to confuse various sorts of persistent or adaptative effects (Engen, 1982) when time series method are used the time lags of residual effects are tractably modelled by MA(1) or MA(2) at the most.

The dark horse in psychophysics is the AR(m) aspect of the response process. It is obvious that if responses are completely uncoupled from their collateral stimulus series then they will either be a random walk or have an AR(m) structure perhaps with periodicities some of which are induced by δt.

As soon as a mixture of stochastic dependence on stimuli and autonomous AR processes exists then various forms of misidentification are possible. The problem can be diagrammed as in Fig. 1:

Figure 1: Direct and indirect lagged stimulus effects on psychophysical responses

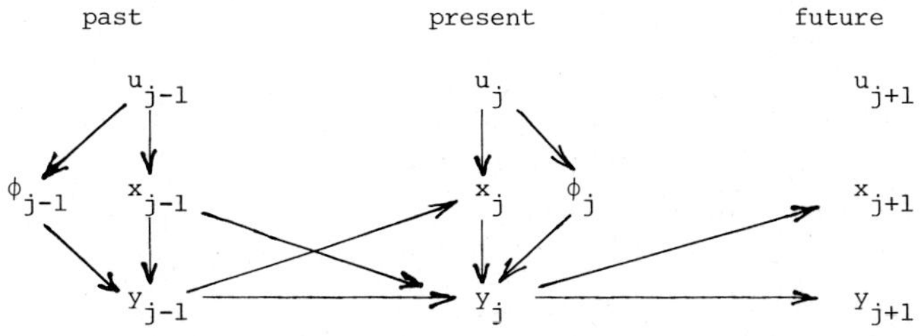

The links in Fig 1 labelled ϕ are the externally-inferred links used in a black box analysis to an input-output psychophysical function. The variables x are state variables, which in psychological theory would be mediated by the representation in stored memory of the events u and y. Hence the observed response y_j can be dependent on previous events at j-k through two pathways (at least);

$$u_{j-1} \text{ to } x_{j-1} \text{ to } y_j$$

or

$$y_{j-1} \text{ to } x_j \text{ to } y_j .$$

It has not been usual separately to try and identify the x, and hence models of psychophysical sequential estimation tasks have looked at the set

$$\{cov(y_j, u_{j-k})\},\ k = 0,1,2,\ldots,8$$

and omitted consideration of the path implicated in

$$\{cov(y_j, y_{j-k})\},\ k = 0,1,2,\ldots,8.$$

There is obviously a confusion, in modelling which does not consider the autoregression of the response sequence y, between the $cov(y_j, u_{j-k})$ and the partial $cov(y_j, u_{j-k}: y_{j-k})$, $k = 1,2,\ldots$.

It is worth saying a little more about what is suspected to go on in psychophysical experiments because it has implications for time series modellers. It is obvious that a system with high $cov(y_j, u_j)$, $\varepsilon > cov(y_j, y_{j-k})$, $\varepsilon \to 0$, and low c in $cov(y_j, u_{j-k}) = c^k$, $c < 1$ will look like a null AR output with strong MA characteristics, y will be an MA filter on u, dying out rapidly. The alternative is that c is quite large, but that the system is two state; call this model M1. We postulate a Markov process with parameter θ;

State A(M1), prior probability of state occupancy θ,

$$cov(y_j, u_{j-k}) = c^k,\ k = 0,1,\ldots,n$$

State B(M1), probability $1 - \theta$,

$$cov(y_j, u_j) < \varepsilon\ .$$

The idea here is that stochastically the observer's responses become locally disconnected from the stimulus sequence and then reconnect. The states x_j may remain connected. This can happen either because of fluctuating attention or because u_j is too small to be detected. Both sorts of models have been extensively considered. The problem is, what happens if we take M1 and misanalyze a realization of it as a case of M2, where M2 is

ARMA(2, 0, 2) on (y, u)?

The answer will depend on θ, and on the Markov transition matrix (compare Jacobs and Lewis, 1983) and if the process is self-excitatory on the values of the input sequence u. This implies in practice that we need ancillary calibration experiments, on the sensitivity and stability of human observers, before too readily deciding on a model. I would declare a preference for a third model, M3:

State A(M3), high prior probability of occupancy

low order ARMA on y, u

State B(M3), low probability and high transition back

into A but input dependent,

cov(u, y) - zero (that is, quasi-random responding).

Another way of expressing this is to say that failure to detect results is guessing and guessing is slightly autoregressive behaviour. The data analyses to be reviewed here are assumed to be compatible with M3 and the lapses into state B treated as an unidentifiable source of bad fit.

METHOD

The data explored here are drawn from a study (Gregson, 1982) in which the experimental design of Table 1 was used. The details of experimental procedures are given in Gregson (1983b). The input series n are randomized at generation on both components independently and are the same for all subjects in all sequences. Only data from the last two sessions of each of 8 subjects have been used; this is a total of 16 sequences each of 56 trials with 2 inputs and 2 outputs on each trial, reanalysed for a total of 5 different models; in all 80 time series analyses using the Genstat computer package, each checked for all autocorrelation spectra, residual autocorrelations, and standard errors on all parameter estimates. This yielded about 40 pages of computer output per subject per session per model, and is not an economical way to conduct routine research but was pursued here precisely because we have no benchmark analyses to refer to.

Table 1: Allocation of subjects to conditions by sessions

		Subjects #					
		1,2	3,4	5,6	7,8	9,10	11,12
Successive sessions	1	A	B	AB	Ba	A	B
	2	B	A	Ba	Ab	Ba	Ab
	3	AB	BA	BA	AB	AB	BA
	4	BA	AB	AB	BA	BA	AB
	5	BA	AB	AB	BA	BA	AB

A,a = amyl acetate, B,b = Bergamote, lower case letters indicate weak background levels only.

RESULTS

The five alternative models are as follows; in each case y_1 refers to the first response given on a trial, but u_1 and u_2 are simultaneously presented in a mixture. These empirical models are denoted by m1, m2, m3 to distinguish them from the previous theoretical discussion of M1, M2 and M3. These m are all cases of 2-input 2-output ARMA single state processes.

m1:

$$y_{1j} = c_1y_{1j-1} + c_2y_{1j-2} + c_3y_{2j} + c_4y_{2j-1} + c_5u_{1j} + c_6u_{2j} + c_7u_{2j-1}$$

$$y_{2j} = c_8y_{2j-1} + c_9y_{2j-2} + c_{10}y_{1j} + c_{11}y_{1j-1} + c_{12}u_{1j} + c_{13}u_{1j-1} + c_{14}u_{2j} .$$

This sort of model was arrived at by successive data analyses on previous extensive experiments, using the Box-Jenkins cyclic approach initially and exploring alternatives to find a simple structure having regard to the convergence of the numerical methods in the Genstat software (a modified form of Ljung and Box, 1979). This model was used in a root locus analysis described in Gregson (1983c). Despite the risk of over-determination and nonconvergence, it was decided to extend the model and also produce versions with first-order differencing. This gives

m2, Δ^0:

$$y_{1j} = c_1y_{1j-j} + c_2y_{1j-2} + c_3y_{2j} + c_4y_{2j-1} + c_5u_{1j} + c_6u_{1j-1} + c_7u_{1j-2} + c_8u_{2j} + c_9u_{2j-1} + c_{10}u_{2j-2}$$

$$y_{2j} = c_{11}y_{2j-1} + c_{12}y_{2j-2} + c_{13}y_{1j} + c_{14}y_{1j-1} + c_{15}u_{1j} + c_{16}u_{1j-1} + c_{17}u_{1j-2} + c_{18}u_{2j} + c_{19}u_{2j-1} + c_{20}u_{2j-2} .$$

m3 has the same structure as m2, but was obtained in a slightly different way. The Genstat program constructs the more complex models from component ARMA terms, as building blocks, and these are given in Table 2.

Table 2: Component ARMA terms of models m2 and m3 used, p, d, q notation

	$m2,\Delta^1$	$m2,\Delta^1$	$m3,\Delta^0$	$m3,\Delta^1$
$y_1 = f(y_1, y_2)$	2 0 1	2 1 1	1 0 1	1 1 1
$y_1 = f(u_1, y_1)$	2 0 2	2 1 2	2 0 2	2 1 2
$y_1 = f(u_2, y_1)$	2 0 2	2 1 2	2 0 2	2 1 2
$y_2 = f(y_1, y_2)$	2 0 1	2 1 1	1 0 1	1 1 1
$y_2 = f(u_1, y_2)$	2 0 2	2 1 2	2 0 2	2 1 2
$y_2 = f(u_2, y_2)$	2 0 2	2 1 2	2 0 2	2 1 2

It will be seen that there is multiple determination of the autoregressive terms in y in this method, m3 is more simple.

Using the autoregressions of the residuals as a test of fit, and obtaining separate fits for the y_1 and y_2 parts of each model, the frequency with which each of the five alternative models give minimum residual solutions in 16 data sets is shown in Table 3.

Table 3: Frequency of best solutions across the set of models

Model	$m1,\Delta^0$		$m2,\Delta^0$		$m2,\Delta^1$		$m3,\Delta^0$		$m3,\Delta^1$	
Session	S4	S5	S4	S5	S4	S5	S4	S5	S4	S5
y_1	4	2	3	3	0	0	1	3	0	0
y_2	5	4	0	0	0	0	1	0	2	4

Table 3 may tentatively be interpreted to mean that overdetermination and overdifferencing does not help the algorithm to converge on a good solution, which is reasonable, and that in a 2-input 2-output psychophysical system it is possible for one sensory dimension, the second one, to be controlled by a different process from that which controls the first. There is a tendency for the stimulus qualities of the first dimension to dominate because of the substances used to generate the odours, but a complication of the sort suggested here is I believe novel to psychophysics though for single input systems both Δ^0 and Δ^1 models have been advanced.

The questions of homogeneity and invariance can be explored by examining some standardized coefficients in the best fit models, over subjects and over sessions.

Coefficients which are known from previous work (Gregson, 1983b) to be dominant in the models have been selected for examination. It is required to know if an affine transformation can relate the values observed over sessions, for a given parameter and for a given subject. It is suspected that subjects are not homogeneous in their stability across sessions, and that some parameters are more stable than others. The set of coefficients used in the 128 cells of Table 4 have overall a unimodal leptokurtic distribution with a small shift down in the mean from session 4 to 5. The distribution of within-subject product-moment correlations between sessions suggests invariance of behaviour in some subjects and not in others. This is unsurprising in olfactory psychophysics but is a warning against pooling data across subjects. Table 5 gives across-session correlations for each subject and each parameter, the d.f. are 6 in each case which means that only larger values could be treated with any confidence.

Table 4: Standardized coefficients under minimum residual solutions

Subject	Session	$y_{1j}=f(u_{1j})$	$y_{1j}=f(u_{2j})$	$y_{2j}=f(u_{1j})$	$y_{2j}=f(u_{2j})$	$y_{1j}=f(y_{1j-1})$	$y_{1j}=f(y_2)$	$y_{2j}=f(y_{2j-1})$	$y_{2j}=f(y_{1j})$
SM	S4	2.84	0.01	-1.53	2.29	-2.32	0.80	-1.37	1.28
	S5	2.56	6.58	-0.55	1.88	-1.50	0.60	-0.26	0.40
DB	S4	1.56	-0.96	-0.82	-0.66	2.00	-2.12	-1.08	-1.15
	S5	4.52	-1.37	1.63	3.01	2.67	-0.23	1.00	0.22
JM	S4	1.34	0	-0.14	3.13	1.15	0.93	49.50	1.20
	S5	3.62	-6.32	0.19	0.30	3.00	-0.61	-0.76	-0.43
MC	S4	2.47	-4.30	-1.71	0.13	0.03	0.76	-6.05	3.00
	S5	0.68	-4.89	-0.87	2.00	3.65	-0.80	-2.08	0
BP	S4	0.89	0.13	2.37	3.74	3.26	-1.32	0.11	-23.33
	S5	0.C6	-0.59	0.07	5.37	0.05	-0.26	-1.44	-0.50
GB	S4	4.52	-0.14	-0.75	2.16	-0.62	-0.38	0.53	-0.77
	S5	5.10	-8.27	-0.61	0.64	0.61	-1.14	2.10	-0.33
AC	S4	1.28	1.00	-0.23	1.78	0.43	0.58	2.57	0.93
	S5	0	-6.77	-2.00	0.38	-29.40	1.40	-0.73	1.27
RJ	S4	0.39	-7.69	-0.31	4.45	-0.13	-0.68	32.33	1.00
	S5	1.27	0.02	-0.40	5.54	-0.08	-0.60	-1.55	-3.86

Table 5: Between session correlations of standardised coefficients for subjects and parameters

Subject	$r_{4,5}$	Parameter	$r_{4,5}$
SM	.447	$^*y_{1j}, u_{1j}$	.559
DB	.740	y_{1j}, u_{2j}	-.156
JM	-.062	y_{2j}, u_{1j}	.129
MC	.611	$^*y_{2j}, u_{2j}$	.378
BP	.321	y_{1j}, y_{1j-1}	.073
GB	.574	y_{1j}, y_{2j}	.228
AC	.307	y_{2j}, y_{2j-1}	-.210
RJ	-.131	y_{2j}, y_{1j}	.019

In Table 5 the two parameters marked with an * are the ones which a priori are expected to be the most stable. The remainder are quite labile from session to session even when they can be significant within a session. This could be due to the process being what I have called M3 but with varying state occupancy between sessions.

The following tentative conclusions are advanced:

(1) It is not in general necessary to difference ARMA models to improve fit in this area of psychophysics, but separate input-output relations could differ in terms of optimal differencing.

(2) The identified models are stable in their larger parameters across sessions for some subjects. The defining characteristics of this subset of stable subjects is not known, it may be related to the dynamics of the feedback loops mediating the stability of the judgments from trial to trial (Gregson, 1983c).

(3) The stability in (2) is up to an affine transformation of the parameters, which implies that loop gain does vary independently of process structure. We have independently shown that this is reasonable by plotting the shifts in root locus diagrams between coupled and uncoupled 2-channel systems on the earlier sessions of the same experiment as used for data here.

(4) The highest stability will occur in the parameters of zero-lag input-output links with "appropriate" connections; that is for

$$(\text{response}_i)_j = f(\text{stimulus}_i)_j.$$

It is this last result which explains why classical psychophysics, which is outside time, works to a first approximation.

This time series analysis approach necessarily opens up new questions to ask about psychosensory processes, about precisely how they are executed and kept reasonably stable. It appears possible that in a multivariate dynamic sequential process some subsystems are much more stable than others.

This research was supported by the Australian Research Grants Scheme in 1982.

REFERENCES

ANDERSON, O.D. (1977). Time series analysis and forecasting: Another look at the Box-Jenkins approach. The Statistician 26, 285-303.

ENGEN, T. (1982). The Perception of Odors. Academic Press, New York.

FOSSARD, A. (1977). Multivariable System Control. North-Holland, Amsterdam.

GEWEKE, J. (1982). Measurement of linear dependence and feedback between multiple time series. Journal of the American Statistical Association 77, 304-314.

GREEN, D. M., LUCE, R. D., and DUNCAN, J. E. (1977). Variability and sequential effects in magnitude production and estimation of auditory intensity. Perception and Psychophysics 22, 450-456.

GREGSON, R.A.M. (1980). Model evaluation via stochastic parameter convergence as on-line system identification. British Journal of Mathematical and Statistical Psychology 33, 17-35.

GREGSON, R. A. M. (1982). Representation of a 2-input, 2-output odour mixture identification task as a multivariate time series. Proceedings of the 2nd Australian Mathematical Psychology Conference, Newcastle, N.S.W.

GREGSON, R. A. M. (1983a). Time Series in Psychology. L. Erlbaum, New Jersey.

GREGSON, R. A. M. (1983b). The sequential structure of odour mixture component intensity judgments. British Journal of Mathematical and Statistical Psychology 36, 132-144.

GREGSON, R. A. M. (1983c). Invariance in time series representations of 2-input, 2-output psychophysical experiments. Submitted for publication.

JACOBS, P.A., and LEWIS, P.A.W. (1983). Stationary discrete autoregressive moving average time series generated by mixtures. Journal of Time Series Analysis 4, 19-36.

KRANTZ, D. H. (1972). A theory of magnitude estimation and cross-modality matching. Journal of Mathematical Psychology 9, 168-199.

LAMING, D. (1982). Differential coupling of sensory discriminations inferred from a survey of stable decision models for Weber's Law. British Journal of Mathematical and Statistical Psychology 35, 129-161.

LJUNG, G. M., and BOX, G. E. P. (1979). The likelihood function of stationary autoregressive-moving average models. Biometrika 66, 265-270.

MARKS, L.E. (1974). Sensory Processes, The New Psychophysics. Academic Press, New York.

PARDUCCI, A. (1963). Range-frequency compromise in judgment. Psychological Monographs 77, 2(565).

PARZEN, E. (1982). Comment. Journal of the American Statistical Association 77, 320-322.

PASSY, J. (1895). Revue générale sur les sensations olfactives. Année Psychologique 2, 363-410.

PETZOLD, P. (1981). Distance effects on sequential dependencies in categorical judgments. Journal of Experimental Psychology: Human Perception and Performance 7, 1371-1385.

TONG, H. and LIM, K. S. (1980). Threshold autoregression, limit cycles and cyclical data. Journal of the Royal Statistical Society, Series B, 42, 245-292.

WAGNER, M. and BAIRD, J. C. (1981). A quantitative analysis of sequential effects with numeric stimuli. Perception and Psychophysics 29, 359-364.

WARD, L. M. (1982). Mixed-modality psychophysical scaling: Sequential dependencies and other properties. Perception and Psychophysics 31, 53-62.

TIME SERIES ANALYSIS: Theory and Practice 7
O.D. Anderson (editor)

VARIANCE CHANGES IN AUTOREGRESSIVE MODEL

P. Baufays and J.-P. Rasson
Mathematics Department, Facultés de Namur, Rempart de la Vierge, 8, B-5000 Namur, Belgium

A new algorithm for estimating the unknown parameters of an autoregressive model, with innovation variance changes at unknown time points, is proposed. This iteratively estimates the autoregressive parameters, given the change points, and the change points given the autoregressive parameters. The procedure's advantage is shown by computer simulations and by reworking a previous case-study.

INTRODUCTION

When fitting an autoregressive (AR) model to economic data, many authors have noticed that the innovation variance, along the series, changes at unknown points (Wichern, Miller and Hsu, 1976; Gulledge and Willis, 1984). Wichern *et al.* proposed a maximum likelihood method for estimating the change points, variances and AR parameters.

This paper describes improvements to this earlier method, aimed at producing an efficient estimation algorithm. Our method consists of iterating two steps. the maximum likelihood estimation (mle) of the change points and variances, given the autoregressive parameters; and the mle of the autoregressive parameters, given the change points and variances.

We give results of simulation experiments which show the advantage of our method, and compare it with that of Wichern *et al.* (1976), for their real data example.

1. THE MODEL

We consider a realization of a univariate AR process of order p :

$$Z_t = \phi_1 Z_{t-1} + \dots + \phi_p Z_{t-p} + A_t \quad , \quad t = -p+1, \dots, 1, \dots, N \tag{1}$$

where (A_t) is a sequence of independent gaussian random variables, with zero mean, for which we assume that the variance changes at k time points, so that

$$\mathrm{Var}(A_t) = \sigma_i^2 \quad , \quad j_i \leqslant t < j_{i+1} \quad . \tag{2}$$

Such an assumption implies that the process is not second order stationary. But, if all the zeros of $P(\xi) = 1 - \phi_1 \xi - \ldots - \phi_p \xi^p$ lie outside the unit circle, its variance is bounded.

In this paper, the order of the process, p, and the number of change points, k, are assumed known. We have to estimate the unknown parameters (AR parameters, variances and change points) from the sample values (z_t).

2. ESTIMATION

We use maximum likelihood, and consider the first p observations as having fixed values.

The loglikelihood function is equal to (up to a constant)

$$L(\underline{\phi}, \underline{\sigma}^2, \underline{j}) = - \sum_{i=1}^{k+1} [(j_i - j_{i-1}) \log \sigma_i^2 + \sum_{j=j_{i-1}}^{j_i - 1} (Z_t - \phi_1 Z_{t-1} - \ldots - \phi_p Z_{t-p})^2 / \phi_i^2] \tag{3}$$

where $\underline{\phi} = (\phi_1, \ldots, \phi_p)$; $\underline{\sigma}^2 = (\sigma_1^2, \ldots, \sigma_k^2)$; $\underline{j} = (j_1, \ldots, j_k)$ and

$j_o = 1$, $j_{k+1} = N + 1$.

Although it is impossible to obtain directly the mle of the parameters, we may compute both

$$L^\star(\underline{\phi}) = \max_{\underline{\sigma}^2, \underline{j}} L(\underline{\phi}, \underline{\sigma}^2, \underline{j}) = L(\underline{\phi}, \underline{\sigma}^2(\underline{\phi}), j(\underline{\phi})) \tag{4}$$

and

$$\tilde{L}(\underline{\sigma}^2, \underline{j}) = \max_{\underline{\phi}} L(\underline{\phi}, \underline{\sigma}^2, \underline{j}) = L(\underline{\phi}(\underline{\sigma}^2, \underline{j}), \underline{\sigma}^2, \underline{j}) . \tag{5}$$

Our algorithm is a two-stage one. Starting with an initial value of $\underline{\phi}$, we update successively $\underline{\sigma}^2$, $\underline{j}$ and $\underline{\phi}$ according to the maximum likelihood principle.

Step 1. The mle of $\underline{\phi}$ given $\underline{\sigma}^2$ and $\underline{j}$ (i.e. $\underline{\phi}^\star = \underline{\phi}(\underline{\sigma}^2, \underline{j})$), is the solution of a linear system of equations

$$\sum_{i=1}^{k+1} (\sum_{t=j_{i-1}}^{j_i - 1} (Z_t Z_{t-m} - \phi_1 Z_{t-1} Z_{t-m} - \ldots - \phi_p Z_{t-p} Z_{t-m})) = 0, \; m = 1, \ldots, p. \tag{6}$$

Contrary to the method of Wichern *et al.*, Step 2 estimates $\underline{j}$ and $\underline{\sigma}^2$ by maximizing $L(\underline{\phi}^\star, \underline{\sigma}^2, \underline{j})$ with respect to $\underline{j}$ and $\underline{\sigma}^2$. We emphasize that this requires the computation of $L(\underline{\phi}^\star, \underline{\sigma}^2, \underline{j})$, not for the $O(N^k)$, but only for $O(N)$, when $k = 1$, and for $O(k N^2)$, when $k \geq 2$, different values of $\underline{j}$.

First, defining

$$e_t(\underline{\phi}^\star) = Z_t - \phi_1^\star Z_{t-1} - \dots - \phi_p^\star Z_{t-p} . \tag{7}$$

We have

$$L(\underline{\phi}^\star, \sigma^2, \underline{j}) = -\sum_{i=1}^{k+1} [(j_i - j_{i-1}) \log \sigma_i^2 + \sum_{t=j_i-1}^{j_i-1} e_t^2(\phi^\star) / \sigma_i^2] \tag{8}$$

so that, for given change points, the mle of the σ_i^2 's are

$$\hat{\sigma}_i^2(\underline{\phi}^\star, \underline{j}) = (\sum_{t=j_{i-1}}^{j_i-1} e_t^2(\phi^\star)) / (j_{i-1} - j_i). \tag{9}$$

Then

$$L(\underline{\phi}^\star, \underline{\hat{\sigma}}^2(\underline{\phi}^\star, \underline{j}), \underline{j}) = -\sum_{i=1}^{k+1} (j_i - j_{i-1}) \log \hat{\sigma}_i^2 - N \tag{10}$$

$$= -S(\phi^\star, j) - N .$$

We compute $\underline{j}^\star$ which minimizes $S(\underline{\phi}^\star, j)$ by means of the Fisher algorithm (Hartigan, 1975).

We define

$$d(i_1, i_2) = (i_2 - i_1 + 1) \log \sigma^2 , \quad 1 \leqslant i_1 \leqslant i_2 \leqslant N \tag{12}$$

with

$$\sigma^2 = (\sum_{t=i_1}^{i_2} e_t^2(\phi^\star)) / (i_2 - i_1 + 1)) \tag{13}$$

and

$$c(i, 1) = d(1, i) , \quad i = 1, \dots, N , \tag{14}$$

$$c(i, m) = \min_{q \leqslant i} (c(q-1, m-1) + d(q, i)) , \; m = 2, \dots, k , \quad i = 1, \dots, N . \tag{15}$$

$id(i, m)$ is an integer q which satisfies (15) . The optimal change points are then given by

$$\hat{j}_k = id(N, k) \tag{16}$$

$$\hat{j}_i = id(\hat{j}_{i+1} - 1 , i) , \quad i = k - 1, \dots, 1 .$$

We remark that $c(N, k)$ is the minimal value of $S(\underline{\phi}^\star, j)$, and so, $L^\star(\underline{\phi}^\star) = -c(N, k) - N$.

The proposed algorithm consists of the iteration of these two steps until the loglikelihood function (lf) value (or $-S(\underline{\phi}^\star, \underline{j}^\star)$ does not increase significantly. This algorithm is naturally convergent because the lf value increases at each step.

3. REMARKS

a) The algorithm needs starting values. As Wichern *et al.* (1976), we propose using the stationary case solution (no variance change) which is generally satisfactory.

b) To avoid undesirable infinite values of the lf, we have to set a minimal distance D (greater than p) between each change point. This is achieved in the algorithm by setting $d(i_1, i_2) = +\infty$ if $|i_1 - i_2| < D$.

c) Our $\underline{j}$ and $\underline{\sigma}^2$ estimates are the maximum likelihood ones when the Z_t 's are independent ($\underline{\phi} = 0$) and when $\underline{\phi}$ is known.

d) One of the advantages of our approach, as compared to the Wichern *et al.* method, is that it describes an efficient method to compute $L^\star(\underline{\phi})$ and therefore allows a fast computation of the overall mle of $\underline{\phi}$ (e.g. by scanning $L^\star(\underline{\phi})$ for each point of a grid on the admissible area for $\underline{\phi}$). As this requires the computation of $S(\underline{\phi}^\star, \underline{j})$ for only $O(kN^2)$ different $\underline{j}$, our algorithm remains attractive for both long series and frequent changes.

e) By using the mle of $\underline{i}$, given $\underline{\phi}$, our algorithm often converges to the overall mle of $\underline{\phi}$, $\underline{\sigma}^2$ and $\underline{j}$.

f) The proposed method may be quite easily modified to deal with step changes in other parameters for AR models (mean and variance of the innovations, AR parameters).

4. SIMULATIONS

Simulations have been performed on the Namur University DEC Computer. Each example is made up of 100 realizations of the same process, for which we apply our method, assuming different numbers of change points. The sample mean and variance of each parameter estimate (for the stationary case and the true number of change points) are given in the Appendix.

The AR parameter estimates are unbiased whatever the assumed number of changes may be; their variance is minimal for the "true" number.

The $\hat{\phi}$ variance decrease, when detecting the change at $t = j_i$, is an increasing function of $r_i = \max(\sigma_i^2/\sigma_{i-1}^2, \sigma_{i-1}^2/\sigma_i^2)$. This means that, as may be expected, a small step change in variance does not greatly affect the estimation of ϕ .

We also notice that the classical formula

$$\mathrm{Var}\ \hat{\phi}_i = (1 - \phi_p^2) / N \qquad (17)$$

seems to remain valid for the true number of changes. With regard to the change points estimates, their sample mean is close to the true value, and their sample standard deviation decreases with r_i . Finally, the variance estimates seem unbiased and their variance can be approximated by $2\,\sigma_i^4 / (j_i - j_{i-1})$.

The procedure generally requires no more than two or three iterations ; moreover, the computing cost remains low : the estimation of all the parameters for one to five change points, for instance, amounts to respectively US $ 0.1 and US $ 1 for series length equal to, respectively, 120 and 360 .

The change points estimated from the initial value of $\underline{\phi}$ are generally close to the final change points estimates.

5. REAL EXAMPLE

Box and Jenkins (1970) fit the ARIMA (0, 1, 1) model

$$z_t - z_{t-1} = \varepsilon_t + 0.1\,\varepsilon_{t-1} \tag{18}$$

to a series of 369 IBM stock prices. They argue that the lack of fit shown by the Box and Pierce test may be due to a change in the moving average parameter. Wichern *et al.* (1976) analyse the first difference of the same series logarithm, for which they propose an AR (1) model with two innovations variance changes. Their method, based on the ml principle, gives the resutls summarized in Table 1.

$\hat{\phi} = 0.13$	$10^4\ \mathrm{Var}(\varepsilon_t)$	$= 0.99$	$t < 180$	$-S(\hat{\phi}, j) = -232$
		$= 0.60$	$t < 235$	
		$= 7.2$	$t \geqslant 235$	

Table 1. The Wichern *et al.* results for IBM data.

These authors also remark that the AR parameter estimate on one portion of the series, defined by the change points, differs significantly from another, although the estimated variances remain quite the same (Table 2).

Period	$\hat{\phi}_i$	$\sigma_i^2\ 10^4$
1 - 179	0.23	0.99
180 - 234	0.13	0.62
235 - 368	-0.02	7.1

Table 2. The Wichern *et al.* results (change of variance and AR parameter).

Our procedure gives for the same series $(\ln (z_t/z_{t-1}))$ the following results (Table 3).

k	$-S(\hat{\phi}, j)$	$\hat{\phi}$	Period	$10^4 \sigma_i^2$
0	-420	0.001	1 - 368	0.32
1	-236	0.13	1 - 234	0.89
			235 - 368	7.2
2	-212	0.12	1 - 234	0.89
			235 - 280	13.
			281 - 368	3.9
3	-207	0.11	1 - 179	.99
			185 - 234	.60
			235 - 280	13.
			281 - 368	3.9
4	-203	0.11	1 - 113	1.0
			114 - 153	0.46
			154 - 234	0.91
			235 - 280	13.
			281 - 368	3.9

Table 3. Results with our method.

The change at $t = 235$, already discovered by Wichern *et al.*, is always found whatever the number of changes chosen. The one at $t = 281$ also seems significant, but the change of Wichern *et al.* at $t = 180$ only appears for $k = 3$.

The choice $k = 2$ seems reasonable, as a greater number of changes does not significantly increase the lf optimal value.

This is much greater when using our procedure, with $j = 2$, than for the Wichern *et al.* method. In fact, the computation of $L^{\star}(\phi)$ (for $\phi = i/100$, $-99 \leqslant i \leqslant 99$) shows that, for each ϕ between -0.12 and 0.25, there exists at least one set of change points which gives a greater likelihood value than those obtained by Wichern *et al.* Furthermore, for each ϕ, the first optimal change points is at $t = 235$ or $t = 237$, and the second between $t = 278$ and $t = 288$. (This show that our initial estimates of the change points are always near to the final ones).

We also notice that the ratios between consecutive variances are greater for our solution than for that of Wichern *et al.*

As often happens with simulated data, our algorithm gives in each case the mle of the parameters. Furthermore, the final solution does not depend on the initial one. Even when starting with $\phi = -1$ or $\phi = 1$, the algorithm converges to the mle; but, in these cases, naturally, the number of iterations is greater (4 or 5) than when starting with the stationarity case solution (2 or 3).

Finally, one may remark that the AR parameter estimate is similar for the two methods; but, if we estimate ϕ for each portion of the series, we get a rather different model. In Table 4, we display both the estimation of ϕ for each portion and the mle when we assume that both the autoregressive parameter and the variance are subject to step changes.

	Period	$\hat{\phi}$
(a)	1 - 234	0.22
	235 - 280	0.02
	281 - 368	-0.12
(b)	1 - 234	0.22
	235 - 278	0.03
	279 - 368	-0.14

Table 4. (a) ϕ estimates for each portion of the series;
(b) ϕ mle when both the variance and ϕ are assumed subject to two changes.

CONCLUSIONS

We have described a new iterative method, attractive for computer use, aimed at estimating the parameters of an AR process for which the variance of the innovation changes at unknown points. Its major advantage is that, for given AR parameters, it computes the ml change points estimates. The reanalysis of a case-study shows improvement on the previous method.

There is now a need for a criterion to determine the number of change points.

REFERENCES

ANDERSON, T.W. (1971). The Statistical Analysis of Time Series. Wiley, New York.

BOX, G.E.P., and JENKINS, G.M. (1970). Time Series Analysis, Forecasting and Control. Holden Day, San Francisco.

GULLEDGE, T.R., and WILLIS, J.E. (1984). The Relationship between Price and Volume of Contracts on the Wheat Futures Market. These Proceedings.

HARTIGAN, J.A. (1975). Clustering Algorithms. Wiley, New York.

WICHERN, D.W., MILLER, R.B., and HSU, D.A. (1976). Changes of Variance in First Order Autoregressive Time Series Models, With an Application. Appl. Statist. 25, 248-256.

APPENDIX. SOME SIMULATION RESULTS

N	ϕ	$\tilde{\phi}$	$V(\tilde{\phi})$	ϕ^*	$V(\phi^*)$	σ^2	σ^{2*}	$V(\sigma^{2*})$	j	j*	SD(j*)
120	.3	.3	.013	.3	.007	.1	.1	6	40	40	2.
						1.	1.	394	80	79	2.5
						.1	.09	4			
120	.3	.3	.016	.3	.008	.1	.1	6	40	40	1.6
						2.	2.1	1600	80	80	1.7
						.1	.1	4			
120	.3	.3	.010	.3	.007	.1	.1	5	40	40	4.3
						.5	.52	120	80	79	3.9
						.1	.1	5			
120	.3	.3	.096	.3	.007	.1	.1	5	40	40	2.
						1.	1.	238			
120	-.3	-.3	.012	-.3	.007	.1	.1	5	70	70	1.2
						2.	2.	1227			
120	.5	.5	.011	.5	.006	.1	.1	4	50	50	1.1
						2.	2.1	3765	80	79	4.9
						.4	.38	1.3			

N	ϕ	$\tilde{\phi}$	$V(\tilde{\phi})$	ϕ^*	$V(\phi^*)$	σ^2	σ^{2*}	$V(\sigma^{2*})$	j	j^*	$SD(j^*)$
120	.6	.6	.011	.6	.008	.1	.1	8	30	30	1.6
	-.2	-.19	.011	-.2	.007	2.	2.1	1577	80	79	4.9
						.4	.38	62			
50	.9	.86	.012	.89	.005	.1	.1	7	30	30	1
						2.	2.	4909			
50	.9	.87	.009	.90	.005	.1	.1	6	30	31	4
						.6	.63	487			
80	.5	.49	.012	.50	.008	.1	.1	7	40	40	6
						.6	.62	211			
80	.1	.1	.014	.1	.009	.1	.1	7	40	40	6
						.6	.6	200			
80	.7	.7	.007	.7	.006	.1	.1	8	40	41	8
						.3	.31	49			

Table 5. $\tilde{\phi}$ is the stationary case estimate of ϕ ;
V(.) denotes the sample variance and SD(.) the sample standard deviation.

TIME SERIES ANALYSIS: Theory and Practice 7
O.D. Anderson (editor)

CHOOSING BETWEEN LINEAR AND THRESHOLD AUTOREGRESSIVE MODELS

Timo Teräsvirta
Research Institute of Finnish Economy, Lönnrotinkatu 4 B, SF-00120 Helsinki 12, Finland

Ritva Luukkonen
Institute of Occupational Health, Haartmaninkatu 1, SF-00290 Helsinki 29, Finland

Tong and Lim (1980) advocate the use of threshold models in certain situations where the linear model is claimed to be inadequate. While a procedure for specifying the dynamics of the threshold model is suggested, no method is given for distinguishing between the linear and threshold models. In this paper, the performances of two model selection criteria, AIC and SBIC when used for this purpose, are investigated. The Monte Carlo experiments indicate that SBIC is clearly the better of these two alternatives.

1. INTRODUCTION

In a recent paper, Tong and Lim (1980) state that "the new era of *practical* non-linear time series modelling is, without doubt, long-overdue" (their italics). To accompany this claim, the authors introduce a family of non-linear models, called threshold autoregressive (TAR) models, and demonstrate their applicability to practical problems by examples.

Even if we accepted the invitation of Tong and Lim to enter the world of non-linear models, we might still at least sometimes be tempted to look back. The theory of linear models is as yet much better developed than that of non-linear models, and many problems are easier to deal with in the linear framework. Weighty arguments in favour of these more complicated models are needed before a model builder might be willing to abandon the more familiar territory of linearity. Therefore, an interesting question is how it could be confirmed that the true model was linear and not threshold autoregressive. In many cases the theory of the phenomenon to be modelled is not very helpful in this respect. If the TAR models are regarded as an alternative in a model building situation, there should be ways of choosing the right family of models on the basis of the evidence contained in the data.

Tong and Lim (1980) employ Akaike's Information Criterion (AIC) (Akaike, 1974) in the specification of TAR models. A natural idea would then be to apply AIC also to choosing between linear and TAR models. This has been proposed by Tong in an unpublished report (Tong, 1979). The Monte Carlo results of this paper demonstrate, however, that AIC is not a suitable criterion for that purpose. They are rather in favour of another criterion considered here, the Schwarz' Bayesian Information Criterion (SBIC), cf. Schwarz (1978). SBIC seems to be much more reliable than AIC in selecting the linear alternative when the true model is linear. It also yields consistent estimates of the dimension of the model. There are also other criteria than SBIC with the last-mentioned property; see e.g. Hannan and Quinn (1979) and Geweke and Meese (1981). In order to keep the exposition brief they are not considered here.

In this paper we first consider TAR models and their specification (Sections 2 and 3). The simulation experiment is described in Section 4 and its results discussed in Section 5. The final remarks are in Section 6.

2. THRESHOLD AUTOREGRESSIVE AND RELATED MODELS

The single-equation TAR model with J regimes can be written as

$$\phi^{(j)}(B)y_t = \beta^{(j)}(B)x_t + \varepsilon_t^{(j)},\ t=1,2,\dots,T,\ E\varepsilon_t^{(j)} = 0,\ \text{var}(\varepsilon_t^{(j)}) = \sigma_j^2,\ j=1,2$$

$$\text{if } c_{j-1} \le u_{t-d} < c_j;\ c_0 = -\infty,\ c_J = \infty,\ d>0,\ j=1,\dots,J. \tag{2.1}$$

In (2.1), u_{t-d} is a random variable, d is the delay parameter, c_j's are fixed threshold values, B is a lag operator, $By_t = y_{t-1}$, and

$$\phi^{(j)}(B) = 1 - \sum_{k=1}^{p_j} \phi_k^{(j)} B^k,\ j=1,2, \text{ and}$$

$$\beta^{(j)}(B) = \sum_{k=0}^{r_j} \beta_k^{(j)} B^k,\ j=1,2.$$

Furthermore, $\text{cov}(\varepsilon_t^{(j)}, \varepsilon_s^{(k)}) = 0,\ t \neq s,\ \forall j, k$.

If $u_t = x_t$, then (y_t, x_t) following (2.1) is called an open-loop threshold autoregressive system (TARSO). Model (2.1) is called a TARSO$(J;p_1,r_1;\dots;p_j,r_j)$ model. If $\beta^{(j)}(B) \equiv 0$, $j=1,\dots,J$, and $u_t = y_t$, we have a self-exciting threshold autoregressive or SETAR$(J,p_1,\dots,p_J)$ model, see Tong and Lim (1980). In the applications published so far, the number of regimes $J=2$.

Consider the following model

$$y_t = D_t \sum_{k=1}^{r_1} \beta_k^{(1)} x_{kt} + (1-D_t) \sum_{k=1}^{r_2} \beta_k^{(2)} x_{kt} + D_t\varepsilon_t^{(1)} + (1-D_t)\varepsilon_t^{(2)}$$

$$E\varepsilon_t^{(j)} = 0,\ \text{var}(\varepsilon_t^{(j)}) = \sigma_j^2,\ \text{cov}(\varepsilon_t^{(j)}, \varepsilon_s^{(k)}) = 0,\ t \neq s,\ j,k = 1,2 \tag{2.2}$$

where $D_t = 0$ if $\sum_{j=1}^{h} \pi_j u_{j,t-d_j} \le 0$, $d_j > 0$, $j=1,\dots,h$, and $D_t = 1$ otherwise. The above model has been discussed by Goldfeld and Quandt (1973), see also Quandt (1982). In order to be able to estimate the $\beta^{(j)}$'s, σ_1^2, σ_2^2 and D_t, $t=1,\dots,T$, Goldfeld and Quandt suggested that D_t be approximated by

$$D_t = (2\pi\sigma^2)^{-1/2} \int_{-\infty}^{\sum_{j=1}^{h} \pi_j u_{j,t-d_j}} \exp\{-\eta^2/2\sigma^2\}\, d\eta. \tag{2.3}$$

Now, a TARSO$(2;0,r_1;0,r_2)$ model is a special case of (2.2) where the x_{kt}'s are lags of the same variable x_t. Furthermore $x_t = u_{1,t}$, $\pi_1 = 1$ and $u_{2,t-d_2} \equiv -1$. In

TARSO models, D_t is supposed to have a degenerate distribution, and π_2 is specified together with the lag structure of the model. Now, the parameters of (2.2) completed with (2.3) can be estimated by the maximum likelihood method, and the linearity of the model can be tested by a LR test, see Goldfeld and Quandt (1973). That is not possible if the approach of Tong and Lim (1980) is applied, as the likelihood function does not meet the necessary regularity conditions.

3. SPECIFICATION OF THRESHOLD AUTOREGRESSIVE MODELS

An inherent feature of a TARSO model is the assumption that the orders of the lag polynomials are unknown a priori. A similar assumption is made by Box and Jenkins (1970) in their treatment of ARMA and transfer function models. Thus they have to be specified from the data, together with the threshold values and the delay parameter. Goldfeld and Quandt (1973) do not have this problem; if their model contains lags, the lag structure is assumed known.

Tong and Lim (1980) have proposed the use of AIC as the main specification criterion of threshold models. Consider a TARSO$(2;p_1,r_1;p_2,r_2)$ model and define a general model selection criterion as

$$MSC(p_j,r_j) = \ln \hat{\sigma}^2(p_j,r_j) + (p_j + r_j + 1)g(T_j) \tag{3.1}$$

where $\hat{\sigma}^2(p_j,r_j) = T_j^{-1} \sum_{t\in j} (\hat{\phi}^{(j)}(B)y_t - \hat{\beta}^{(j)}(B)x_t)^2$ and T_j is the efficient number of observations in regime j. Setting $g(T_j) = 2T_j^{-1}$ in (3.1) yields AIC whereas $g(T_j) = T_j^{-1} \ln T_j$ corresponds to SBIC. Let the threshold value c be fixed at the sample $100q^{th}$ percentile ξ_q of x. Define

$$AIC(\hat{p}_j,\hat{r}_j) = \min_{0\le p_j\le P_j;\,0\le r_j\le R_j} h_j AIC(p_j,r_j)\ , \quad j = 1,2\ .$$

where $h_j = T_j/T$, $j = 1,2$. Then

$$AIC(\xi_q) = AIC(\hat{p}_1,\hat{r}_1) + AIC(\hat{p}_2,\hat{r}_2) \tag{3.2}$$

is the AIC value of the whole model when the threshold lies at ξ_q. Tong and Lim choose $\Xi = \{\xi_{0.3}, \xi_{0.4}, \xi_{0.5}, \xi_{0.6}, \xi_{0.7}\}$ and compute (3.2) for all values of Ξ keeping the delay parameter d fixed. The model corresponding to the minimum of (3.2) over Ξ is selected among the models with the same delay parameter. The above exercise is repeated for other values of $d_i \in \mathcal{D} = \{d_1,\ldots,d_m\}$ and this way $d, c, p_1, r_1, p_2, r_2, \sigma_1^2$ and σ_2^2 can be estimated. In practice it seems that the delay parameter is only a minor problem in the specification. Quite often there is one obvious alternative suggesting itself so that the others are easily excluded.

The specification procedure outlined in Tong and Lim (1980) does not contain any proviso for the possibility that the true model is linear although the use of AIC was suggested in Tong (1979). Since this criterion is used for the specification of the threshold model it would be easy to compute its value also for a set of linear models and compare the results. The minimum value of this extended set of alternatives would then indicate the final model.

It is well-known that AIC does not estimate the dimension of the model consistently when a sequence of nested models is considered. The asymptotic probability of choosing too large a model remains positive. This has been shown in connection with AR models (Shibata, 1976), ARMA models (Hannan, 1980) and finite distributed lag models (Geweke and Meese, 1981). The same is true for polynomial distributed lag models when the lag length is determined first and the degree of polynomial thereafter (Teräsvirta and Mellin, 1983). For a rather general treatment of the problem, see Kohn (1983). In TARSO models, the situation is more complicated; the alternatives do not necessarily form a sequence of nested hypotheses. We are able to show that if the alternatives are nested, then AIC has a tendency to overestimate the dimension of the model. In this context overestimation means selecting a threshold model with a positive probability even asymptotically when the true model is linear. See also Section 4 and the appendix.

On the other hand, SBIC estimates the dimension consistently in all cases of the nested models mentioned above. Nevertheless, the asymptotic properties may mean little in a customary application, but the differences in the asymptotic behaviour of AIC and SBIC do motivate their small sample comparison. Besides, recent simulations with a finite distributed lag model and a polynomial distributed lag model indicate that SBIC can be superior to AIC in small samples as well, see Geweke and Meese (1981) and Teräsvirta and Mellin (1983).

4. SIMULATION EXPERIMENT

In order to compare the performance of AIC and SBIC, when testing the linearity of assumed threshold models, we have carried out a simulation study. TAR models are usually applied to situations where the output variable displays cyclical variation. This is of course a rather superficial observation; more discussion on the nature of this cyclical variation can be found in Tong and Lim (1980). However, a linear model with this property has been constructed here. It is

$$(1 - 0.8B)y_t = (1 + B + B^2)x_t + \varepsilon_t \qquad (4.1)$$

where $\varepsilon_t \sim N(0,1)$, $cov(\varepsilon_t, \varepsilon_s) = 0$, $s \neq t$. Furthermore

$$x_t = (1 - 0.5B)(1 - 0.9B^4)\zeta_t \qquad (4.2)$$

where $\zeta_t \sim N(0,1)$, $cov(\zeta_t, \zeta_s) = 0$, $s \neq t$. From (4.1) and (4.2) it is seen that the output variable contains cyclical variation. Since this is a preliminary study, we economised the computations. The whole spectrum of threshold models was not scanned through in search of the best non-linear alternative. The choice was first limited to the family

$$(1 - \phi_1^{(j)}B)y_t = \mu^{(j)} + (\beta_0^{(j)} + \beta_1^{(j)}B + \beta_2^{(j)}B^2)x_t + \varepsilon_t^{(j)},\ E\varepsilon_t^{(j)} = 0,$$

$$var(\varepsilon_t^{(j)}) = \sigma_j^2,\ cov(\varepsilon_t^{(j)}, \varepsilon_s^{(k)}) = 0,\ t \neq s,\ j,k = 1,2,$$

$$j = 1 \text{ if } x_{t-4} \leq c;\ j = 2 \text{ if } x_{t-4} > c. \qquad (4.3)$$

Note that (4.1) is nested in (4.3). Thus the whole specification procedure described in the preceding section was not carried out since $p_1 = p_2 = 1$ and $r_1 = r_2 = 2$ were fixed. A limited experiment showed that if d was considered unknown, it received a value four in more than nine cases out of ten. It was then permanently given that value. The threshold was specified by using the set Ξ of sample percentiles as in Tong and Lim (1980), and the model with the smallest AIC or SBIC value was selected. This obviously favours the selection of the linear model. If a larger set of combinations (p_1,r_1,p_2,r_2,Ξ) had been checked instead of merely $(1,2,1,2,\Xi)$, there would have been a positive probability of finding a TARSO model with a still lower AIC or SBIC value. On the other hand, the degrees of the rational distributed lag in (4.1) are assumed known and are not varied during the experiment.

Suppose the true linear model is nested in the threshold model. Assume furthermore that the threshold is given in advance. Then it can be shown that, asymptotically, SBIC chooses the linear model with probability one. When AIC is applied to the same problem, the probability of choosing the threshold model remains positive as the number of observations increases. This is shown in the appendix. The result indicates that SBIC has an edge over AIC at least in large samples.

The effective sample sizes in the Monte Carlo experiment were 50,100,150 and 500, respectively. The number of trials in each experiment was 400.

5. RESULTS

Restricting the family of threshold models in advance as in (4.3) has the effect that both AIC and SBIC often yield the same threshold value. This is because the number of parameters in the penalty function remains unchanged throughout. Minor differences arise from the fact that, as the penalty function of AIC contains the factor 2, this is replaced by ln T_j in SBIC. The distributions of the threshold values are shown in Figures 1 and 2. An interesting observation is that either the 30th or 70th percentile is chosen in more than half of the trials. This tendency is more pronounced with SBIC than AIC.

The results of the performance of the two model selection criteria are given in Table 1. They demonstrate that AIC is not a reliable criterion for detecting linearity when the alternative is a TARSO model. Even for the largest sample size (T = 500), the relative frequency of erroneously choosing the TARSO model is 0.4. SBIC does not perform well either, when the number of observations is small, but is fairly satisfactory and clearly better than AIC for T = 100. A tentative

Figure 1. The distribution of threshold values in the Monte Carlo experiment (400 trials) when AIC was used in the specification

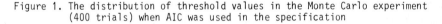

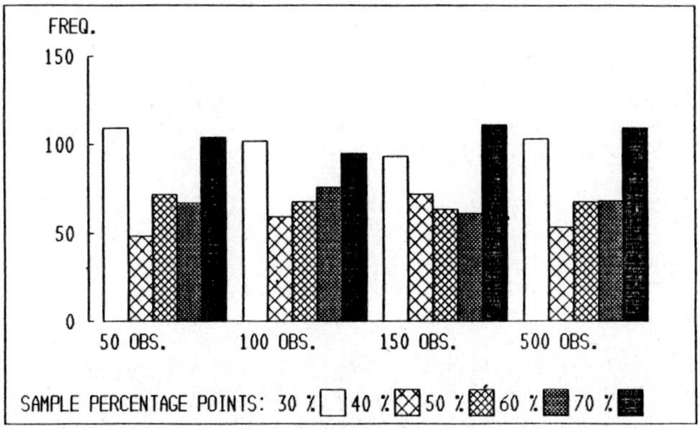

Figure 2. The distribution of threshold values in the Monte Carlo experiment (400 trials) when SBIC was used in the specification

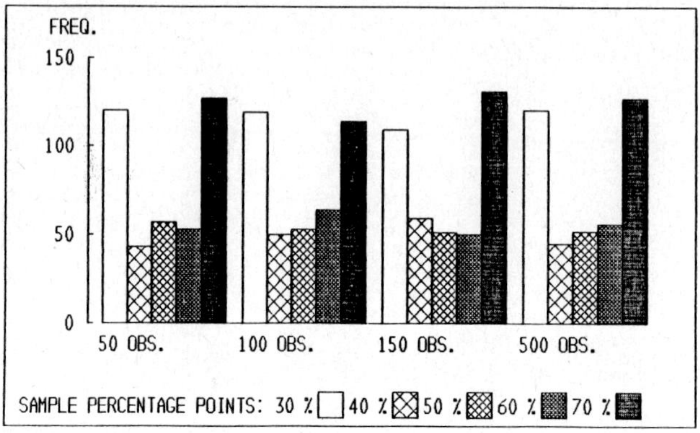

Table 1. The observed relative frequencies of choosing the TARSO model when the true model is (4.1)

Number of observations	Model selection criterion	
	AIC	SBIC
50	0.573	0.438
100	0.460	0.090
150	0.423	0.040
500	0.395	0.008

conclusion is that the model builder would be well advised to prefer SBIC to AIC in the specification of threshold models.

A logical question is what are the reasons for the mediocre performance of the model selection criteria AIC and SBIC in small samples. Why is the fit improved so dramatically when two regimes are assumed while in reality there is only one? One conspicuous detail in the applications of Tong and Lim (1980) is that the residual variances of the piecewise linear models are quite different. At least in ecological applications, it might be feasible to think that the white noise driving the system would have constant power without switching according to the regime. If changes in the variance of the disturbances were not assumed, would this affect the detection of linearity in small samples?

To investigate this possibility we constructed a restricted TARSO (RTARSO) model by setting $\sigma_1^2 = \sigma_2^2$ in (4.3). The simulations were repeated by using the RTARSO model as the alternative to the linear model instead of (4.3). The results are in Table 2. AIC still has a tendency of choosing the threshold model rather frequently, even in large samples. On the other hand, SBIC performs very well already when T = 50. When $T \geq 100$, the observed frequency of erroneously choosing a RTARSO model is zero. We may thus conclude that the error variance is a crucial parameter in the specification of TARSO models. Note that assuming $\sigma_1^2 = \sigma_2^2$, however, affects the model selection criteria but not the estimation of the parameters of the threshold model.

Table 2. The observed relative frequencies of choosing the RTARSO model when the true model is (4.1)

Number of observations	Model selection criterion	
	AIC	SBIC
50	0.355	0.023
100	0.263	0
150	0.283	0
500	0.238	0

The models in this experiment were nested. Therefore, we also computed the values of the likelihood ratio (LR) statistic using threshold values from the specification procedure. Supposing (incorrectly) that the LR statistic has a large sample χ^2 distribution under H_0 (linearity), a theoretical 0.05 significance level was used to test this hypothesis. The LR test fared somewhat better than AIC but was inferior to SBIC in every experiment with one exception. It rejected the linear model in 38 per cent of all cases, when the alternative was (4.3) and $T = 50$. On the other hand, Goldfeld and Quandt (1973) reported that the LR test works reasonably well in their experiments in which the likelihood function satisfies the regularity conditions.

6. FINAL REMARKS

This paper has investigated checking linearity in connection with the specification of TARSO models using model selection criteria. Of course, we are not suggesting that other possible techniques do not exist. An ultimate practical check would be post-sample prediction which has been applied to some examples in the paper of Tong and Lim (1980). The technique for obtaining these predictions in TAR models has been discussed for instance by Tong (1982).

In the Monte Carlo experiment we in fact also computed fifteen one-step ahead predictions for each trial. The average prediction accuracy of linear models measured with the RMSE was not substantially higher than that of the threshold models. When the RMSE of fifteen one-step ahead forecasts was the sole criterion of choice between linear and TARSO models, the relative frequency of incorrectly selecting the non-linear model was slightly less than 0.3. The number of observations used in estimation was either 100 or 150. The differences between SBIC and AIC were minimal, as these criteria were now only applied for locating the threshold value in the TARSO model.

APPENDIX. Asymptotic properties of AIC and SBIC when these criteria are applied to choosing between the linear and TAR models

Consider a TAR model (2.1) with $J = 2$ and $\phi^{(j)}(B) \equiv 1$, $j = 1,2$. It can be written in matrix form as

$$\begin{bmatrix} y_1 \\ y_2 \end{bmatrix} = \begin{bmatrix} X_1 & 0 \\ 0 & X_2 \end{bmatrix} \begin{bmatrix} \beta_1 \\ \beta_2 \end{bmatrix} + \begin{bmatrix} \varepsilon_1 \\ \varepsilon_2 \end{bmatrix}, \quad E\varepsilon = 0, \ \text{cov}(\varepsilon) = \text{diag}(\sigma_1^2 I_{T_1}, \sigma_2^2 I_{T_2}) \tag{A1}$$

where observation (y_t, x_t) belongs to the first submodel if $x_{t-d} \leq c$ and, otherwise, to the second submodel. Let X_j be a $T_j \times p$ matrix, $\mathrm{rank}(X_j) = p$, and assume c fixed and known. When model selection criteria are used to distinguish between the linear and TAR models the latter will be chosen if

$$\mathrm{MSC}(p+1) - \mathrm{MSC}(2p+2) \geq 0 \ . \tag{A2}$$

From (3.1) it follows that (A2) is equivalent to

$$\ln\hat{\sigma}^2 + (p+1)g(T) - \{h_1 \ln\hat{\sigma}_1^2 + h_2 \ln\hat{\sigma}_2^2 + (p+1)(h_1 g(T_1) + h_2 g(T_2))\}$$

$$= \ln \frac{\hat{\sigma}^2}{\hat{\sigma}_1^{2h_1}\hat{\sigma}_2^{2h_2}} + (p+1)\{g(T) - h_1 g(T_1) - h_2 g(T_2)\} \geq 0 \tag{A3}$$

where $\hat{\sigma}^2 = T^{-1}(y - Xb)'(y - Xb)$, $\hat{\sigma}_j^2 = T_j^{-1}(y_j - X_j b_j)'(y - X_j b_j)$,

$$b = (X'X)^{-1}X'y, \ y = (y_1', y_2')', \ X = \mathrm{diag}(X_1, X_2), \ b_j = (X_j'X_j)^{-1}X_j'y_j \ ,$$

$$h_j = T_j/T, \ j = 1,2 \ .$$

The logarithm in (A3) is 2/T times the log-likelihood ratio $LR(\hat{\sigma}^2, \hat{\sigma}_1^2, \hat{\sigma}_2^2)$ when the null hypothesis $\beta_1 = \beta_2$, $\sigma_1^2 = \sigma_2^2$ is tested against (A1). Thus, asymptotically under the null hypothesis,

$$2LR(\hat{\sigma}^2, \hat{\sigma}_1^2, \sigma_2^2) = T \ln \hat{\sigma}^2 \hat{\sigma}_1^{-2h_1} \hat{\sigma}_2^{-2h_2} \sim \chi^2(p+1).$$

The large sample probability of choosing the TAR model when the true model is linear is then

$$q = \Pr\{2LR(\hat{\sigma}^2, \hat{\sigma}_1^2, \hat{\sigma}_2^2) \geq T\,(p+1)(h_1 g(T_1) + h_2 g(T_2) - g(T))\} \ . \tag{A4}$$

Using a result in Rao (1965, p. 78 (iv)), (A4) can be approximated from above and we obtain

$$q \leq T^{-1}(h_1 g(T_1) + h_2 g(T_2) - g(T))^{-1} \ . \tag{A5}$$

Consider SBIC so that $g(k) = k^{-1} \ln k$. Then the r.h.s. of (A5) converges to zero as $T \to \infty$. When AIC is employed for the model selection, the r.h.s. of the probability inequality in (A4) converges to $2(p+1)$ as $T \to \infty$. Thus the probability q remains positive even asymptotically. These asymptotic results are also valid for TARSO models containing lags of the independent variable if the threshold value is assumed known because the disturbances of the model are white noise, see e.g. Harvey (1981, pp. 48-49). In the numerical example of this paper, $\lim_{T\to\infty} q$ for AIC reaches the value .075.

A guess based on that example could be that the corresponding value is even higher if the threshold is not fixed in advance. In fact, we repeated our simulations assuming c = 0 throughout. Then, for T = 500, the relative frequency of erroneously choosing the TAR model by using AIC was 0.148 which is much closer to the asymptotic value than the corresponding value in Table 1.

REFERENCES

AKAIKE, H. (1974). A new look at the statistical model identification. *IEEE Transactions on Automatic Control* AC-19, 716-723.

BOX, G.E.P. and G.M. JENKINS (1970). *Time series analysis, forecasting and control*. San Francisco: Holden-Day.

GEWEKE, J.F. and R. MEESE (1981). Estimating regression models of finite but unknown order. *International Economic Review* 22, 55-60.

GOLDFELD, S.M. and R.E. QUANDT (1973). The estimation of structural shifts by switching regressions. *Annals of Economic and Social Measurement* 2, 475-485.

HANNAN, E.J. and B.G. QUINN (1979). The determination of the order of an autoregression. *Journal of Royal Statistical Society* B 41, 190-195.

HANNAN, E.J. (1980). The estimation of the order of an ARMA process. *Annals of Statistics* 8, 1081-1081.

HARVEY, A.C. (1981). *The econometric analysis of time series*. Deddington: Philip Allan.

KOHN, R. (1983). Estimation of minimal subset dimension. *Econometrica* 51, 367-376.

QUANDT, R.E. (1982). Econometric disequilibrium models. *Econometric Reviews* 1, 1-63.

RAO, C. R. (1965). *Linear statistical inference and its applications*. New York: Wiley.

SCHWARZ, G. (1978). Estimating the dimension of a model. *Annals of Statistics* 6, 461-464.

SHIBATA, R. (1976). Selection of the order of an autoregressive model by Akaike's information criterion. *BIOMETRIKA* 63, 117-126.

TERÄSVIRTA, T. and I. MELLIN (1983). Estimation of polynomial distributed lag models. University of Helsinki, Department of Statistics, Research Report No 41.

TONG, H. (1979). An introduction to threshold autoregressive models, unpublished manuscript.

TONG, H. (1982). A note on using threshold autoregressive models for multi-step-ahead prediction of cyclical data. *Journal of Time Series Analysis* 3, 137-140.

TONG, H. and K.S. LIM (1980). Threshold autoregression, limit cycles and cyclical data. *Journal of Royal Statistical Society* B 42, 245-292 (with Discussion).

TIME SERIES ANALYSIS: Theory and Practice 7
O.D. Anderson (editor)

RANDOM OSCILLATIONS MODELLING

Christine Jacob

Laboratoire de Biométrie du Centre de Recherches Zootechniques,
F78350 Jouy-en-Josas, France

The intuitive idea of a cyclic trajectory (or, more generally, of a stochastic cycle) is defined mathematically. The object of these definitions is to directly characterize the random oscillations of each trajectory of the process, without using definitions of a deterministic type which concern either a series from which the noise has been eliminated (limit cycle) or the averaged process behaviour (mean, autocorrelations, spectra). We also define the stability and self-exciting properties of a stochastic cycle. We then construct and study a general time series model for such a cycle.

1. INTRODUCTION

The object of this paper is to try to study "random oscillations" in discrete time series directly. Up to now, these oscillations have generally been modelled by either the periodic regression model or the autoregressive model. The periodic regression model fits a periodic function to the series and studies the characteristics (amplitudes and frequencies of the different harmonics) of this periodic function. In this model, the residuals represent the deviation of the random oscillations from the deterministic rhythm. This model is adequate if the rhythm is assumed exogeneous, that is, imposed on the biological system by external conditions : for example, the annual rhythms of plants and animals are due to the earth's annual revolution around the sun.

If the rhythm is endogeneous, that is, created by the biological system itself - which is the case in most biological rhythms - then, in order to make the process random memory explicit, nonlinear autoregressive models with limit cycles are used (Tong, 1980 ; Ozaki, 1980). The limit cycle is supposed to represent the oscillations of the system. But this method has important drawsbacks, mainly :

1) In order to measure the quality of the fitting, we have to calculate the deviations between the random oscillations and the limit cycle ; which, for this model, are not represented by the residuals. So, how can these deviations be calculated ?

2) If the deterministic difference equation has 2 limit cycles, which limit cycle is supposed to best represent the random oscillations ?

There exist intermediate models such as the periodic linear autoregressive models (Jones and Brelsford, 1967 ; Pagano, 1978), which are linear autoregressive models with periodic coefficients. These explain the random memory of processes with a periodic probabilistic structure such as exogeneous rhythms ; but they cannot explain endogeneous rhythms - in particular self-exciting cycles (see § 3.2.).

So we tried to model the random oscillations themselves without using methods of a deterministic type (periodic regression, limit cycle). First we have to define mathematically the intuitive idea of random oscillations, and then we look for models satisfying the definition. We restrict ourselves here to endogeneous rhythms, so we study non linear autoregressive models. We will show that, generally, classical models cannot have random oscillations, even if they have a limit cycle. Then, we will have to define a new general model for such oscillations.

2. DEFINITIONS

The oscillations we are studying are constant period oscillations. That is, their periods are independent of the process state. These oscillations are the stochastic equivalent, in discrete time, of limit cycles - solutions of equations $\ddot{x} + f(x)\dot{x} + \alpha x = 0$, where the damping force $f(x)\dot{x}$ is non linear and the restoring force αx is linear (Ozaki, 1980 ; Pavlidis, 1973). In the deterministic case, the process returns to the same value after a time lag equal to the period d, and the sequence $(x_1,\dots, x_d)$ defines the cycle. In the stochastic case, the process does not necessarily return to quite the same value : it moves inside an interval C_j, and the sequence $(C_1,\dots, C_d)$ will define a stochastic cycle (figure 1).

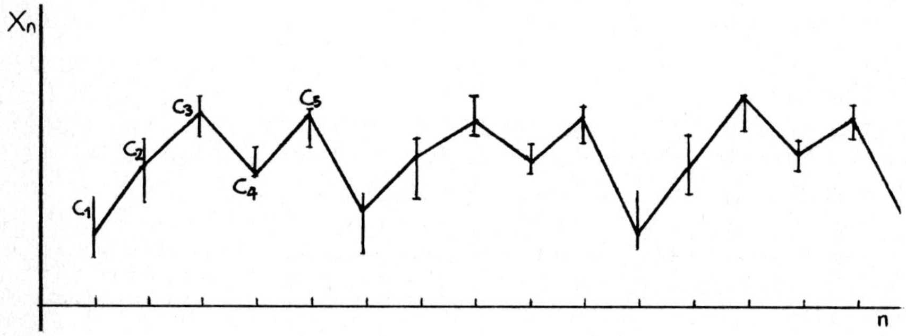

Fig. 1 : Cyclic trajectory. The cycle is a defined by the sequence (C_1,C_2,C_3,C_4,C_5)

We suppose $(X_n)_{n\in\mathbb{N}}$ is a process taking values in some arbitrary space $(\mathcal{X},\mathcal{F})$, and has a probability measure $P_n(X_1,\dots, X_n)$ for all n. We define a trajectory cycle, a stochastic cycle and a markovian cycle in the following way.

Definition 2.1 : Definition of a trajectory cycle

A trajectory of the process X_n is cyclic (or periodic) if there exist d sets $C_1,\dots, C_d \in \mathcal{F}$ $(d>1)$ such that :

i) two of them (at least) are disjoint

ii) the observations sequences

(A) $(X_{1+(\ell-1)d},\dots, X_{\ell d}) \in (C_1,\dots, C_d) \quad \forall \ell \in \mathbb{N}$

iii) d is the smallest such positive integer.

When the C_j are each reduced to a single point, this defines a deterministic periodic trajectory.

Definition of a stochastic cycle

We suppose here that the initial instant of the mathematical process is also that for the studied biological phenomenon. Otherwise, the following condition D which concerns the initial law of the process has no meaning.

Definition 2.2 : if there exist d sets $C_1,\ldots, C_d \in \mathcal{F}$ such that :

i) 2 of them (at least) are disjoint (if $d \geq 2$)

ii) (B) $P_{n+1}(X_{n+1} \in C_{jo+n+1} | X_1 = x_1,\ldots, X_n = x_n) = 1$
$\forall (x_1,\ldots, x_n) \in (C_{jo+1},\ldots, C_{jo+n})$, $\forall n \geq 1$

where the sequence $(C_{jo+1},\ldots, C_{jo+n})$ is defined as the periodic sequence

$C_{jo+1}, C_{jo+2},\ldots, C_d, C_1,\ldots, C_{jo+1},\ldots, C_d,\ldots, C_1,\ldots, C_j$, with $n = kd-j_o+j$,

iii) d is the smallest such positive integer

then the sequence $C = (C_1,\ldots, C_d)$ is a stochastic cycle for the process if $d>1$, and a singular set if $d=1$.

This definition means that if the initial state X_1 belongs to the cycle, then the trajectory is cyclic.

If, in addition, X_n is h-markovian, that is $\mathcal{X}_n = (X_n, X_{n-1},\ldots, X_{n-(h-1)})$ is markovian, then :

$$(B) \Longleftrightarrow \begin{cases} (C_1)\ P_{n+1}(\mathcal{X}_{n+1} \in \mathcal{C}_{jo+n+1} | \mathcal{X}_n = x) = 1 \quad \forall\ x \in \mathcal{C}_{jo+n},\ \forall n \geq h \\ \quad \text{where } \mathcal{C}_j \text{ is defined as the sequence } = (C_j \times C_{j-1} \times\ldots\times C_{j-h+1}) \\ (D)\ P_{m+1}(X_{m+1} \in C_{jo+m+1} | X_1 = x_1,\ldots, X_m = x_m) = 1 \\ \quad \forall\ x_1,\ldots, x_m \in C_{jo+1},\ldots, C_{jo+m},\ \forall m < h \end{cases}$$

This last condition (D) concerns the initial law of the process.

Because of the stationarity property, we will restrict ourselves here to markov chains with temporally homogeneous transition probabilities. And because the initial instant of the measured process is not always the initial instant of the biological phenomenon itself, we will no longer assume (D).

So, suppose now $(\mathcal{X}_n)_{n\in\mathbb{N}}$ is a markovian chain on $(\mathcal{X}^h, \mathcal{F}^h)$ with transition probabilities $P(x, B) = P(\mathcal{X}_{n+1} \in B | \mathcal{X}_n = x)$, and x_o denotes the initial state. The definition of a markovian cycle is then :

Definition 2.3 : definition of a markovian cycle

A sequence $(\mathcal{C}_1,\ldots, \mathcal{C}_d)$ of set $\mathcal{C}_j \in \mathcal{F}^h$ is a markovian cycle of length d (if $d>1$) or a singular set (if $d=1$) for the process $\mathcal{X}_n$ if

i) 2 of them (at least) are disjoint (if $d \geq 2$)

ii) (C_2) $P(x, \mathcal{C}_{jo+n+1}) = 1\ \forall\ x \in \mathcal{C}_{jo+n}$, $\forall n$, where
$\mathcal{C}_j = (C_j \times C_{j-1} \times\ldots\times C_{j-(h-1)})$

iii) d is the smallest such positive integer.

This last relation can be written in an equivalent way according to :

$$(C_3)\quad \begin{cases} P(\mathcal{X}_{n+1} \in \mathcal{C}_{j+1} \mid \mathcal{X}_n = x,\ \phi_n = j \ (\mathrm{mod}\ d)) = 1 \quad \forall\, x \in \mathcal{C}_j,\quad \forall\, n \\ P(\mathcal{X}_{n+1} \in \mathcal{C}_{j+1} \mid \mathcal{X}_n = x,\ \phi_n = k \ (\mathrm{mod}\ d)) = 0 \quad \forall\, x \in \mathcal{C}_j,\quad k \neq j,\ \forall\, n \\ \phi_{n+1} = \phi_n + 1 \\ \phi_0 = j_0 \ \text{ with } \ j_0 \in (1,\dots,d). \end{cases}$$

The system runs cyclically through the sets of states $\mathcal{C}_1,\dots,\ \mathcal{C}_d,\ \mathcal{C}_1,\dots$ -the cycle has d different phases which are sequentially $(1,2,\dots,d)$. The variable ϕ_n represents the markovian memory of the cycle phase at instant n.

In order to have cycles as narrow as possible (minimal cycles), we must add the following condition : If the process has two cycles $(\mathcal{C}_1,\dots,\mathcal{C}_d)$ and $(\mathcal{C}'_1,\dots,\mathcal{C}'_{d'})$, then they are completely disjoint, that is $\bigcup_{j=1}^{d} \mathcal{C}_j \cap \bigcup_{j=1}^{d'} \mathcal{C}'_j = \emptyset$. If we do not use the phase memory, the relation $P(x, \mathcal{C}_{j+1}) = 1\ \forall\, x \in \mathcal{C}_j$, $\forall\, j\ (\mathrm{mod}\ d)$, defines a "stationary" cycle of length d if and only if the $\mathcal{C}_j$ are all disjoint. Orey's definition (1971), as well as the cyclically moving classes defined by Doob (1953), correspond to that case. Otherwise, suppose $\mathcal{C}_k \cap \mathcal{C}_m \neq \emptyset$; then, if $\mathcal{X}_n \in \mathcal{C}_k \cap \mathcal{C}_m$, we must have $\mathcal{X}_{n+\ell} \in \mathcal{C}_{k+\ell} \cap \mathcal{C}_{m+\ell}$, $\forall\, \ell = 1,2\dots$. So, from time n, the cycle $(\mathcal{C}_j)_{j=1,\dots,d}$ can be reduced to the intersections $(\mathcal{C}_{k+\ell} \cap \mathcal{C}_{m+\ell})_{\ell=1,\dots,d}$, if those are assumed to be disjoint and not empty. The assumption that the $\mathcal{C}_j$ are all disjoint is a very restrictive condition. It becomes more restrictive with smaller h and shorter sampling interval Δ between two measuring instants. In the limit, as Δ tends to 0, when $h = 1$, the C_j must tend to a single point, $\{x_j\}$. This implies that this assumption prevents us from defining a stochastic cycle in continuous time (with the C_j not reduced to a single point), as would be analogous to the Orey's definition in discrete time.

Definition 2.4 : markovian cycle properties

Here, d is any positive integer ; and, for $d = 1$, the cycle is just a singular set.

2.4.1. Stochastic limit cycle and stability property.

Suppose $\mathcal{C} = (\mathcal{C}_1,\dots,\mathcal{C}_d)$ is a markovian cycle for the process $\mathcal{X}_n$. $\mathcal{C}$ is an almost sure (a.s.) stochastic limit cycle if there exists x_0 such that the probability $P(\mathcal{X}_n \in \mathcal{C} \mid \mathcal{X}_0 = x_0)$ tends to 1 when n tends to ∞. This assumption can be too strong ; so we also define a $(1-\varepsilon)$ a.s. limit cycle when the limit of $P(\mathcal{X}_n \in \mathcal{C} \mid \mathcal{X}_0 = x_0)$ as n tends to ∞ is greater or equal to $1-\varepsilon$, where ε is a small positive constant. The set $\{x_0 \in \mathcal{X} : \lim_n P(\mathcal{X}_n \in \mathcal{C} \mid \mathcal{X}_0 = x_0) = 1$ (resp. $\geq 1-\varepsilon$)$\}$ is called the stability domain of the cycle. If this set is a neighbourhood of the cycle (for the topology of $\mathcal{X}^h$), the cycle is called stable. This definition is the stochastic equivalent of Tong's definition for deterministic limit cycles (Tong and Lim, 1980).

2.4.2. Self-exciting cycle

A self-exciting cycle is a cycle created by the biological system itself and not modified overall by the random innovation. The cycle $\mathscr{C}$ is a self-exciting cycle if and only if

i) the process from which the innovation has been eliminated, has a deterministic limit cycle $x = (x_1, \ldots, x_d)$ having the same period d as the stochastic cycle.

ii) $(x_1, \ldots, x_d) \in (\mathscr{C}_1, \ldots, \mathscr{C}_d)$ and $E(\mathscr{X}_{n+1} | \mathscr{X}_n = x_j) = x_{j+1} \quad \forall\ j \pmod d$.

3. MODELS FOR MARKOVIAN CYCLE

3.1. If the process has no innovation, the general model for the markovian process $\mathscr{X}_n$ is

(M1) $X_{n+1} = f(\mathscr{X}_n)$

with a given initial law for $\mathscr{X}_0$. Then the process $\mathscr{X}_n$ has a cycle $\mathscr{C}$ if either f has a limit cycle x, or f has a "chaotic" (or "pseodo") periodicity. In the first case, $\mathscr{C}_j = \{x_j\}$; and, in the second, the C_j are not reduced to one point. (If $\mathscr{X} = \mathbb{R}$, then C_j is an interval of $\mathbb{R}$).

3.2. If the process has an innovation, the usual autoregressive model is

(M2) $X_{n+1} = f(\mathscr{X}_n) + \varepsilon_{n+1}$

with ε_{n+1}, a stationary white noise independent of $\mathscr{X}_n$, which takes values in Ω_ε . Then, the only possible cycles for this model presuppose that $\Omega_\varepsilon \subset \bigcap_j \bigcap_{x \in \mathscr{C}_j} C_{j+1} - f(x)$ where $C_{j+1} - f(x)$ denotes the set $\{c : c+f(x) \in C_{j+1}\}$.

Proof : An arbitrary sequence $(\mathscr{C}_1, \ldots, \mathscr{C}_d)$ $(d>1)$, where at least 2 of them are disjoint, is a cycle, if $\mathscr{X}_n$ satisfies condition (C_3), that is,

$$\begin{aligned} P(\mathscr{X}_{n+1} \in \mathscr{C}_{j+1} | \mathscr{X}_n = x, \phi_n = j(d)) &= P(X_{n+1} \in C_{j+1} | \mathscr{X}_n = x, \phi_n = j(d)) \\ &= P(\varepsilon_{n+1} \in C_{j+1} - f(x)) \\ &= 1 \quad \forall\ x \in \mathscr{C}_j, \quad \forall\ j \pmod d. \end{aligned}$$

This is possible if and only if $\Omega_\varepsilon \subset \bigcap_j \bigcap_{x \in \mathscr{C}_j} C_{j+1} - f(x)$. This condition is not possible if $\Omega_\varepsilon = (-\infty, +\infty)$ (Tong and Wu, 1982, were aware of this condition) and is possible when Ω_ε is a finished interval, if $\bigcap_j \bigcap_{x \in \mathscr{C}_j} C_{j+1} - f(x) \neq \emptyset$. This last condition is a very restrictive one. So we are led to build another general model for random oscillations. Intuitively, in order to have random oscillations, ε_{n+1} will have to be chosen in such a way that X_{n+1} will take a value in C_{j+1}, if $\mathscr{X}_n$ was in $\mathscr{C}_j$.

That is, the innovation will depend in a certain manner on the value of $\mathscr{X}_n$. In biology, it can be interpreted as a "regulation noise". A general model satisfying definition 2.3, that is, which admits a markovian cycle is :

$$(M3)\quad \begin{cases} X_{n+1} = f(\mathcal{X}_n) + \sum_j 1_{\mathcal{C}_j}(\mathcal{X}_n)\, 1_j(\phi_n (\text{mod } d))\, \varepsilon_{n+1,j+1} \\ \phi_{n+1} = \phi_n + 1 \\ \mathcal{X}_o \in \mathcal{C}_{jo} \; ; \; \phi_o = j_o \; ; \; j_o \in (1,\ldots, d). \end{cases}$$

$1_{\mathcal{C}_j}(\mathcal{X}_n)$ and $1_j(\phi_n(\text{mod } d))$ are indicator functions ; when $\mathcal{X}_n$ is in $\mathcal{C}_j$ and $\phi_n = j(\text{mod } d)$, the innovation is $\varepsilon_{n+1,j+1}$. The $\varepsilon_{n+1,j+1}$ are stationary white noises, taking values in $\Omega_{j+1} = \bigcap_{x \in \mathcal{C}_j} C_{j+1} - f(x)$, and are assumed to be independent for all n and j, and independent of $\mathcal{X}_n$. The stationarity of the $\varepsilon_{n+1,j+1}$ implies the stationarity of the transition probabilities of the process satisfying this model ; and $(\mathcal{X}_n, \phi_n)$ is a markovian process.

It is interesting to note that, for a given initial phase ϕ_o, the probabilistic structure of the innovation is periodic, which implies a periodic structure for the process $\mathcal{X}_n$ as well. The trajectories will then be not only periodic in "mean" (the series running cyclically through the sets $\mathcal{C}_1,\ldots, \mathcal{C}_d, \mathcal{C}_1,\ldots$), but also in "variance" (the lengths of the $\mathcal{C}_j$ as well as the variances of the $\varepsilon_{n,j}$ being not necessarily all the same). Processes with a periodic probabilistic structure have been modelled for example by means of linear autoregressive models with periodic coefficients. In these models, the oscillations are not created by the system itself, as in model (M3), but are created by the periodicity of the model. Moreover the noise structure of these models does not allow the oscillations to stay in a bounded cyclic band such as $(C_1,\ldots, C_d, C_1,\ldots)$. For these two reasons, model (M3) is the most suitable for the modelling of self-exciting cycles.

In the following, we suppose $\mathcal{X} = \mathbb{R}$, and that the C_j and Ω_j are intervals of $\mathbb{R}$: $C_{j+1} = (a_{j+1}, b_{j+1})$ and $\Omega_{j+1} = (a_{j+1} - f(x)_m, b_{j+1} - f(x)_M)$, where $f(x)_m = \min_{x \in \mathcal{C}_j} f(x)$ and $f(x)_M = \max_{x \in \mathcal{C}_j} f(x)$.

A particular case of this model is when $d = 1$, $C_1 = (-\infty, +\infty)$ and f is bounded. Then $\Omega_1 = (-\infty, +\infty)$ and the model is reduced to the usual aperiodic model (M2) (with $\Omega_\varepsilon = (-\infty, +\infty)$). The probabilistic structure of this model has been studied by Doukhan and Ghindes (1980). Another particular case is when $C_j = \{x_j\}$ and $\varepsilon_{n+1,j+1} = \{x_{j+1} - f(x_j)\}$. Then the trajectories are the deterministic cycle $\{x_1,\ldots, x_d\}$ which is the limit cycle of f if $f(x_j) = x_{j+1}$ $(j = 1,\ldots, d)$.

Remark on the existence of the model

For a given stochastic process $\mathcal{X}_n$, supposed to satisfy the model (M3), the form of the function f depends on the noise, or conversely, the noise depends on f. This can be easily seen in the previous particular case.

i) if the Ω_j are known, f will exist if $\mathcal{L}(C_j) \geq \mathcal{L}(\Omega_j)$ $(j = 1,\dots, d)$, where $\mathcal{L}(I)$ notes the length of the inverval I. That means that the range of $\varepsilon_{n,j}$ values must not be greater than the length of C_j.

ii) if f is known, Ω_{j+1} will exist if $\mathcal{L}(C_{j+1}) \geq f(x)_M - f(x)_m$.

A more general model would include several cycles and the possibility for the process to begin outside the cycle :

$$(M4)\quad \begin{cases} X_{n+1} = f(\mathcal{X}_n) + \sum\limits_{k=1}^{K} \sum\limits_{j=1}^{dk} 1_{\mathcal{C}_j^k}(\mathcal{X}_n)\, 1_j(\phi_{n,k}(\text{mod } d))\varepsilon_{n+1,j+1,k} \\ \qquad + 1_{(\cup \mathcal{C}^k)^c}(\mathcal{X}_n)\, \varepsilon_n \\ \phi_{n+1,k} = \phi_{n,k} + 1 \quad \forall\ k = 1,\dots, K \\ \mathcal{X}_0 \in \mathcal{C}_{j_0}^k \ ;\ \phi_{0,k} = j_0 \end{cases}$$

where ε_{n+1} is a white noise taking values in $(-\infty, +\infty)$, independent of $\mathcal{X}_n$, and independent of the $\varepsilon_{n+1,j+1,k}$. The K cycles are assumed completely disjoint. In most applications, the model (M3) will be sufficient. So, except for the stability property study, we will restrict our attention to this model.

4. STUDY OF THE CYCLE PROPERTIES

4.1. Stability property in the model (M4) (K=1)

Lemma 1. If the process $\mathcal{X}_n$ is geometrically ergodic up to the entry time into the cycle, then the cycle is an a.s. stable limit cycle in the model (M4), and its stability domain is $\mathcal{X}^h$.

Proof : We have to prove that $P(\mathcal{X}_n \in \mathcal{C} | \mathcal{X}_0 = x_0) \xrightarrow[n \to \infty]{} 1 \quad \forall\ x_0 \in \mathcal{C}^c$. Denote

$$P_{x_0}(\mathcal{X}_n \in \mathcal{C}) = P(\mathcal{X}_n \in \mathcal{C} | \mathcal{X}_0 = x_0) = P_{x_0}(\mathcal{X}_n \in \mathcal{C} \cap \mathcal{X}_{n-1} \in \mathcal{C}) + P_{x_0}(\mathcal{X}_n \in \mathcal{C} \cap \mathcal{X}_{n-1} \notin \mathcal{C}).$$

$$P_{x_0}(\mathcal{X}_n \in \mathcal{C} \cap \mathcal{X}_{n-1} \in \mathcal{C}) = \sum_{k=1}^{d} P_{x_0}(\mathcal{X}_n \in \mathcal{C} \cap \mathcal{X}_{n-1} \in \mathcal{C} \cap \phi_{n-1} = k)$$

$$= \sum_k \int P(\mathcal{X}_n \in \mathcal{C} | \mathcal{X}_{n-1} = x,\ \phi_{n-1} = k)\ g_{x_0,n-1,k}(dx)$$

where $g_{x_0,n-1,k}(dx)$ denotes the probability law of $(\mathcal{X}_{n-1}, \phi_{n-1})$ conditional to the starting value x_0. The right member of the previous equation is also

$$\sum_k \int_{\mathcal{C}_k \cap \{\phi_{n-1}=k\}} P(\mathcal{X}_n \in \mathcal{C}_{k+1} | \mathcal{X}_{n-1} = x,\ \phi_{n-1} = k)\ g_{x_0,n-1,k}(dx) = P_{x_0}(\mathcal{X}_{n-1} \in \mathcal{C})$$

because of the cycle property (def. 2.3). We use now the fact that $\mathcal{X}_n$ is markovian up to the entry time in the cycle :

$$P_{x_0}(\mathcal{X}_n \in \mathcal{C} \cap \mathcal{X}_{n-1} \notin \mathcal{C}) = \int_{\mathcal{C}^c} P(\mathcal{X}_n \in \mathcal{C} | \mathcal{X}_{n-1} = x)\ \frac{g_{x_0,n-1}(dx)}{P_{x_0}(\mathcal{X}_{n-1} \notin \mathcal{C})} \cdot P_{x_0}(\mathcal{X}_{n-1} \notin \mathcal{C})$$

where $g_{x_0,n-1}(dx)$ is the density law of $\mathcal{X}_{n-1}$ conditional on the initial state

x_o. Denote $p_n = P_{x_o}(\mathcal{X}_n \in \mathcal{C})$ and

$$q_n = \int_{\mathcal{C}^c} P(\mathcal{X}_n \in \mathcal{C} | \mathcal{X}_{n-1} = x) \frac{g_{x_o,n-1}(dx)}{P_{x_o}(X_{n-1} \notin \mathcal{C})}.$$

If we suppose $\mathcal{X}_n$ ergodic up to the entry time into the cycle, then

$$\frac{g_{x_o,n-1}(dx)\, 1_{\mathcal{C}^c}(dx)}{P_{x_o}(\mathcal{X}_{n-1} \notin \mathcal{C})}$$

which is the density law of $\mathcal{X}_{n-1}$, restricted to $\mathcal{C}^c$ and conditional on the starting point x_o, tends to an invariant density law when n tends to ∞. Then q_n tends to a certain q. So we have

$$p_n = p_{n-1} + q_n(1 - p_{n-1}) = p_{n-1}(1-q_n) + q_n$$

and writing $q_n = q + \alpha_n$, where $\alpha_n \to 0$ when $n \to \infty$, we get

$$\begin{aligned} p_n &= p_{n-1}(1-q) + q + \alpha_n(1 - p_{n-1}) \\ &= p_{n-2}(1-q)^2 + q(1-q) + q + \alpha_{n-1}(1-q)(1 - p_{n-2}) + \alpha_n(1 - p_{n-1}) \\ &\vdots \\ &= p_{n-(n-1)}(1-q)^{n-1} + q \sum_{i=0}^{n-2} (1-q)^i + \sum_{m=1}^{n} \alpha_{n-m+1}(1-q)^{m-1} (1 - p_{n-m}) \\ &= (1-q)^{n-1} (p_1-1) + 1 + \sum_{\ell=0}^{n-1} \alpha_{n-\ell}(1-q)^{\ell} (1 - p_{n-\ell-1}) \end{aligned}$$

using the definition of the model, $q = \int_{\mathcal{C}^c} P(\varepsilon_n \in C - f(x)) \frac{g_{x_o,n-1}(dx)}{P_{x_o}(X_{n-1} \notin \mathcal{C})}$.

$C - f(x) \subset (-\infty, +\infty)$, the space of ε_n. Then $P(\varepsilon_n \in C - f(x)) > 0$ for all $x \in \mathcal{C}^c$, which implies that $(1-q) < 1$. Then $(1-q)^{n-1} \to 0$, when n tends to ∞,

and the quantity $|\sum_{\ell=0}^{n-1} \alpha_{n-\ell}(1-q)^{\ell} (1-P_{n-\ell-1})| \leq \min(q, 1-q)(1-(1-q)^n)/q$, $\to[\min(q,1-q)]/q < 1$ So this quantity has a limit which is not necessarily equal to 0 ; which implies that p_n does not necessarily tend to 1. For that, the α_j must satisfy certain assumptions. For example, if we suppose that $|\alpha_j| < a\rho^j$ where $0 < \rho < 1$ (geometric ergodicity), then

$$\begin{aligned} |\sum_{\ell=0}^{n-1} \alpha_{n-\ell}(1-q)^{\ell}(1-p_{n-\ell-1})| &\leq a \sum_{0}^{n-1} \rho^{n-\ell} (1-q)^{\ell} \\ &= a\, \rho^n \sum_{0}^{n-1} \left(\frac{1-q}{\rho}\right)^{\ell} \\ &= \begin{cases} a\, \rho^n . n & \text{if } \rho = 1 - q \\ a\, \rho \dfrac{((1-q)^n - \rho^n)}{1-q-\rho} & \text{if } \rho \neq 1 - q \end{cases} \end{aligned}$$

which tends to 0 - and then $p_n \to 1$ with a convergence rate function of q (which depends on x_o). Thus the stability property comes from both the fact that we have, at every time, the same positive probability q of entering into the cycle, and from a particular ergodicity. (Note : sufficient conditions for ergodicity can be found in Tweedie's paper (1975), for general markov chains, or in Doukhan and Ghindes's thesis (1980) for nonlinear autoregressive models.).

4.2. Self-exciting cycle

Lemma 2. If $\mathfrak{X} =]\alpha,\beta[\subset \mathbb{R}$ (with α and β finished or not) and if f is continuous on $[\alpha,\beta]$, then a necessary and sufficient condition for a self-exciting cycle is $E\,\varepsilon_{n,j} = 0$, $\forall\, j = 1,\dots, d$. $(d \geq 1)$.

Proof: We have to prove that def. 2.4.2 (i and ii) is equivalent to $E\,\varepsilon_{n,j} = 0$. We prove here that $E\,\varepsilon_{n,j} = 0$ implies i (or rather, that non i implies $E\,\varepsilon_{n,j} \neq 0$, for some j). The other implications are immediate.

Suppose $h = 1$ (the proof is similar for $h > 1$) ; and suppose that f has neither a deterministic cycle nor singular point (cycle of period 1) except perhaps in α or β. So, as f is continuous, f must satisfy either (1) or (2) that follow :

$f(x) > x \quad , \quad \forall\, x \in\,]\alpha,\beta[\qquad (1)$

$f(x) < x \quad , \quad \forall\, x \in\,]\alpha,\beta[\quad . \qquad (2)$

Suppose (1), for example. Then, if 0 belongs to Ω_j, for all j, $X_{n+1} = f(\mathfrak{X}_n)$ belongs to $\mathfrak{X}$ (3), which implies that $f(\mathfrak{X}) \subset \mathfrak{X}$ (4). The relations (1) and (4) imply that $f(\beta) = \beta$, while the relations (1) and (3) imply that $X_{n+1} > X_n$. Moreover f is continuous. Then, as f has no singular point except β, f satisfies :

$$f(x) \underset{x \uparrow x_0}{\not\longrightarrow} x_0, \quad \forall\; x_0 \neq \beta.$$

So, $(X_n)_n$ is an increasing sequence such that the difference $X_{n+1} - X_n$ tends to 0 as n tends to ∞ only when X_n tends to β. So X_n tends to β as n tends to ∞ whatever the initial state x_0. But, if x_0 belongs to C_{j_0}, then $X_{kd} = f^{(kd)}(x_0)$ belongs also to C_{jo}, for all k, because of the cycle property. This implies that β belongs to C_{jo}, for all j_0. This result is incompatible with the assumption that at least two different C_j are disjoint. So we cannot suppose that 0 belongs to Ω_j for all j, which implies that we cannot suppose $E\,\varepsilon_{n,j} = 0$, for all j.

This last condition, $E\,\varepsilon_{n,j} = 0$, for all j, means that such an innovation does not maintain the cycle. For example, if f is linear, the only possible cycles suppose at least one Ω_j entirely positive (or negative). This Ω_j compensates for the natural damping of the oscillations.

5. ESTIMATION OF THE CYCLE

We will assume here that f and d are known or already estimated. We do not need the model to estimate the cycle. The model will be used to estimate the order h of the autoregression, which allows us to build the $(\mathscr{C}_j)_j$ from the $(C_j)_j$. Suppose we have an infinite length cyclic series with no experimental error. The only noise is the regulation noise of the system. Then, if the C_j are closed intervals of $\mathbb{R}$,

$$C_j = \{\inf_k X_{j+kd}, \sup_k X_{j+kd}\} \tag{5.1}$$

where k is such that $j+kd \geq T$, the time of entry into the cycle.

On the other hand, if the observations include an error term, the model will be from instant T

$$X_{n+1} = f(\mathscr{X}_n) + \Sigma 1_{\mathscr{C}_j}(\mathscr{X}_n)\ 1_j\ (\phi_n(\text{mod } d))\ \varepsilon_{n+1,j+1}$$

$$Y_{n+1} = X_{n+1} + \eta_{n+1}$$

where Y_{n+1} is the observation and η_{n+1} is the unknown experimental error. We have to estimate the $(C_j)_j$ through the series $(Y_n)_n$. We can do that either by smoothing the series $(Y_n)_n$ by means of a moving average for example, and then estimating the C_j as in the non-errored case (5.1), either by smoothing the C_j built directly from the $(Y_n)_n$, or by smoothing the cyclic continuous band built from $(C_1, \ldots, C_d)$. This last estimation can be done by cutting the series according to intervals of length d, superposing all these pieces of series and then smoothing the resulting observed band, using a non parametric estimation in 2 dimensions.

In every case, the smoothing partially eliminates the experimental error, but can change the shape of the theoretical cycle. However the last method has three main advantages : i) it does not depend too much on d which can be very big. The first two methods estimate d different parameters whereas the last one estimates the cyclic band as a whole ; ii) it is the only possible method in continuous time ; iii) it is also the only method if the sampling is not at regular time intervals.

We suppose in this paragraph that T is known and the series infinite ; although, in practice, that is rarely the case the case. Then we can find the C_j by a sequential method of estimation. Suppose that $N = m+kd$ is the length of the series, and that the sequential intervals of estimation are of the form $I_k = [m+kd, N]$, where $k = 0,1,\ldots, K-2$. Then the estimates of the C_j will be the values $\hat{C}_{j,k_0}$, such that the criterion $\mathscr{S}_{k_0} = \sum_{j=1}^{d} \mathscr{L}(C_{j,k_0}) / \sum_{j=1}^{d} \mathscr{L}(C_{j,k_{0-1}})$ is significantly less than 1, while $\mathscr{S}_{k_{0+1}}$ is not significantly less than 1.

In the following example, we will not use such a criterion ; but, rather, do empirical estimations -for we have not yet studied its probability law.

6. EXAMPLE - THE CANADIAN LYNX DATA

We have applied this cyclic model to the series of the annual records of the number of canadian lynx trapped in the years 1821 to 1934 in the Mackenzie River district of North-West Canada. This series, which consists of 114 data points, presents random oscillations (fig. 2). These are created by the biological system itself, that is, probably by a predator-prey system. Many authors (see, in particular, Campbell and Walker, 1977 ; Tong, 1977; Tong and Lim, 1980 ; Haggan and Ozaki, 1981 & 1982) have tried to model this series. Here, we try to explain the observed self exciting random oscillations with a cyclic nonlinear autoregressive model of the type (M3), assuming d and f known. The average period of the oscillations is about 9.5 years. In fact, the five first cycles have an average period of about nine years and the six others, a period of ten years. We have chosen d=10. We have first estimated the $(C_j)_j$ from a smooothing of the $(C_j)_j$ calculated from the data. For this smoothing, we have used an ordinary moving average of length 3. Because of the variation of the period value, the resulting band was very wide ; so, we have then restricted our estimation to the six last cycles (64 years) (fig. 2).

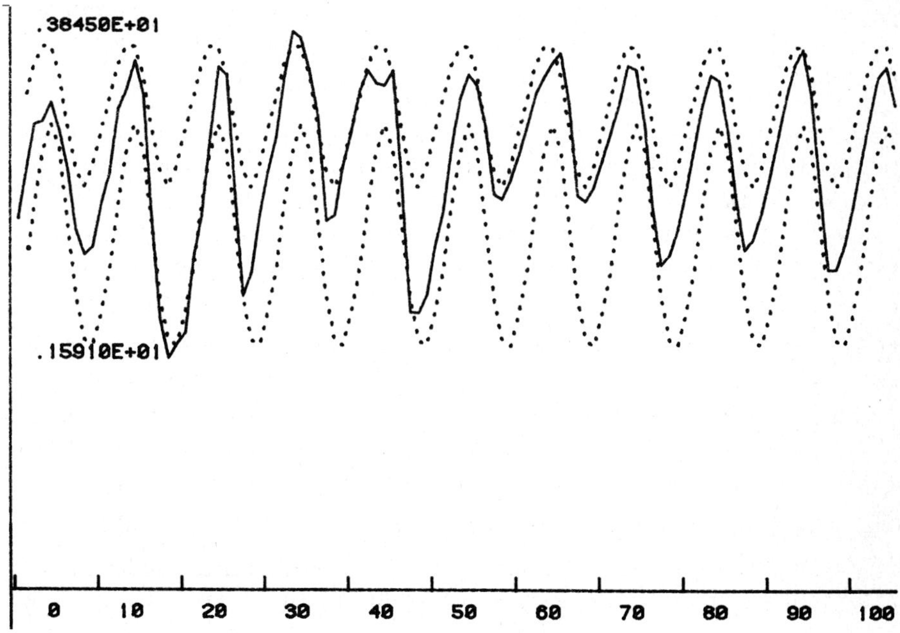

Fig. 2. The observed series (114 data points from 1821 to 1934) , with the estimated cyclic band calculated from the last 64 data points (dotted).

We have then estimated the unknown function f by means of a Threshold model (TAR). As Tong (1980) pointed out, this model has the advantage of approximating any general function f provided that f is uniformly continuous in the cycle. But Ozaki (1981) claimed that TAR cannot give a good approximation to non linear random vibrations for two reasons. The first is that the introduction of constant terms is not natural in vibrations because 0 must be a singular point. The second is that the TAR fitted for the lynx data (Tong 1980) has no limit cycle for certain initial values. But these two reasons are not really valid. First, we can find TAR models with a zero singular point and a limit cycle. For example

$$X_{n+1} = 0.5 - 0.2\,X_n \quad \text{if } X_n < 0$$
$$X_{n+1} = -\,0.8\,X_n \quad \text{if } X_n \geq 0$$

has the following limit cycle : $\{-0.4762,\ 0.5952\}$. Second, the stability domain of a limit cycle generally consists only of a part of the space ; and, certainly, this is the case in Biology.

For the estimation of f, we have supposed that the variances of the $\varepsilon_{n,j}$ were all the same, which implies that the AIC procedure is the same in the cyclic model (M3) as in the usual model (M2). We used Tong's normalized AIC (Tong and Lim, 1980) for TAR models. NAIC was minimized automatically according to Rosenbrock's algorithm (Kuester and Mize, 1973). The value of NAIC was -3.2396 and the associated fitted model was the following :

$$X_n = 0.55482 + 1.1253\,X_{n-1} - 0.17800\,X_{n-2} - 0.04059\,X_{n-3} - 0.10811\,X_{n-4}$$
$$+ 0.17325\,X_{n-5} - 0.32182\,X_{n-6} + 0.28476\,X_{n-7} - 0.073248\,X_{n-8} + \varepsilon_n^1, \quad X_{n-2} \leq 3.3858$$
$$X_n = 2.3836 + 1.5116\,X_{n-1} - 1.1857\,X_{n-2} - 0.093339\,X_{n-3} + \varepsilon_n^2, \quad X_{n-2} > 3.3858$$

where var $\varepsilon_n^1 = 0.015776$; var $\varepsilon_n^2 = 0.068419$. The estimated threshold function has a limit cycle of period 10 years (fig. 3).

We then simulated this TAR model with two gaussian white noises ε_n^1 and ε_n^2, and the cyclic TAR model with truncated gaussian white noises $\varepsilon_{n,j}^1$ and $\varepsilon_{n,j}^2$ with the same variances as ε_n^1 and ε_n^2 . These variances are very small. However the TAR series is not cyclic according to def. 2.3 (fig. 4), whereas the cyclic TAR series looks quite similar to the observed series (fig. 5).

The residual variance of the difference between the observed series and the simulation is much less for the cyclic TAR (0.18951) than for the TAR (0.56245). Moreover the autocorrelations of the cyclic TAR simulation is cyclic and quite similar to the observed autocorrelations, which is not so for the TAR simulation (fig. 6, 8, 9). The damping of the oscillations that we see in the figures is only due to the usual biased estimator we have used. (See the graphic plot of the unbiased estimator in fig. 7).

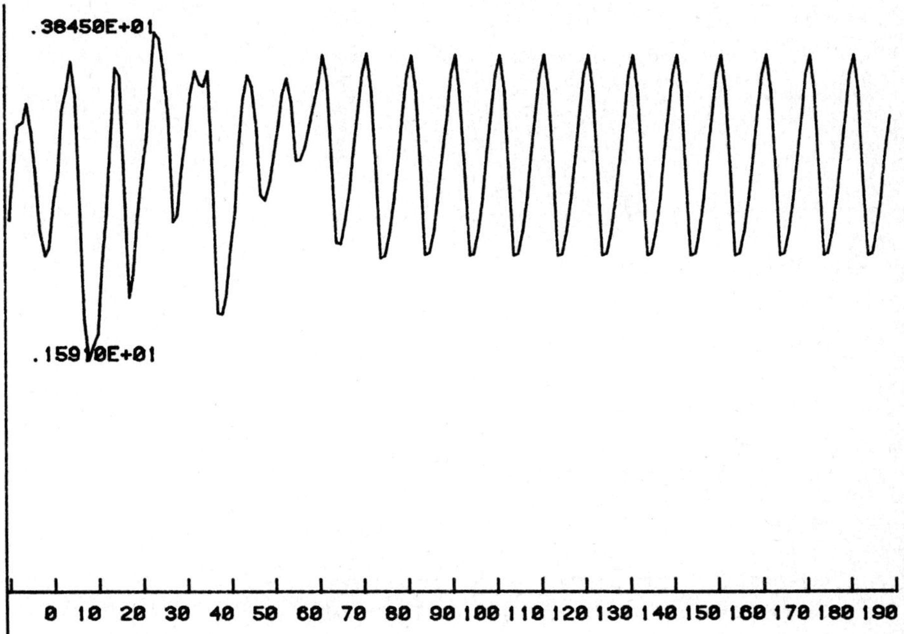

Fig. 3. The last 64 points from the observed series followed by the deterministic limit cycle of the TAR function.

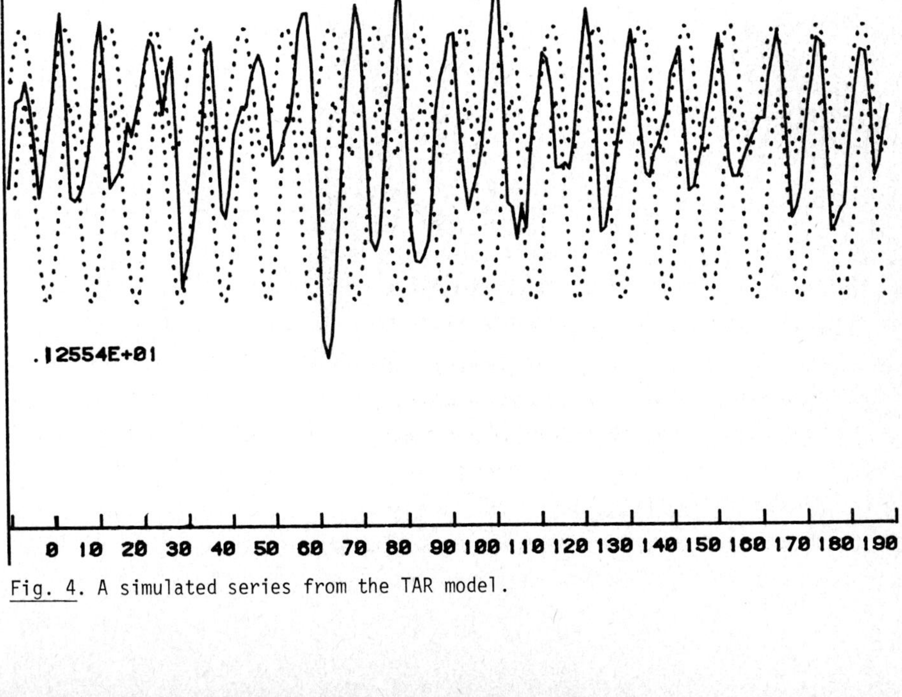

Fig. 4. A simulated series from the TAR model.

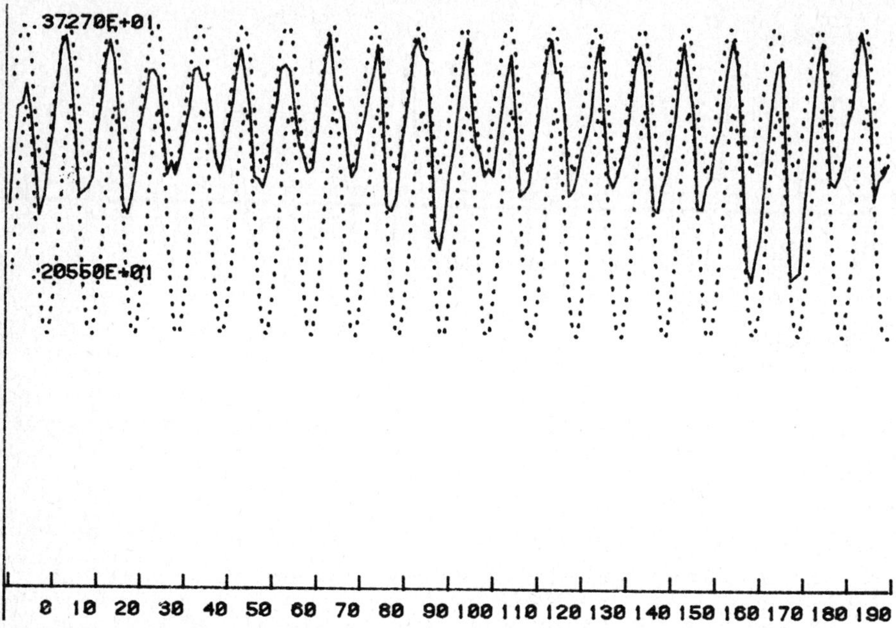

Fig. 5. A simulated series from the cyclic TAR model.

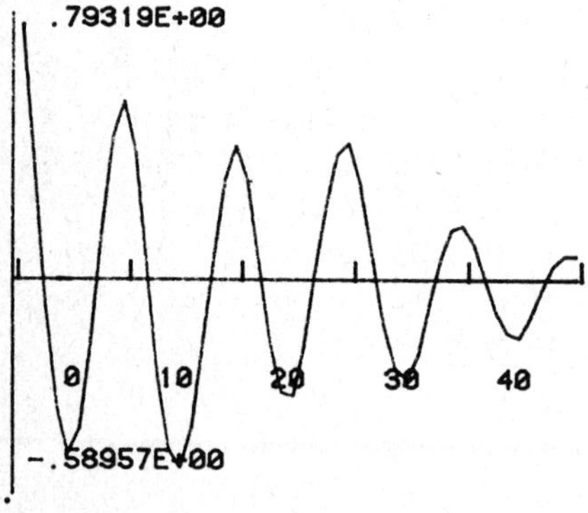

Fig. 6. The autocorrelations of the observations (biased estimator)

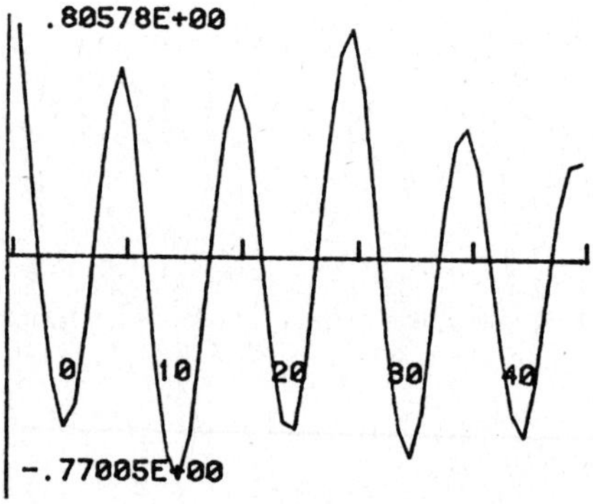

Fig. 7. The autocorrelations of the observations (unbiased estimator)

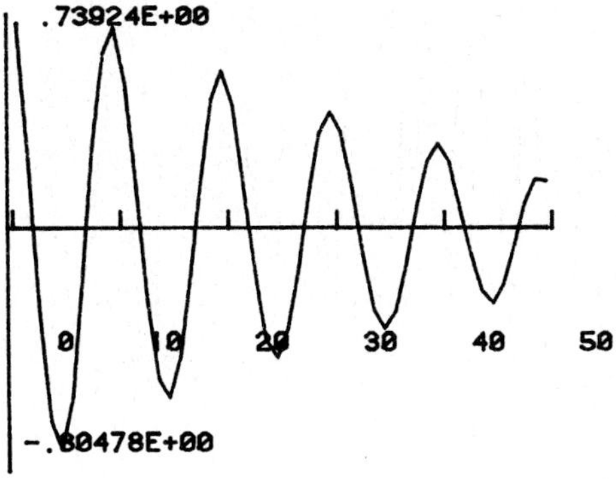

Fig. 8. The autocorrelations of the cyclic TAR simulation (biased estimator)

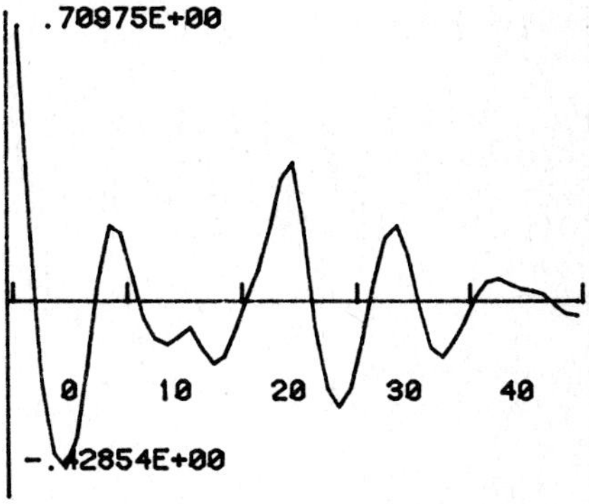

Fig. 9. The autocorrelations of the TAR simulation (biased estimator)

Note. The function f could be estimated using other types of models like, for example, non linear threshold models (Ozaki,1981) or exponential models (Ozaki, 1982). But it would not change the conclusion which depends essentially on the noise structure.

7. CONCLUSION

The estimation of the cycle $\mathcal{C}$ allows us to make long-lead forecasts, which take into account the system regulation noise, for we know that the future values must belong to the stochastic cycle of the process. Tong and Wu's forecasts (1982) also used the periodicity of the memory of the process, but in a different apparent manner. However, their predictions are found to be cyclic according to def. 2.3 (fig. 10). The cyclic band $(C_1, \ldots, C_d)$ is then a generalisation of their method which unfortunately cannot be applied to models other than threshold ones. Moreover, the cyclic model allows us to simulate cyclic trajectories. For example, this can be used to study the structural stability of the stochastic cycle ; that is, to study the changes or the vanishing of the cycle when f is slightly changed.

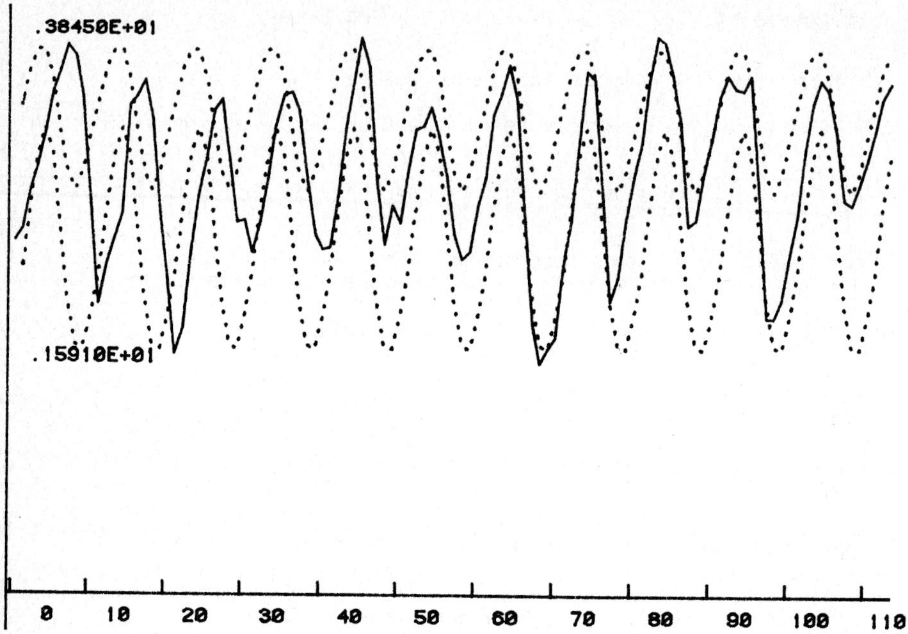

Fig. 10. The last 64 points from the observed series followed by Tong and Wu's predictions (42 values)

The author is grateful to Dr J. Demongeot for his fruitful comments.

REFERENCES

CAMPBELL, M.J. and WALKER, A.M. (1977). A survey of statistical work on the Mackenzie River series of Annual Canadian Lynx trappings for the years 1821-1934, and a new analysis. *J. Roy. Statist. Soc. A 140*, 411-431.

DOOB, J.L. (1953). *Stochastic processes*. New-York, Wiley.

DOUKHAN P. and GHINDES M. (1980). Etude des processus $X_n = f(X_{n-1}) + \varepsilon_n$. Thèse de 3ème cycle. Université Paris Sud, Département de Mathématique, Orsay, France.

GUMOWSKI, I. and MIRA, C. (1980). *Dynamique chaotique*. Editions Cepadues, Toulouse.

HAGGAN, V. and OZAKI, T. (1981). Modelling nonlinear random vibrations using an amplitude-dependent autoregressive time series model. *Biometrika 68*, 189-196.

HAGGAN, V. and OZAKI, T. (1980). Amplitude-dependent exponential AR model fitting for non linear random vibrations. In *Time Series*. Ed. O.D. Anderson, North Holland, Amsterdam & New-York, 57-71.

JONES, R.H. and BRELSFORD, W.M. (1967). Time series with periodic structure. *Biometrika 54*, 403-408.

KUESTER, J.L. and MIZE, J.H. (1973). *Optimization techniques with fortran.*

McGraw Hill Book Company.

OREY, S. (1971). Lecture notes on limit theorems for Markov chain transition probabilities. Van Nostrand Reinhold Mathematical Studies, London.

OZAKI, T. (1980). Non linear time series models for non linear random vibrations. *J. Appl. Prob.* *17*, 84-93.

OZAKI, T. (1981). Non linear threshold autoregressive models for nonlinear random vibrations. *J. Appl. Prob.* *18*, 443-451.

OZAKI, T. (1982). The statistical analysis of pertubed limit cycle processes using nonlinear time series models. *J. Time Series Analysis* 3, 29-41.

PAGANO, M. (1978). On periodic and multiple autoregressions. *Annals of Statistics* *6*, 1310-1317.

PAVLIDIS, T. (1973). *Biological oscillators : their mathematical analysis*. Academic Press, New-York & London.

TONG, H. (1977). Some comments on the canadian lynx data. *J. Roy. Statist. Soc.* A *140*, 432-436.

TONG, H. and LIM, K.S. (1980). Threshold autoregression, limit cycles and cyclical data. *J.R. Statist. Soc.* B *42*, 245-292.

TONG, H. (1980). A view on non linear time series building. In *Time Series*. Ed. O.D. Anderson, North Holland, Amsterdam & New-York, 41-56.

TONG, H. and WU, Z.M. (1982). Multi-step ahead forecasting of cyclical data by threshold autoregression. In *Time Series Analysis : Theory and Practice 1*. Ed. O.D. Anderson, North-Holland, Amsterdam & New-York, 733-753.

TWEEDIE, R.L. (1975). Sufficient conditions for ergodicity and stationarity of markov chains on a general state space. *Stoch. Proc. & Appl.* *3*, 385-403.

TIME SERIES ANALYSIS: Theory and Practice 7
O.D. Anderson (editor)

USING STATE DEPENDENT MODELS FOR PREDICTION OF TIME SERIES WITH MISSING OBSERVATIONS

Phillip A. Cartwright
Department of Economics and Division of Research, College of Business Administration, University of Georgia, Athens, Georgia 30602, USA

In this paper, extensions of the state dependent model developed by Priestley to allow prediction of observations for time series having blocks of missing observations are considered. Alternative models are applied to the statistical analysis of data and the numerical results are discussed.

I. INTRODUCTION

In recent years, nonlinear time series models have come to play an important role in applied time series analysis. To date, the broadest class of models developed in the literature is the class of state dependent models (S.D.M.) initially presented by Priestley (1980, 1982), and extended by Cartwright and Newbold (1983). The purpose of this paper is to present preliminary work on the problem of extending the S.D.M. for predicting time series having blocks of missing observations. Section II provides a brief description of the S.D.M., i.e., procedures for estimating the model and generating forecasts are reviewed. Section III considers problems of estimation and forecasting with S.D.M.s when the time series of interest has a block of missing observations, i.e., the series is missing several observations in a row. Section IV presents the results of applications of the model to the Wölfer sunspot series under alternative procedures and model selection criteria. In order to test model performance several observations are deleted from the series. The paper concludes with a discussion of model performance.

II. THE MODEL

A detailed description of the S.D.M. has been given by Priestley (1980, 1982). Presentation of the model in its most general form is provided below. The ARMA(p,q) S.D.M. is described by

$$\begin{aligned} &X_t + \phi_1(x_{t-1})X_{t-1} + \dots + \phi_p(x_{t-1})X_{t-p} \\ &= \mu(x_{t-1}) + \varepsilon_t + \theta_1(x_{t-1})\varepsilon_{t-1} + \dots + \theta_q(x_{t-1})\varepsilon_{t-q} \end{aligned} \tag{1}$$

The process characterized by the ARMA scheme in (1) is governed by the level parameter, μ, the AR coefficients, $\phi_1, \dots, \phi_p$, and a set of MA coefficients,

$\theta_1, \ldots, \theta_q$. Each of the coefficients vary with the current state of the process captured by $x_{t-1} = \{\varepsilon_{t-q}, \ldots, \varepsilon_{t-1}, X_{t-p}, \ldots, X_{t-1}\}$.

Assuming that the coefficients are smooth functions of the state vector, the model may be fit to data without the imposition of further conditions on the functions relating the coefficients to the state vector. The general model for the level parameter, the autoregressive coefficients and the moving average coefficients may be written as

$$\mu(x_t) = \mu(x_{t-1}) + \Delta x_t' \alpha^{(t)} \tag{2}$$

$$\theta_n(x_t) = \theta_n(x_{t-1}) + \Delta x_t' \beta_n^{(t)} \tag{3}$$

$$\phi_n(x_t) = \phi_n(x_{t-1}) + \Delta x_t' \gamma_n^{(t)} \tag{4}$$

where $\{\alpha^{(t)}\}$, $\{\beta_n^{(t)}\}$, $\{\gamma_n^{(t)}\}$ are gradients defined in terms of deviations of the process from the current state. The gradients are permitted to wander in the form of a random walk, and the model bends so as to minimize the distance between X_t and the predicted value $\hat{X}_t$. The random walk model for the gradients is

$$B_t = B_{t-1} + V_t \tag{5}$$

where $B_t = \{\alpha^{(t)}; \beta_q^{(t)}, \ldots, \beta_1^{(t)}; \gamma_p^{(t)}, \ldots, \gamma_1^{(t)}\}$. V_t is a sequence of independent matrix random variables with each matrix assumed to be MVN $\sim (0, \Sigma_V)$. The observation and state equations are

$$X_t = H_t \theta_t + \varepsilon_t \tag{6}$$

and

$$\theta_t = F_{t-1}\theta_{t-1} + W_t \tag{7}$$

respectively. H_t is a $1 \times (p+q+1)^2$ vector, $H_t = (1, \varepsilon_{t-1}, \ldots, \varepsilon_{t-q}, -X_{t-1}, \ldots, -X_{t-p}, 0, 0, \ldots, 0)$, θ_t is a $(p+q+1)^2$ column vector, $\theta_t = (\mu_t, \theta_{1,t}, \ldots, \theta_{q,t}, \phi_{1,t}, \ldots \phi_{p,t}, \alpha^{(t)}, \beta_1^{(t)}, \ldots, \beta_q^{(t)}, \gamma_1^{(t)}, \ldots, \gamma_p^{(t)})^t$, and F_{t-1} is a $(p+q+1)^2 \times (p+q+1)^2$ transition matrix

$$F_{t-1} = \left[\begin{array}{c|c} I_{(p+q+1)} & I_{(p+q+1)} \otimes \Delta x_{t-1}' \\ \hline 0 & I_{(p+q)(p+q+1)} \end{array}\right] \tag{8}$$

The elements $\Delta x_{t-1}' = \{(\varepsilon_{t-1} - \varepsilon_{t-2}), \ldots, (\varepsilon_{t-q} - \varepsilon_{t-q-1}), (X_{t-1} - X_{t-2}), \ldots, (X_{t-p} - X_{t-p-1})\}$, and the vector W_t is given by

$$W_t = (0, \ldots, 0, v_{1t}, v_{2t}, \ldots, v_{(p+q)(p+q+1)t}), \tag{9}$$

where the first (p+q+1) elements are zero and $v_{1t}, v_{2t}, \ldots,$ are the elements in the columns of V_t of equation (5).

Applying the Kalman algorithm (Kalman, 1963) directly to (6) and (7) yields the recursion given by (10) - (14).

$$\hat{C}_t = \hat{P}_t - \hat{K}_t(H_t\hat{P}_tH_t' + \hat{\sigma}_\varepsilon^2)K_t' \tag{10}$$

$$\hat{P}_t = F_{t-1}\hat{C}_{t-1}F_{t-1}' + \hat{\Sigma}_W \tag{11}$$

$$\hat{K}_t = \hat{P}_tH_t'(H_t\hat{P}_tH_t' + \hat{\sigma}_\varepsilon^2)^{-1} \tag{12}$$

$$\hat{\theta}_t = F_{t-1}\hat{\theta}_{t-1} + \hat{K}_t(X_t - H_tF_{t-1}\hat{\theta}_{t-1}) \tag{13}$$

$$\hat{X}_t = H_t\hat{\theta}_t \tag{14}$$

In equations (10) - (14), P_t is the variance-covariance matrix of the one-step prediction errors of θ_t, i.e., $P_t = E[(\theta_t - F_{t-1}\hat{\theta}_{t-1})(\theta_t - F_{t-1}\hat{\theta}_{t-1})']$, $C_t = E[(\theta_t - \hat{\theta}_t)(\theta_t - \hat{\theta}_t)']$, the variance-covariance matrix of $(\theta_t - \hat{\theta}_t)$, K_t is the Kalman gain vector, and σ_ε^2 is the residual variance.

At time $t = t_0$ the system requires start-up values for $\hat{\theta}_{t_0}$ as well as the variance-covariance matrix $\hat{C}_{t_0}$ and $\hat{\sigma}_\varepsilon^2$. These may be obtained by estimating a standard linear model using an initial stretch of the data. The recursion may be started midway into the initial stretch. The initial values of the gradients may be set to zero. Values must be chosen for Σ_W so as to minimize prediction error, and these may be estimated using a grid search procedure (Cartwright and Newbold, 1983).

The output of the fitting algorithm gives estimated values of θ_t based upon information given by the value of X at period t. From the estimates of the parameters available at period t, and given knowledge of X_{t-1}, estimated values of X_t are computed. The algorithm may be used for the purpose of generating forecasts by making use of the information captured by the estimate of θ_t and the most recent observation, X_t. The appropriate forecasting equation is

$$\hat{X}_{t+1} = \hat{\mu}_t + \hat{\phi}_{1,t}X_t + \ldots + \hat{\phi}_{p,t}X_{t-p} + \hat{\theta}_{1,t}\hat{\varepsilon}_{t-1} + \ldots + \hat{\theta}_{q,t}\hat{\varepsilon}_{t-q} \, . \tag{15}$$

III. THE S.D.M. WITH MISSING OBSERVATIONS

The problem of estimating a time series model for series with missing data has received a considerable amount of attention. Recently, the missing observation problem has been analyzed by Ansley and Kohn (1983), Ljung (1982), Dunsmuir (1981),

Harvey (1981), and Jones (1980). In these studies, the authors have addressed the problem within the maximum likelihood framework. Given the complexity of the general S.D.M., and thus, of the likelihood function, the adaptation of procedures suggested in recent papers to use with S.D.M.s is not immediate. Despite its ad hoc nature, an immediate solution is to replace missing observations with one-step forecasts. However, when a block of data is missing, the problem is complicated in that (1) the information about the past contained in the state vector fades away, i.e., $\theta_{t+k|t}$ approaches zero as $k \to \infty$, and the variance-covariance matrix of the one-step prediction errors of θ_t returns to the initial state variance-covariance matrix, and (2) for series having bounded values, in the absence of appropriate constraints on model behavior, the model may generate predicted values outside the permissible set. In fact, this problem may arise in the application of S.D.M.s to data series without missing observations.

A solution to the first complication is to rewrite the algorithm given by equations (10) - (14) so as to replace missing values by one-step forecasts. For example, consider a block of missing observations occurring midway into a series ($t = \frac{n}{2}$ for even n and t = 1, ..., n) and extending over k time periods ($k = 1, \ldots, m$, $m < n$). Let X^o_{t+k} denote a missing observation at period t+k. Replace X^o_{t+k} by $\hat{\hat{X}}_{t+k}$, the one-step forecast of X_{t+k} generated on the basis of information available at period t+k-1. The forecast of the next missing observation, X^o_{t+k+1}, is generated based upon information available at period t+k, and so forth, i.e., the forecast $\hat{\hat{X}}_{t+k}$ is taken as part of the information set. Following period t+m, the final period associated with the occurrence of a missing observation, the algorithm uses X_{t+m+1} and all remaining available observations at t+m+1, t+m+2, ..., n.

The second problem is more difficult. The procedure given by Cartwright and Newbold (1983) for fitting S.D.M.s to data involves implementing a grid-type search over the diagonal elements of Σ_v given by equation (5) so as to achieve the best possible fit to the data in the sense of mean squared error. The rate of adjustment of the model parameters is governed by the magnitude of the diagonal elements of Σ_v relative to σ^2_ε, e.g., if the estimated values of Σ_v are large relative to $\hat{\sigma}^2_\varepsilon$, the model parameter estimates are permitted to change rapidly.

As indicated above, for series having boundaries on the set of possible values, constraints may have to be specified in order to prevent the model from generating unreasonable predictions as outcomes from the fitting and/or forecasting algorithms. Such unacceptable model performance may be realized for cases involving no missing observations, however, the problem is likely to be more prevalent when forecasts are used to generate values for blocks of missing observations. In addition, for series characterized by a considerable level of period-to-period variability, the problem is more severe. The models yielding the best fit for these series are generally characterized by large ratios of the diagonal elements of $\hat{\Sigma}_v$ to $\hat{\sigma}^2_\varepsilon$. As a result, while estimates of parameter values may be admissible, and the model fit may be good in the sense of mean squared error, the likelihood that the predicted values of the series will be unacceptable increases. Again, the problem becomes more prevalent when forecasts are used to replace missing observations.

If best fit is the only criterion for model selection, as a first option, the analyst may decide to ignore predictions deemed to be unreasonable. A second option is to impose reasonable constraints on the values of the predictions, i.e., for series having blocks of missing observations, an S.D.M. may be fit to data in the following way. Fit the S.D.M. using the procedure discussed by Cartwright and Newbold (1983). All observations may be used to fit the model, however, a missing value code should be entered for appropriate periods and these should be replaced by one-step forecasts. Restrictions on values of the predictions of X_t at period t and one-step forecasts may be imposed. A third alternative requires reestimation of the diagonal elements of $\hat{\Sigma}_v$ under appropriate restrictions, i.e., restrictions may be placed on the variances of the random walk parameters, hence constraining the adjustment process. A fourth possibility involves specification of restrictions of the state vector θ_t, and the variance-covariance matrix, P_t. In theory, the restrictions may be stochastic or non-stochastic.

The first option seems unacceptable, particularly for cases involving missing observations, as unacceptable forecasts may be used to replace missing values. If the analyst is committed to using the minimum mean squared error criterion for model selection, the second alternative seems appropriate. It is indicated below that choice of the third option may be more reasonable than either the first or second, yet this option may require that the analyst accept a substantial reduction in model fit. Moreover, it is possible that the grid search process will not locate values of the diagonal elements consistent with permissible forecast values. Providing the restrictions are correct, i.e., the restrictions incorporate all useful information relating to the process, it seems reasonable to expect implementation of restrictions on the state vector, θ_t, and variance-covariance matrix, P_t, will result in superior model performance, yet the specification of appropriate restrictions may not be a trivial matter.

Conceivably, given a sufficiently long data series, the procedures indicated could be implemented for series with blocks of missing data of varying lengths and series with more than one block of missing data. However, as a practical matter, due to the increased risk of forecast error as larger sets of missing observations are replaced by one-step forecasts, the procedures become less viable as the length and number of missing blocks increases.

IV. APPLICATION

In order to evaluate the procedures outlined in Section III, the S.D.M. is applied to 100 observations from the Wölfer sunspot series. The annual observations for the well-known series date from 1770 to 1869 and are given in Box and Jenkins (1976) as Series E. Future research should evaluate model performance using other time series, however, the sunspot data are selected for this preliminary study as the series is characterized by substantial variability, and the data are bounded from below, i.e., only non-negative predictions are permissible. Ten consecutive observations for periods 1850 through 1859 are deleted and these periods are identified as having missing observations. Following the procedures outlined in Section III, an AR(2) S.D.M. is fit to the sunspot series. No restrictions are imposed on the model or predicted values. Start-up values for the state vector, $\hat{\theta}_{t_0}$, the variance-covariance matrix $\hat{C}_{t_0}$ and the residual variance $\hat{\sigma}^2_{\varepsilon}$ are obtained using the first 30 observations to estimate a standard linear model. Start-up values are shown in Table 1.

Table 1. Start-up Values for AR(2) Model

$\hat{\mu} = 22.468$	$\hat{\sigma}^2_{\varepsilon} = 415.820$	$\hat{\sigma}^2_{\phi_2} = .022$	$\hat{\sigma}_{\phi_1\phi_2} = -.017$
$\hat{\phi}_1 = 1.345$	$\hat{\sigma}^2_{\mu} = 52.976$	$\hat{\sigma}_{\mu\phi_1} = -.208$	
$\hat{\phi}_2 = -.723$	$\hat{\sigma}^2_{\phi_1} = .021$	$\hat{\sigma}_{\mu\phi_2} = -.337$	

The minimum MSE is measured over the last 70 periods for which observations are given, and it is approximately zero. The minimum is achieved by setting the variances of the random walk terms equal to 100,000.0. The fact that the diagonal elements of $\hat{\Sigma}_V$ are large relative to $\hat{\sigma}^2_{\varepsilon}$ indicates that the model parameters are being permitted to change rapidly.

Over the periods for which observations are available, the predictions generated from the fitting algorithm do not violate the requirement that values be non-negative in any meaningful statistical sense. At period 41, the observation $X_t = 0$, and the prediction generated by the model is $\hat{X}_t = -.0001$. Four of the forecasts for periods with missing observations are negative, and these values are significantly different from zero. The forecasts, $\hat{X}_t$, for periods designated as having missing values and the actual observations for those periods are shown in Table 2.

Table 2. Forecasts of Missing Values
for Unrestricted and Restricted AR(2) Models

Period	Observation, X_t	Forecast, $\hat{\hat{X}}_t$	Forecast, $\hat{\hat{X}}_t^r$
81	66.0	38.93	38.92
82	64.0	-16.77	0.00
83	54.0	-48.90	0.00
84	39.0	-50.38	29.14
85	21.0	-28.33	68.28
86	7.0	2.40	99.05
87	4.0	27.18	111.09
88	23.0	37.47	104.23
89	55.0	32.75	86.01
90	94.0	18.72	66.66

The results given for the unrestricted model provide motivation for specifying restrictions on the predictions generated by the model or on model behavior. First, restrictions are placed directly on the forecasts. The following set of constraints is imposed on the forecasts, $\hat{\hat{X}}_t^r$, where the superscript r denotes a value subject to the restrictions

$$\hat{\hat{X}}_t^r = \begin{cases} 0 \text{ if } \hat{\hat{X}}_t < 0. \\ \hat{\hat{X}}_t \text{ if } \hat{\hat{X}}_t \geq 0. \end{cases} \tag{16}$$

One-step forecasts are generated under the restrictions given by equation (16). The lower constraint is binding for seven periods. The MSE measured over the last 70 observation periods is 323.55. The forecasts, $\hat{\hat{X}}_t^r$, for periods 81 through 90 are shown in Table 2. The lower constraint is binding in periods 82 and 83. The forecast error is quite large, but all values are permissible.

It is of interest to note the forecasting performance of the model over the periods immediately following the block of missing observations. The one-step forecasts and actual observations for periods 91 through 95 are shown in Table 3. Except for the value reported for period 92, the model appears to perform reasonably well in the sense of forecast error.

Table 3. Forecasts for AR(2) Restricted Model

Period	Observation, X_t	Forecast, $\hat{X}_t^r$	Forecast, $\hat{X}_t^{r'}$	Forecast, $\hat{X}_t^{r''}$
91	96.0	54.30	41.85	41.81
92	77.0	149.90	109.11	108.46
93	59.0	29.51	50.95	50.94
94	44.0	49.04	40.92	40.86
95	47.0	37.32	34.05	34.01

Following the discussion in Section III, the analyst might choose to estimate the diagonal elements of Σ_v, i.e., the variances on the random walk parameters, subject to the constraint that predicted values generated from the fitting and forecasting algorithms be in a permissible set. Implementation of this procedure may, however, be inconsistent with the model selection criterion of best fit. In order to test this procedure, the AR(2) model is reestimated subject to the single constraint that the predicted values be non-negative. By placing the constraint directly on the variances of the random walk parameters rather than on the predicted values, the model adjustment process is constrained. The estimated values of the variances are zero. As anticipated, the model fit is not nearly as good as that under the set of restrictions given by equation (16). The MSE measured over the last 70 observation periods is 230.67. This contrasts sharply with the near perfect fit associated with the model when restrictions are placed on the predicted values.

One-step forecasts are generated using the model with restrictions placed on the diagonal elements of Σ_v. The MSE over the last 70 observations is 247.89. This compares favorably with the value of 323.55 recorded for the performance of the model under the previous restrictions. The forecasts and actual values for the periods 81 through 90 are shown in Table 4, and in order to facilitate comparison, the values for periods 91 through 95 are shown in Table 3. The r' superscript on the forecast indicates the forecast value is subject to the restriction placed on the diagonal elements of Σ_v. In terms of forecast error over the ten periods with missing observations, and the five following periods, the model outperforms the model with the constraints on the forecast values. Notice, however, that the prediction generated by the model for period 83 is quite close to the lower bound, and the estimated variances on the random walk parameters given in equation (5) are set to zero. Given the specification of the parameters in the model, it is possible that the grid search may fail to locate variance estimates consistent with the restriction(s) imposed. In such an event, the analyst may consider the second alternative discussed, or consider the fourth alternative, i.e., specify restrictions on the elements of the state vector. This latter alternative is discussed below.

Table 4. Forecasts of Missing Values for Restricted AR(2) Model

Period	Observation, X_t	Forecast, $\hat{X}_t^{r'}$	Forecast, $\hat{X}_t^{r''}$
81	66.0	54.36	54.36
82	64.0	18.17	18.16
83	54.0	.41	.40
84	39.0	4.21	4.19
85	21.0	23.60	23.59
86	7.0	47.77	47.79
87	4.0	66.31	66.30
88	23.0	73.14	73.12
89	55.0	68.03	68.00
90	94.0	55.44	55.41

It has been indicated that restricting the coefficients of the state vector is likely to result in superior model performance relative to the alternatives considered thus far, provided the restrictions are correct. For series exhibiting behavior captured by reasonably stable coefficients, the analyst might restrict the coefficients of the state vector, θ_t, and the variance-covariance matrix P_t to assume the values computed for the previous period, i.e., $\hat{\theta}_{t|t} = \hat{\theta}_{t|t-1}$ and $\hat{P}_{t|t} = \hat{P}_{t|t-1}$. Implementation of this procedure over a long block of missing observations amounts to starting the Kalman recursion at the other end, and, for series exhibiting state-dependency or nonlinear behavior, e.g., the oil discovery series modelled by Cartwright and Newbold (1983), the specification of restrictions is considerably more difficult. More research is required in order to fully address the problems relating to the application of restrictions on parameters.

Predictions for observations missing from the sunspot series are generated under the restriction that the estimates of θ_t and P_t be equal to the values computed in the last period prior to the occurrence of a missing observation. In addition, on the basis of the results discussed above, the diagonal elements of Σ_v are set to zero. The values for the restricted parameters are shown in Table 5. Forecasts, $\hat{X}_t^{r''}$, are shown in Table 3 and 4. The values of $\hat{X}_t^{r''}$ and $\hat{X}_t^{r'}$ are not significantly different. This is to be expected given that the diagonal elements of Σ_v are set equal to zero indicating the absence of state dependency. For this case the coefficients of the state vector are quite stable, thus justifying the restrictions. For series characterized by state dependency it is likely that more complex restrictions would be required. The MSE over the last 70 observation periods is 247.34. This is not significantly different from the value computed for the case in which restrictions are placed only on the diagonal elements of Σ_v.

Table 5. Values for Restricted Parameters
for Periods with Missing Observations, t = 81, 82, ..., 90.

$\hat{\mu} = 18.024$	$\hat{P}_{11} = 4.527$	$\hat{P}_{22} = .009$
$\hat{\phi}_1 = 1.402$	$\hat{P}_{12} = \hat{P}_{21} = .021$	$\hat{P}_{23} = \hat{P}_{32} = -.009$
$\hat{\phi}_2 = -.793$	$\hat{P}_{13} = \hat{P}_{31} = -.011$	$\hat{P}_{33} = .009$

V. CONCLUSIONS

This paper presents preliminary work on extending the S.D.M. for prediction of time series with blocks of missing observations. The paper indicates that one-step forecasts may be used for periods with missing observations, however, restrictions on the predictions or on model behavior may be necessary in order to prevent generation of inadmissible forecasts. Three types of restrictions are implemented in this paper. First, restrictions are placed directly on the forecast values, second, estimates of the variances on the random walk parameters are constrained, i.e., restrictions are placed on the model adjustment process, and third, exact restrictions are placed on the elements of the state vector and the variance-covariance matrix. The preliminary evidence indicates the later two procedures are preferable, however, placement of restrictions on the diagonal elements of Σ_v may not result in model behavior consistent with the model selection criterion of best fit. Moreover, it is conceivable that despite restrictions on the variances, unacceptable forecasts may be generated. At this point, it appears that future research efforts would be best spent focusing on the specification of restrictions on the state vector, θ_t, and variance-covariance matrix, P_t.

ACKNOWLEDGEMENTS

The author wishes to thank Paul Newbold and R. Carter Hill for valuable suggestions on earlier versions of the paper. In addition, he wishes to thank anonymous referees for helpful comments.

REFERENCES

ANSLEY, C. and KOHN, R. (1983). Exact Likelihood of Vector Autoregressive - Moving Average Process with Missing or Aggregated Data. Biometrika 70, 275-278.

BOX, G.E.P. and JENKINS, G. (1976). Time Series Analysis: Forecasting and Control. Revised Ed. Holden-Day.

CARTWRIGHT, P.A. and NEWBOLD, PAUL (1983). A Time Series Approach to the Prediction of Oil Discoveries. In Time Series Analysis: Theory and Practice 4, Proceedings of the Conference held at Cincinnati, August 1982. Ed.: O.D. Anderson, North-Holland Publishing Co., 265-289.

DUNSMUIR, W. (1981). Estimation for Stationary Time Series when Data are Irregularly Spaced or Missing. In Applied Time Series Analysis II. Ed.: D. Findley, Academic Press, 609-650.

JONES, R.H. (1980). Maximum Likelihood Fitting of ARMA Models to Time Series with Missing Observations. Technometrics 22, 389-395.

HARVEY, A. (1981). The Kalman Filter and Its Applications in Econometrics and Time Series Analysis. Paper presented at 6 Symposium uber Operations Research, Augsburg, September, 1981.

KALMAN, R.E. (1963). New Methods in Wiener Fitting Theory. In Proceedings of the First Symposium on Engineering Applications of Random Function Theory and Probability. Ed.: John L. Bogdanoff and Frank Kozin, John Wiley and Sons, Inc., 270-388.

LJUNG, G.M. (1982). The Likelihood Function for a Stationary Gaussian Autoregressive-Moving Average Process with Missing Observations. Biometrika 69, 265-268.

PRIESTLEY, M.B. (1980). State-Dependent Models: A General Approach to Non-Linear Time Series Analysis. Journal of Time Series Analysis 1, 47-71.

PRIESTLEY, M.B. (1982). On the Fitting of General Non-Linear Time Series Models. In Time Series Analysis: Theory and Practice 1. Proceedings of the International Conference held at Valencia, Spain, June 1981. Ed.: O.D. Anderson, North-Holland Publishing Co., 717-731.

TIME SERIES ANALYSIS: Theory and Practice 7
O.D. Anderson (editor)

EXTENSIONS OF FORECASTING METHODS FOR IRREGULARLY SPACED DATA

D.J. Wright
Faculty of Administration, University of Ottawa, Ottawa K1N 9B5, Canada

The aim of this paper is to describe computationally efficient forecasting procedures or irregularly spaced data, and to analyze and compare their behaviour on examples of such data. The aim of computational efficiency is achieved by making the procedures sequential. A sequential regression procedure is described and extensions of the Brown and Holt methods are given which require the same order of magnitude of computational resources as the regular versions of those methods.

1. INTRODUCTION

This paper addresses a topic which is of great practical importance but which has received little attention in the literature. In practice, most data series contain irregularities in their time spacing whereas most methods for analysis and forecasting assume regular spacing. Two recent reviews of the state of the art in time series analysis and forecasting, Fildes (1979) and Newbold (1981), do not mention any methods for dealing with unequally spaced data.

The work reported here arose out of a study involving the forecasting of prices given in tenders. Part of the data set was the prices given in previous tenders whose dates were at quite irregular time spacings. This is one example of a completely irregular time series. Other examples occur when data points are missing from an otherwise regular series. The data may have been removed to save space in the compilation of published statistics or may simply be unavailable. Many summaries are published of Government statistics, for instance, in which older data are reported less frequently than recent data. Another situation is where the frequency of data collection changes so that, for instance, a quarterly series becomes a monthly one. Unequal time spacings also occur when data points have been deliberately removed during the analysis of a series because there is reason to regard them as outliers.

In this paper we give three methods that are highly efficient computationally. This computational efficiency is achieved by making the methods sequential. Normally, sequential methods are implicitly associated with regularly spaced data. For instance, Kendall (1975) gives sequential regression equations and Gilchrist (1976) includes time discounting in them. We give a discounted sequential regression procedure for irregularly spaced data. The work of Chambers (1971) and Gragg, Leveque and Trangenstein (1979) on this topic in the statistical literature has not been applied in the context of forecasting. The other two methods we give are extensions of the Brown method of double exponential smoothing and of the Holt method. Recent work on irregularly spaced time series is available but it is restricted to

---

This research was supported by National Sciences and Engineering Research Council of Canada Grant Number A8604.

the stationary case and the methods are non-sequential: Dunsmuir (1981), Lambe (1982).

The paper is organised as follows. Section 2 and 3 give the extensions of the Brown and Holt methods and Section 4 gives the sequential regression method. The performance of the methods is then analysed from the point of view of computational efficiency in Section 5 and from the point of view of forecasting accuracy in Section 6. Conclusions regarding choice of method for a given data type are presented in Section 7.

2. EXTENSION OF THE BROWN METHOD FOR IRREGULAR DATA

For a regular series of data points $x_o,\ldots,x_t$ occurring at times $0,\ldots,t$, regular exponential smoothing is based on a weighted average in which the point x_i is weighted by an amount a^{t-i}, for a smoothing parameter a, $0 < a < 1$. In the case of an irregularly spaced series $X_o,\ldots,X_n$ of data points at times $t_o,\ldots,t_n$, we extend exponential smoothing by using a weighting factor

$$a^{t_n - t_i} \tag{1}$$

for the point X_i, to give a corresponding weighted average. The resulting extension of single exponential smoothing is given by Wright (1983 a). The singly smoothed estimate, $\bar{X}_n$, is given by equations (2),...,(6):

$$\bar{X}_n = (1-V_n)\bar{X}_{n-1} + V_n X_n \tag{2}$$

where

$$V_n = 1/W_n \tag{3}$$

$$W_n = b_n W_{n-1} + 1 \tag{4}$$

$$b_n = a^{t_n - t_{n-1}} \tag{5}$$

with initial condition

$$W_o = 1/(1-a^q) \tag{6}$$

where q is the average time spacing of the data. We note that the variable V_n plays the same role in (2) as the smoothing parameter, a, does in regular single exponential smoothing.

The extension of the Brown (1963) method of double exponential smoothing is also given in Wright (1983 a), based on the doubly smoothed estimate $\bar{\bar{X}}_n$:

$$\bar{\bar{X}}_n = (1-V_n)\,\bar{\bar{X}}_{n-1} + V_n\bar{X}_n. \tag{7}$$

The current smoothed estimate at time t_n is:

$$\hat{\mu}_n = 2\bar{X}_n - \bar{\bar{X}}_n \tag{8}$$

and the current estimate of the slope at time t_n is:

$$\beta_n = \frac{W_n}{Q_n}(\bar{X}_n - \bar{\bar{X}}_n) \tag{9}$$

where Q_n is obtained from the recurrence relation:

$$Q_n = b_n Q_{n-1} + (t_n - t_{n-1})\,(W_n - 1). \tag{10}$$

A forecast of X_{n+1} at some future time $t_{n+1} > t_n$, is, therefore:

$$\hat{X}_{n+1} = \hat{\mu}_n + (t_{n+1} - t_n)\,\hat{\beta}_n. \tag{11}$$

The recurrence equations (2), (7), (10) have initial conditions:

$$\bar{X}_o = X_o - \beta_o Q_o / W_o \tag{12}$$

$$\bar{\bar{X}}_o = X_o - 2\beta_o Q_o / W_o \tag{13}$$

$$Q_o = qa^q/(1 - a^q)^2 \tag{14}$$

where β_o is our initial estimate of the slope of the data series.

Equations (2),...,(14) constitute an extension of the Brown method of double exponential smoothing to irregularly spaced data. We call this the IDES procedure (Irregular Data using Exponential Smoothing).

3. EXTENSION OF THE HOLT METHOD FOR IRREGULAR DATA

The Holt method of forecasting regularly spaced data is based on fitting a local linear trend line to the recent data points. The parameters of the line are adjusted according to single exponential smoothing equations. In our case of data points $X_o,\dots,X_n$ occurring at times $t_o,\dots,t_n$, we fit a line $X_i = L_n - M_n(t_n - t_i) + e_i$ where L_n is the level of the trend line at time t_n, M_n is its slope, and e_i is the fitting error. The equations for L_n and M_n are obtained in Wright (1983(c)) as:

$$L_n = (1-V_n)\,(L_{n-1} + (t_n - t_{n-1})M_{n-1}) + V_n X_n \tag{15}$$

$$M_n = (1-V_n)\,M_{n-1} + V_n(L_n - L_{n-1})/(t_n - t_{n-1}) \tag{16}$$

with initial conditions $L_o = X_o$, $M_o = \beta_o$. Here we have used the same variable V_n in (15) and (16). Different variables could be used corresponding to Holt's two smoothing parameters via equations (3), (4), (5). However we have not found this necessary in practice.

A forecast of X_{n+1} at some future time, t_{n+1}, is given by:

$$L_n + (t_{n+1} - t_n)M_n. \tag{17}$$

The extension of Holt's method to deal with missing data points consists therefore of equations (15), (16), (17), (3), (4), (5) and (6). We call this the IDUHO procedure (Irregular Data Using Holt's method).

4. SEQUENTIAL DISCOUNTED REGRESSION FOR IRREGULAR DATA

The sequential regression procedure uses the same discount factors (1) as used in the extension of the Brown method. The line

$$X_i = U + rt_i + e_i \tag{18}$$

is fitted by minimising the sum of squares of discounted errors:

$$\sum_{i=o}^{n} e_i^2\, a^{t_n - t_i}.$$

The procedure, derived in Wright (1983 b), is as follows. The estimates of U and r are:

$$U = (C_n D_n - B_n E_n) Z_n \tag{19}$$

$$r = (A_n E_n - B_n C_n) Z_n \tag{20}$$

where

$$Z_n = (A_n D_n - B_n^2)^{-1} \tag{21}$$

$$A_n = b_n A_{n-1} + 1 \qquad A_o = 1 \tag{22}$$

$$B_n = b_n B_{n-1} + t_n \qquad B_o = t_o \tag{23}$$

$$C_n = b_n C_{n-1} + X_n \qquad C_o = X_o \tag{24}$$

$$D_n = b_n D_{n-1} + t_n^2 \qquad D_o = t_o^2 \tag{25}$$

$$E_n = b_n E_{n-1} + t_n X_n \qquad E_o = t_o X_o \tag{26}$$

$$b_n = a^{t_n - t_{n-1}}. \tag{27}$$

A forecast of X_{n+1} at a future time, t_{n+1}, is given by:

$$U + rt_{n+1}. \tag{28}$$

Equations (19),...,(28) constitute a sequential discounted regression procedure for irregularly spaced data. We call this the IDRE procedure (Irregular Data using Regression).

5. COMPUTATIONAL EFFICIENCY

A major element in the requirements for computation time is equation (5) or (27) in which a number is raised to a power. This is necessary for each of the three methods when there is general irregularity in the time spacing between data points. However, if there is some pattern in this irregularity, as is the case for missing points from an otherwise regular time series, it is not necessary to evaluate equations (5) and (27) every time a data point is added. Instead, some numbers

$$b'_s = a^{s+1} \tag{29}$$

corresponding to (5) and (27) when there has just been a gap of s missing data points, can be stored at the start of the analysis for some reasonable range of values s = 1,...,S. (The time spacing when there is no missing data is assumed to be scaled to 1.)

We therefore have two types of data to delineate in assessing the relative computational efficiency of the methods. First, general irregularity in the time spacing and, second, missing data points from an otherwise regular series. The second case must also be subdivided into two, depending on whether there has been a gap in the series immediately prior to the data point being looked at. If there has not, then $t_n - t_{n-1} = 1$, so that equations (10), (15) and (16) can be simplified.

The computational requirements are given in Exhibit 1 for each of our methods, IDES, IDUHO and IDRE, with the requirements for the regular methods of Brown, Holt and Regression given alongside for comparison.

For regularly spaced data with points missing, there is a clear hierarchy in terms of computation time. This is important in applications involving a large number of forecasts e.g. in inventory control. The storage requirements are less important and are, in any case, much the same for all methods, except for the Brown and Holt methods, which require half the space of the others.

Exhibit 1

Computational requirements for analysis of each data point. The total computation time is based on Raising to a power being equivalent to 10 Multiplications or Divisions and a Multiplication or Division being equivalent to 10 Additions or Subtractions. The first of the three figures gives the requirement for general irregularly spaced data. The second is after a gap in an otherwise regular series. The third is for a regular series with gaps, where no gap has occured immediately prior to the point being analysed.

	IDES			IDUHO			IDRE			Brown	Holt	Regr
Multiplications	6	6	5	3	3	2	13	13	13	3	2	13
Divisions	2	2	2	2	2	1	2	2	2	1	0	2
Additions & substractions	11	11	10	10	10	8	8	8	8	8	8	8
Raising to a power	1	0	0	1	0	0	1	0	0	0	0	0
Total Computation Time	191	91	80	160	60	38	258	158	158	48	28	158
Storage	12	12	12	9	9	9	13	13	13	6	5	12

For a series with general irregularity in the time spacing between data points, the processing time is substantially affected by raising a number to a power. The ranking of the methods in terms of computation speed is the same as in the case of missing points from an otherwise regular series, but the relative differences between the methods are less pronounced.

6. FORECASTING PERFORMANCE

We now investigate the forecasting behaviour of our methods on various types of data. In order to obtain results of general applicability, we use time series obtained from random number generators. The most basic such series is the global linear trend:

$$X_i = c + dt_i + e_i \tag{30}$$

where the e_i are uniformely distributed random numbers. We also analyse series with local linear trends obtained from (30) by changing the parameters c and d at certain times t_k, in such a way as to maintain continuity of $c + dt_k$. This gives us two types of series. In addition we have the two types of time spacing: general irregularity and missing points from an otherwise regular series. In order to analyse the behaviour on the full range of possible situations, we investigate 3 different proportions of data points missing:10%, 20% and 40% which we designate low, medium and high percentages of missing data. These are obtained by removing points at random from the series. This also corresponds to low medium and high degrees of irregularity in the time spacing. We therefore maintain comparability between the missing data and the general irregularity case by choosing 3 levels of irregularity in the latter case: low, medium and high, whose standard deviations equal those of the series with 10%, 20% and 40% of missing points. The time intervals for the generally irregularly spaced data are obtained from random number generators with means and standard deviations equal to those of the time intervals in regular series from which 10%, 20% and 40% of points are missing.

Exhibit 2

Mean square forecast errors, $\overline{E^2}$, over 50 time units for a global trend with different percentages of points missing, (a) 10%, low, (b) 20%, medium, (c) 40%, high.

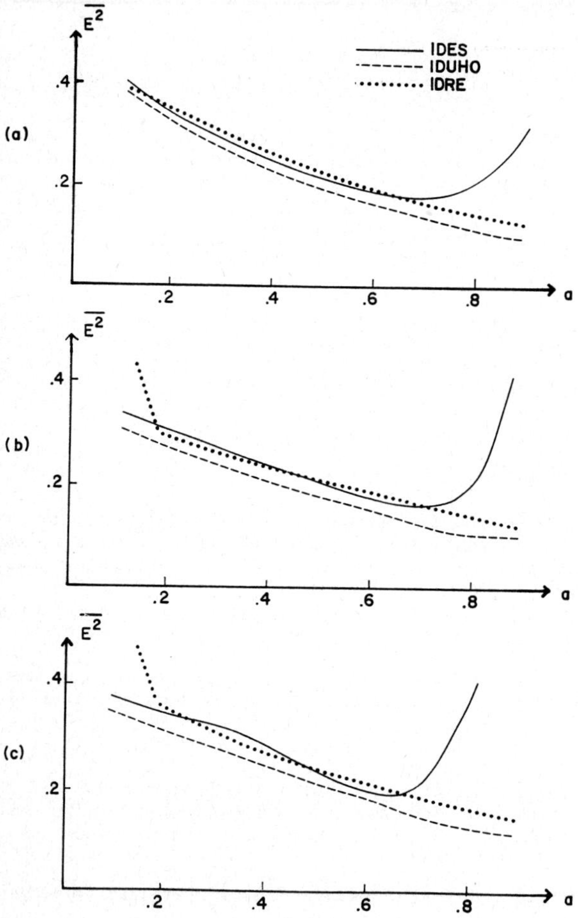

Exhibit 3

Mean square forecast errors, $\overline{E^2}$, over 50 time units for local trends with different percentages of points missing: (a) 10%, low, (b) 20%, medium, (c) 40%, high.

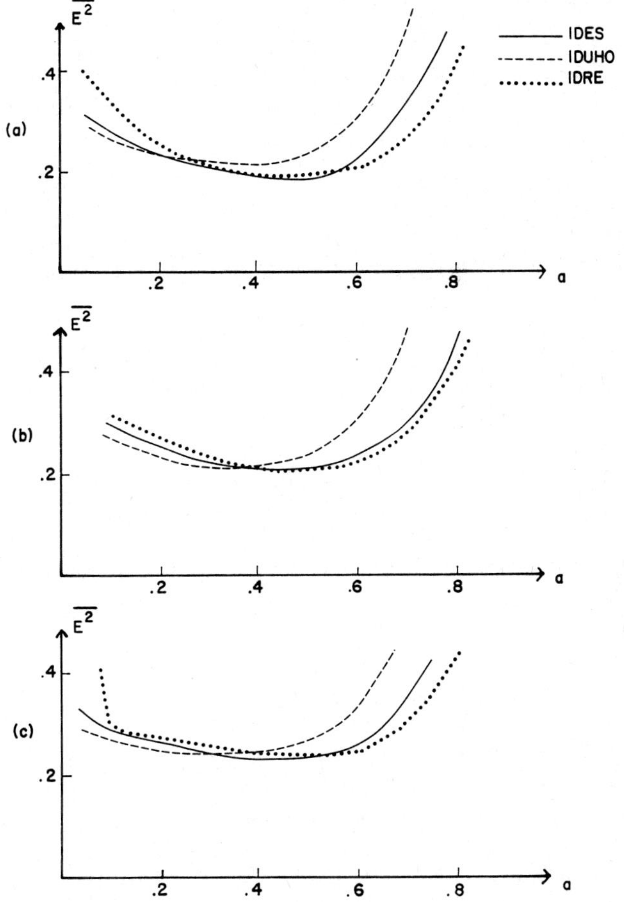

The results are given in Exhibits 2, 3, 4, 5 where the mean square forecast error is plotted against the smoothing parameter, a, for the 3 forecasting methods, IDES, IDUHO and IDRE. The absolute error, relative absolute error and relative square error were also used as criteria in addition to the square error. Sample results are given in Exhibit 6. These Exhibits usually give us a clear indication of the appropriate range of values of smoothing parameter to chose for a given data type, but care should be taken in using them to choose one method in preference to another. The data has been generated from a linear model (30) so as to be appropriate to our forecasting methods. Therefore once the smoothing parameter is chosen appropriately, each of the methods should produce a mean square forecast error equal to the variance of e_i in (30). Any small residual differences in mean square error between the methods should not be regarded as significant. In place of investigating whether one method is more accurate then another, we therefore consider their relative sensitivities to changes in the smoothing parameter. This is important since the smoothing parameter which gives minimum mean square error for one series may well be slightly different for another. Sensitivity to choice of smoothing parameter can be overcome if the forecasting environment allows for trial runs with different parameters. This process of trial runs can be automated, as in existing commercial packages using regular smoothing and regression methods, so as to determine the optimal parameter.

The method of making the forecasts has been chosen differently for the different types of time spacing. All forecasts are short term, which is the situation for which the methods are designed. For the missing data case, all the forecasts are one step ahead. For the general irregularity case, the forecasts are to the next data point. This has been done to illustrate the effect of time horizon on forecast accuracy. As the degree of irregularity increases, the forecast accuracy decreases in the case of general irregularity, Exhibits 4 and 5, but not in the case of missing points, Exhibits 2 and 3. This increase can therefore be attributed to the larger average time horizon of the forecasts rather than to there being less data.

Exhibit 2 shows clearly that, for data series with missing points and a global trend, a high value of a = 0.7 to 0.9 gives the least mean square forecast error for IDUHO and IDRE. The same result appears in Exhibit 4 for general irregularity in the time spacing with the IDES method becoming progressively more unsuited to this data as the degree of irregularity in time spacing increases. The poor performance of IDES with high values of a is a general feature of the method, Wright (1983 a).

In the case of data with local trends, Exhibit 3 shows that, with missing data points, a broad range of smoothing parameters a = 0.2 to 0.5, give similar low values to the mean square error with very little sensitivity to variation in a for any forecasting method. In the case of general irregularity, Exhibit 5, the position is more complex. The best choice of smoothing parameter is different for different methods and for different degrees of irregularity in time spacing. Moreover the IDUHO method is more sensitive to the choice of smoothing parameter than are the other methods for low and medium irregularity.

These results using square error as a measure of forecast accuracy have been presented in detail because of the widespread use of this criterion. Other measures are also important and results were obtained using absolute error, relative absolute error and relative square error. The ranges of smoothing parameters for which the methods display low sensitivity to these criteria

Exhibit 4

Mean square forecast errors, $\overline{E^2}$, over 50 time units for a global trend with general irregularity in the time spacing (a) low, (b) medium, (c) high.

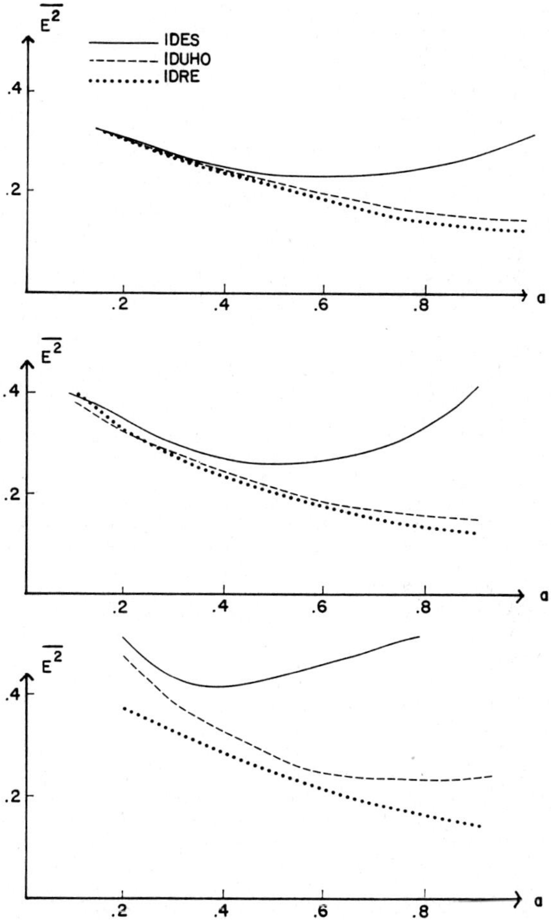

Exhibit 5

Mean square forecast error, $\overline{E^2}$, over 50 time units for a local trend with general irregularity in the time spacing (a) low, (b) medium, (c) high.

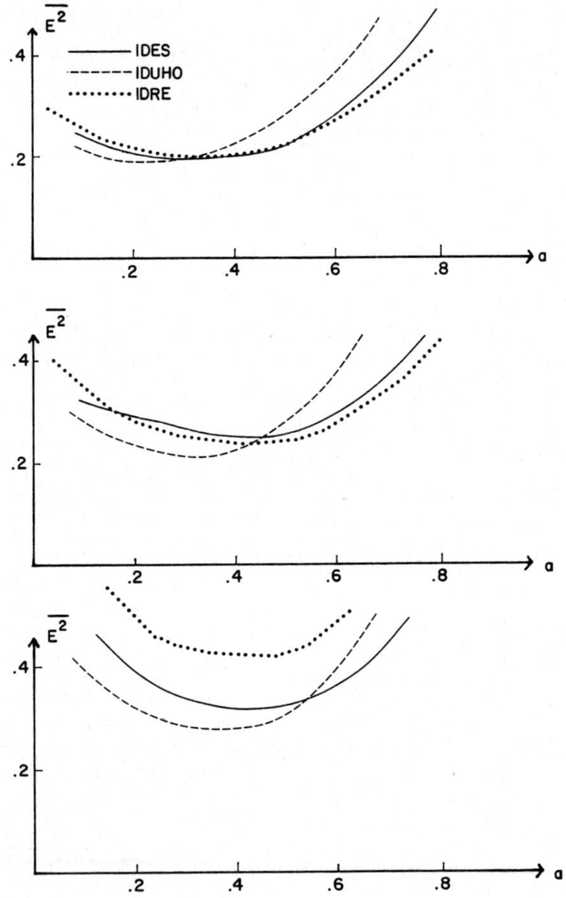

Exhibit 6

Mean relative square forecast error, $\bar{R}^2$, over 50 time units for a local trend with different percentages of points missing. (a) 10%, low, (b) 20%, medium, (c) 40%, high.

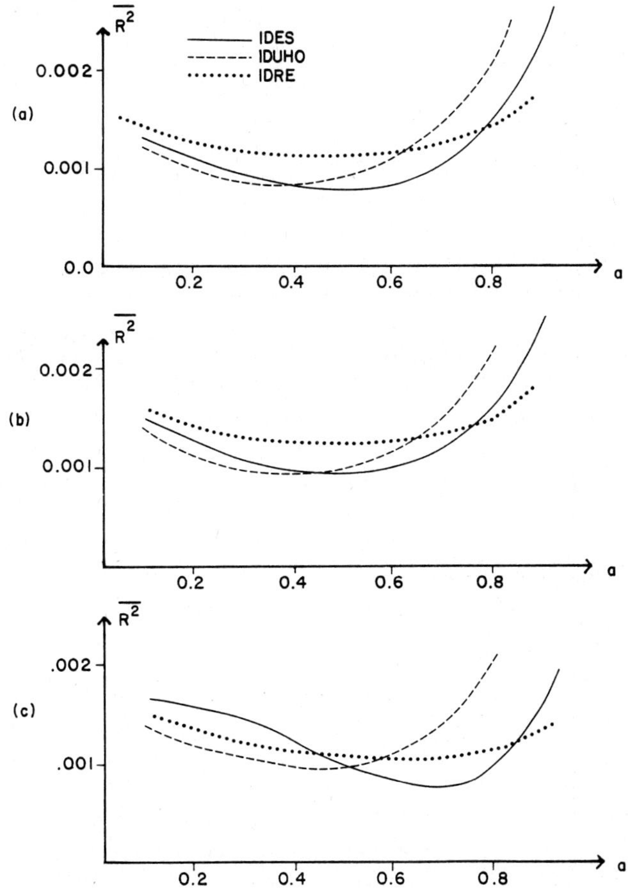

are very similar to those obtained above for square error. An exception is the case of relative square error on local trends with missing data. Becuase of space restrictions only these exceptional results are presented here in Exhibit 6. They are noteworthy for two reasons. Firstly the regression based procedure, IDRE (which is designed to minimise square error), performs poorly when evaluated according to relative square error. Secondly the ranges of insensitivity change as the number of missing data points increases.

CONCLUSIONS

Computationally efficient methods for the analysis of irregularly spaced time series are lacking in the literature despite the frequent occurrence of such data in practice.

Three methods of dealing with this situation are proposed in this paper and their performance is analysed in tests involving a total of 6921 forecasts.

Exhibit 7

Choice of forecasting method and smoothing parameter appropriate to different types of time series and forecasting environment.

FORECAST ENVIRONMENT	TIME SERIES	GENERAL IRREGULARITY		MISSING POINTS	
		GLOBAL TREND	LOCAL TRENDS*	GLOBAL TREND	LOCAL TRENDS
TRIALS ON a	COMPUTATION SPEED IMPORTANT	IDUHO a = 0.7 - 0.9	IDUHO a = 0.2 - 0.5	IDUHO a = 0.7 - 0.9	IDUHO a = 0.2 - 0.5
	NOT	IDUHO IDRE a = 0.7 - 0.9	IDUHO a = 0.2 - 0.5 IDES, IDRE a = 0.2 - 0.7	IDUHO IDRE a = 0.7 - 0.9	IDES IDUHO IDRE a = 0.2 - 0.5
NO TRIALS	IMPORTANT	IDUHO a = 0.9	IDES a = 0.4	IDUHO a = 0.9	IDUHO a = 0.35
	NOT	IDHUO IDRE a = 0.9	IDES IDRE a = 0.4	IDUHO IDRE a = 0.9	IDES IDUHO a = 0.35

* The ranges of smoothing parameters given in this column can be reduced if one knows the degree of irregularity of the time spacing: Exhibit 5.

In order to present a method and choice of smoothing parameter for a given time series, various factors must be taken into account. Firstly three features concerning the series are important: a) whether it has a global trend or a sequence of local trends, b) whether the irregularity in time spacing is of a general nature or is due to missing points from an otherwise regular series, c) the degree of irregularity of the time spacing. This third factor is only relevant when the series has local trends and general irregularity. Secondly two features of the forecasting environment must be taken into account i) whether it is possible to do a few trial runs to overcome sensitivity of the method to the value of smoothing parameter, ii) whether computation speed is important in obtaining forecasts. The choices of methods and smoothing parameters appropriate to each of these combinations is given in Exhibit 7, based on square error as the measure of forecasting performance. Other criteria lead to similar results with certain exceptions, e.g. Exhibit 6.

REFERENCES

BROWN, R.G. (1963). Smoothing, Forecasting and Prediction of Discrete Time Series. Prentice Hall, Englewood Cliffs, N.J.

CHAMBERS, J.M. (1971). Regression Updating. J. Amer. Stat. Assoc. 66, 744-748.

DUNSMUIR, W. (1981). Estimation for Stationary Time Series when Data are Irregularly Spaced or Missing. In Applied Time Series Analysis II, Ed: D.F. Findlay, Academic Press, New York, 609-648.

FILDES, R. (1979). Quantitative Forecasting - The State of the Art: Extra-polative Models. J. Opl. Res. Soc. 30, 691-710.

GILCHRIST, W. (1976). Statistical Forecasting. Wiley, New York.

GRAGG, W.B., LEVEQUE, R.J. and TRANGENSTEIN, J.A. (1979). Numerically Stable Methods for Updating Regressions. J. Amer. Stat. Assoc. 74, 161-168.

KENDALL, M.G. (1975). Time Series. Griffin, London.

LAMBE, T.A. (1982). A Bayesian View of Forecasting by Exponential Smoothing. TIMS/ORSA Conference Paper, San Diego.

NEWBOLD, P. (1981). Some Recent Developments in Time Series Analysis. Internat. Stat. Rev. 49, 53-66.

WRIGHT, D.J. (1983a). Forecasting Irregularly Spaced Data: An Extension of Double Exponential Smoothing. Computers and Industrial Engineering forthcoming.

WRIGHT, D.J. (1983b). Sequential Forecasting Methods for Irregularly Spaced Data. TIMS/ORSA Conference, Chicago.

WRIGHT, D.J. (1983c). Missing Data in Time Series: Forecasting without Interpolation. International Symposium on Forecasting, Philadelphia.

TIME SERIES ANALYSIS: Theory and Practice 7
O.D. Anderson (editor)

A CLOSER LOOK AT A PROOF OF TIME-REVERSIBILITY

John Brode
Department of Mathematics, University of Lowell,
Lowell, MA 01854, U.S.A.

Weiss (1975) has related time-reversibility to the Gaussian distribution. Weiss' theorem can be extended to any stationary process by a theorem of Dynkin. This extension, however, as also the original theorem, can only be proved if we make an explicit assumption as to the nature of infinity. As we are not obliged to make this assumption, we are not obliged to accept any results that depend on our having made such an assumption.

INTRODUCTION

In a paper presented at the 1983 ITSM in Nottingham, A.J. Lawrence used a theorem of Weiss (1975) on time-reversibility. The implication seemed to be that time-reversibility was a Gaussian property. But surely, any symmetric, stable process is time-reversible. This is important to know, as stable processes are becoming increasingly important. At the 12th Conference on Stochastic Processes and their Applications given by the Bernoulli Society at Cornell University in July 1983, one of every eight papers was on stable processes.

This paper will try to clarify this problem by looking more closely at Weiss' theorem which in turn is based on a theorem of Rao (1966). These theorems are quite correct--that is not the question. In fact, a theorem of Dynkin will allow us to extend Weiss' result. However, their application is fraught with danger. They represent a clear case of the potential for interaction between post-1963 mathematical logic and the set theory implicit in the above theorems. In short, the extension of a countable index set, T, of times to the continuum is tricky. The results of such an extension will depend on the basic assumptions made. But there are many conflicting assumptions all of which are logically consistent with each other. We are not obligated to accept any particular one of these possible assumptions. Therefore, the results of one or another assumption may well vanish under a different assumption--a 'proof' may turn out to be no more than a passing whim!

WEISS' THEOREMS

First let us state a definition for time-reversibility (taken from Weiss, 1975):

DEFINITION: A stationary process X(t) is time-reversible if, for every n and every t[1], ..., t[N], [X(t[1]), X(t[2]), ..., X(t[N])] and [X(t[N]), X(t[N-1]), ..., X(t[1])] have the same joint probability distributions.

Weiss states that this definition applies both when T = R and T = N. He then proves the following theorem:

THEOREM 1: Let X(t) be a stationary Gaussian process, ie., all finite dimensional distributions of X(t) are multivariate normal. Then X(t) is time-reversible.

and then its converse:

THEOREM 2: Let X(t) be a discrete time, stationary, mixed auto-regressive moving-average process given by:

$$X(t) = \sum^{N} a(i)X(t-i) + \sum^{M} b(i)\mathcal{E}(t-i)$$

with $[\mathcal{E}(i)]$ an independent identically distributed sequence of non-degenerate random variables and with $\mathcal{E}(t)$ independent of X(t-s) for s > 0. Let N ≠ 0, or if N = 0, let the case b(k) = b(M-k) for k=0,1,...,M and the case b(k) = -b(M-k) for k=0,1,...,M be excluded. If X(t) is time-reversible, then $\mathcal{E}(t)$ is normally distributed.

This theorem would seem to imply that only Gaussian processes are time-reversible.

RAO'S THEOREM

Weiss depends on a paper by Rao (1966) for the proof of theorem 2. Rao shows that if a p-dimensional random variable X can be represented as X = AF and X = BG, where A and B are fixed matrices and F and G are vectors of independent non-degenerate random variables. Then X is a Gaussian random variable whenever there is no column of A that is a multiple of a column of B. Weiss takes A as given, F = G = X, and B = HA, where H is the identity matrix with its columns arranged in reverse order. BY is, therefore, the time-reversal of AY. Thus by Rao's theorem, if the two representations are equivalent, and no column of A is a multiple of a column of B, then X is Gaussian.

As might be expected, Rao's work is extremely solid. The conclusion is irrefutable. This would seem, therefore, to preclude the possibility of a non-Gaussian stable time-reversible process.

AN INTERPRETATION

Let us rephrase Rao's theorem to bring out a key element in his result. His theorem states that only a Gaussian process can be exactly represented by two linear transformations on random variables projected onto a p-dimensional space where p is finite. It is, of course, well-known that the Gaussian distribution is the only symmetric, stable distribution that can be exactly represented by a finite set of statistics. In fact, this is what Rao uses (Lemma 5) in the proof of his theorem. Can we redefine time-reversible to avoid this finite character?

A GENERALIZATION

Feller generalized the diffusion process by defining its local character. At any point, the process can be completely defined by the derivatives of its functional form. (These derivatives are not to be taken as ordinary derivatives, but their precise definition need not detain us here.) The local character, however, does imply time-reversibility as the derivatives are not directional. (This has been proved in a theorem presented by R. Holley showing that time reversibility is equivalent to the self-adjointness of the infinitessimal generator of the process. Self-adjointness of the infinitessimal generator exists for all symmetric, stable processes with characteristic exponent $1 < \alpha < 2$ -- Brode, 1981.) These processes with local character are called Feller-Dynkin processes for which we have the following theorem due to Dynkin (see Williams, 1979, p. 122):

THEOREM: Let T be a $[\mathcal{F}_t]$ stopping time, then, for $\forall \mu \in Pr(E')$, and $\forall \xi \in F_b(\mathcal{F})$, we have:

$$E^{\mu}[\theta_T \xi | \mathcal{F}_{T+}] = E^{X(T)}[\xi] \quad \text{a.s. } (P^{\mu}).$$

This theorem implies that any Feller-Dynkin process is, essentially, Gaussian (ie., it is a Wiener process). Thus, we have clearly extended Weiss' results to continuous time processes. Note, however, the use of E'. This is the Alexandrof-Hopf one-point compactification of the underlying Polish space (essentially a topological space that is locally compact, σ-compact, and metrizable--the real line is the most widely used example of such a space). Thus the passage to the continuum does not seem to avoid the problem we were left with in the previous section.

THE ALEXANDROF-HOPF COMPACTIFICATION

This basic one-point compactification reads as follows:

THEOREM: Any locally compact space X can be embedded in another compact space X', having just one more point than X, in such a way that X is dense in X' (ie., the closure of X is X').

Let us prove this theorem so as to be quite sure of what is implied by it. We will label ω an element that is distinct from any element contained in X. Let [0] be the class of all open sets in X such that its complement in X is compact (ie., that $0^c = X-0$ is compact). Note that X itself is a member of 0 since its complement is the null-set which is compact. Let X' be the set X to which we have adjoined ω. Any set in X' will be called open if:

1) it does not contain ω and is open in X; or
2) it does contain ω and its intersection with X is a member of [0].

X' is defined in terms of its open sets in such a way as to ensure that the relative topology of X' coincides with that on X.

Let us take [V] to be a family of open sets that covers X'. (At this point, we would come in a proof of this theorem to the following phrase that I quote from Yosida, 1978, p. 5):

'Then there must be some member of [V] of the form $0 \cup [\omega]$, where $0 \in [0]$.'

Given our definition of [0], 0^c must be compact as a member of X and can be covered by a finite number of open sets. Thus X' can be covered by this finite number of open sets plus O^c. Q.E.D.

Let us return to the phrase cited from Yosida. What is behind this statement? Obviously, ω can not be adjoined to any open set in X. If X were the real line, it would not do to adjoin ω to the open interval (0,1). The complement of (0,1) in this X is not compact. This will in fact be the case for any open interval in X whose end points are finite numbers or even simply countable numbers. Does any interval exist which satisfies the requirements of the theorem? The answer is yes--provided we can use Zorn's lemma:

ZORN'S LEMMA: Let [V] be a partially ordered set such that any totally ordered subset of [V] has an upper bound. Then [V] has a maximal element. (adapted from Bourbaki, 1970, III.2.4. See also Moore, 1982, pp.224ff.)

and we can 'name' the upper bound. Note that Zorn's lemma implies (or is equivalent to) the axiom of choice (see Jech, 1978,p.40). We must be able to choose the maximal element from the essentially infinite set. Further we must have an upper bound. We may assume that ω is just this upper bound. With these two assumptions (the axiom of choice and the existence of the upper bound ω), Zorn's lemma can be applied and the compactification required is proved.

BUT IS IT A PROOF?

Unfortunately, the answer to the title of this section is no. This 'no' then cascades down through the other proofs cited above back to Weiss. Can this be right? Yes, though to explain this exactly is beyond the scope of this paper. We see here the importance of keeping up with recent developments in mathematical logic (the interested reader might wish to look at Jech, 1978, Kunen, 1980, or Rudin, 1975. Since 1963, logicians have proved the following seemingly contradictory statements:

ZF and the axiom of choice form a consistent model
ZF and the negation of the axiom of choice form a consistent model

where ZF stands for the usual Zermelo-Fraenkel axioms of set theory (assumed consistent amongst themselves). Are such nonsensical results possible? Yes--they come respectively from Goedel's work and from Cohen's application of forcing (Jech, 1978, pp. 108, 493, and 527).

Note that even if we decide not to drop the axiom of choice, we still do not get the desired result unless we are able to assume a least upper bound. The axiom of choice is widely used in topology, functional analysis, and other areas of mathematics. Amongst others, we would lose the Hahn-Banach continuation theorem. The assumption of a least upper bound, equivalent to the continuum hypothesis, is much less vital, and could be dropped with few repercussions outside of this compactification theorem.

The point to be aware of here, is that, essentially, you get what you assume. The various assumptions used to get the results listed above have all been shown to be independent of the Zermelo-Fraenkel axioms of set theory. By independent, we mean that we are not obligated either to make or not to make any of these assumptions.

Thus, any 'results' coming from such an assumption or its negation should be carefully labelled. The 'proof' is no more valid than the assumption. Since for many assumptions it has been rigorously proved that we can not establish their validity from within ZF, the 'proof' is no more than the 'whim' behind the assumption.

REFERENCES

BOURBAKI, N. (1970), Théorie des ensembles, Paris: Hermann.

BRODE, J. (1981), On Estimating the Trajectories of Stochastic Processes with Infinite Variance. In Time Series Analysis (Proceedings of the International Conference held at Houston, Texas, August 1980). Eds: O.D. Anderson and M.R. Perryman, North-Holland, Amsterdam and New York, pp. 43-62.

JECH, T. (1978), Set Theory, Academic Press, New York.

KUNEN, K. (1980), Set Theory: An Introduction to Independence Proofs, North-Holland, Amsterdam and New York.

LAWRENCE, A.J. (1983), Directionality in Time Series. Presented to the International Time Series Meeting held at Nottingham University, UK, April 1983.

MOORE, G.H. (1982), Zermelo's Axiom of Choice: Its Origins, Development, and Influence, Springer Verlag, Berlin.

RAO, C.R. (1966), Characterisation of the Distribution of Random Variables in Linear Structural Relations, Sankhya Series A 28, 251-60.

RUDIN, M.E. (1975), Lectures on Set Theoretic Topology, American Mathematical Society, Providence, R.I.

WEISS, G. (1975), Time-reversibility of Stochastic Processes, J Appl Prob 12, 831-6.

WILLIAMS, D. (1979), Diffusions, Markov Processes, and Martingales, Vol. 1: Foundations, Wiley, New York.

YOSIDA, K. (1978), Functional Analysis, Springer Verlag, Berlin, 5th ed.

TIME SERIES ANALYSIS: Theory and Practice 7
O.D. Anderson (editor)

A METHOD FOR APPROXIMATING THE MEAN-VALUE OF NONSTATIONARY RANDOM DATA

George Treviño[1]
CHIRES Associates, Inc., Box 412, Las Cruces, NM 88004, USA

By "nonstationary" is meant data whose mean-value, $\mu(t)$, is a definite function of time, in contrast to "stationary" data where $\mu(t)$ is constant. The method is formulated by first assuming that while the functional form of $\mu(t)$, $0 \le t \le T$, is in general some polynomial in t, $\mu(t) = \sum_{n=0}^{\infty} a_n t^n$, this can be approximated linearly if T is "small enough." The defining slope and intercept are then determined from the available data. The slope is computed following Treviño (1982), while the intercept is computed using a "shifted-average" scheme, the application of which is new.

1. INTRODUCTION

Perhaps the most widely used method of approximating the mean-value, μ, of (continuous) random data is the so-called "simple-average" method, or some attendant variation of it; the fundamental mathematical form of this method is

$$\underline{\mu}(T) = T^{-1} \int_0^T x(\xi)d\xi \tag{1}$$

where $x(\cdot)$ represents the data which, as indicated, is itself defined over some finite time domain $(0,T)$. In this form the underscored variable denotes a random variable, random because $x(\cdot)$ in the integrand of Equ. (1) typically represents only one record of a complete ensemble of such records, the complete ensemble necessarily describing the total physical phenomena being analyzed. The unfortunate feature of the simple-average method is that it yields statistically valid results only when the measured phenomena is both stationary, i.e. its mean-value does not vary in time, and ergodic, i.e. the phenomena is such that one sample record of it almost always takes on (nearly) all modes of statistical behavior of every other possible sample record of the process, and therefore, that the given sample is indeed "representative" of the total physical phenomena. And although the joint requirements of stationarity and ergodicity are rarely (if ever) satisfied by real physical processes, it is often common practice to simply overlook this unsatisfactory behavior in any experimentally obtained data and to tacitly assume that the available data satisfies both criteria, and subsequently proceed accordingly; the underlying motivation for pursuing this rather dubious procedure is undoubtedly the fact that if it is assumed otherwise, viz. that the data is assumed to be nonstationary, there currently exist (scarcely) few known methods for accurately approximating the total time-dependence of the mean-value of the appropriate data. It is therefore the specific purpose of this note to somewhat alleviate this state of affairs.

2. APPROXIMATION OF $\mu(t)$

The mean-value of nonstationary data can in the most general case be represented by some polynomial function in t , viz. as

$$\mu(t) \simeq \sum_{n=0}^{\infty} a_n t^n \, , \qquad 0 \leq t \leq T \, ;$$

if, however, the given data interval is "small" enough the mean-value can for practical purposes be adequately approximated linearly, viz. as

$$\mu(t) \simeq \mu_1 t + \mu_0$$

where μ_1 and μ_0 are constants. A method for statistically approximating the "trend", μ_1 , from available data is given by Treviño (1982), and the intercept parameter, μ_0 , can be well-approximated by first forming the random variable

$$\underline{\mu}(t,T) = T^{-1} \int_0^T x(t+\xi)d\xi - \mu_1 T/2 \tag{2}$$

and then using both the available data and the value of T to compute μ_0 ; this is accomplished by selecting an initial value of t and then numerically determining $\underline{\mu}(t,T)$, say for example via a computer scheme. Note that in such a computation the "expected value" of $\underline{\mu}(t,T)$ is in fact equal to $\mu_1 t + \mu_0$, indicating that $\underline{\mu}(t,T)$ in Equ. (2) is an <u>unbiased</u> estimate of $\mu(t)$. It is also <u>consistent</u> if

$$E(t) = \langle \underline{\mu}(t,T)^2 \rangle - \langle \underline{\mu}(t,T) \rangle^2$$

is equal to zero for all t . In terms of the (stationary) covariance function, $C(t_1,t_2) = \langle \{x(t_1) - \langle x(t_1)\rangle\}\{x(t_2) - \langle x(t_2)\rangle\}\rangle = C(t_2 - t_1)$, E(t) expresses as

$$E(t) = T^{-2} \int_0^T C(t+\xi_1, t+\xi_2)d\xi_1 d\xi_2$$

$$= T^{-2} \int_0^T C(\xi_2 - \xi_1)d\xi_1 d\xi_2 \tag{2a}$$

and the conditions for E(t) to approximate zero for all t are then that $C(\cdot)$ itself approach zero "rapidly enough" over the (square) domain defined by (0,T) , and that T be "large" when compared against the square-root of the integral

$$\int_0^T C(\xi_2 - \xi_1)d\xi_1 d\xi_2 \, .$$

The first condition is likely to be satisfied for almost all physical phenomena for which the integral scale is not overly large while the second condition can be satisfied by simply gathering an excessive amount of data (cf. Eqs. (6) and (9) of Treviño (loc. cit.)).

This approach toward computing μ_0 is fundamentally different from the approach earlier formulated by Treviño in that this approach specifically utilizes the "shifted-average" integral. The appealing consequence of the present approach is that the statistical approximations to both μ_1 and μ_0 are simultaneously unbiased and consistent while in the previous approach the approximation to μ_0 was unbiased only and not necessarily consistent. Clearly, the extension formulated herein is on firmer statistical footing.

3. ILLUSTRATIVE EXAMPLE

The formulated approach was applied toward the determination of the mean-value of atmospheric velocity data obtained in a boundary-layer experiment undertaken by the Global Atmospheric Research Program-Atlantic Tropical Experiment (GATE), results of which were reported by Pennell and LeMone (1974). In particular, the analyzed data represents measurements of the "cross-wind" velocity, $v(x) = v(Ut) \sim v(t)$, gathered by flying an instrumented aircraft through the atmosphere (in the "x-direction") at a constant velocity $\underset{\sim}{U}$ at a given horizontal level. A preliminary statistical analysis of the data was first undertaken by evaluating the random variable.

$$\underline{\hat{\mu}}(t,\Delta T) = \Delta T^{-1} \int_0^{\Delta T} v(t + \xi)d\xi \tag{3}$$

for varying values of t and a constant value of $\Delta T < T$. The results of this initial evaluation are depicted in Figure 1. The "trend" depicted in Figure 1

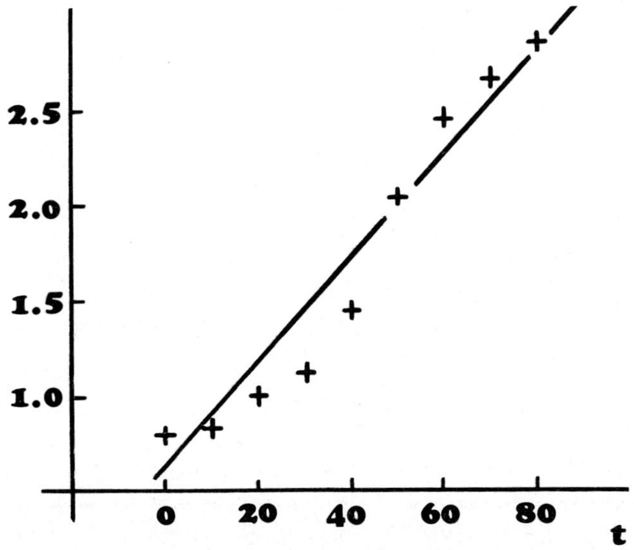

Figure 1: Plot of $\underline{\hat{\mu}}(t,\Delta T)$ vs. t for ΔT = 40-sec and varying values of t (in 10-sec increments). The "original" data represents "cross-wind" velocity, v (in m/sec), obtained during a two-minute (120-sec) data run. The illustrated straight line is "fitted" to the depicted data points by numerically computing μ_1 (μ_1 = 0.027) and then "visually" passing a straight line with the given slope through the points.

is essentially "representative" of the exact trend present in $\langle v(t)\rangle$ for the indicated time domain, a qualitative consequence which can be arrived at by purely theoretical computations. For example, if in the indicated domain $\langle v(t)\rangle = \mu(t) = \mu_1 t + \mu_0$, then $\langle\underline{\mu}(t,\Delta T)\rangle = \mu_1 t + \mu_0 + \mu_1\Delta T/2$; note that $\langle\underline{\mu}(t,\Delta T)\rangle$ is biased but that it is biased by the constant amount, $\mu_1\Delta T/2$, while the slope of the mean-value is unaffected by the averaging (integration) process of Equ. (3). Graphically, this means that if a straight line of slope μ_1 were to be somehow "fitted" to the data represented by Figure 1 this straight line would be parallel to the particular straight line which is the exact mean-value required; the amount of (vertical) "shift" between the two straight lines is then essentially equal to $\mu_1\Delta T/2$. Similarly, if the mean-value is quadratic in t , i.e. if $\langle v(t)\rangle = \mu(t) = \mu_2 t^2 + \mu_1 t + \mu_0$, then Equ. (3) provides

$$\hat{\underline{\mu}}(t,\Delta T) = \mu_2 t^2 + (\mu_2\Delta T + \mu_1)t + \mu_0 + \mu_1\Delta T/2 + \mu_2\Delta T^2/3$$

showing that this result is also biased but that the quadratic term, $\mu_2 t^2$, is left unchanged and will therefore appear directly in an appropriate graph of $\hat{\underline{\mu}}(t,\Delta T)$ vs. t . The straight line that was "graphically" (i.e. visually) fitted to the data points of Figure 1 is

$$\hat{\underline{\mu}}(t,\Delta T) \simeq 0.03t + 0.10 \qquad (4)$$

and since $\{\mu_1\Delta T/2\} \simeq 0.60$ this line rather suggests a mean-value,

$$\mu(t) \simeq 0.03t - 0.50 \qquad (5)$$

the "exact" mean-value from Equ. (2) is

Figure 2: Normalized covariance function, $\{C(\tau)/C(0)\}$, plotted versus the time-lag, τ , where C(0) = 0.41. The integral scale, Λ , is represented by the time-lag width, $\tau \simeq$ 5-sec , of the "rectangle" whose altitude is unity.

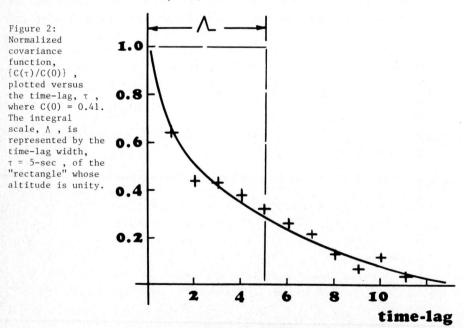

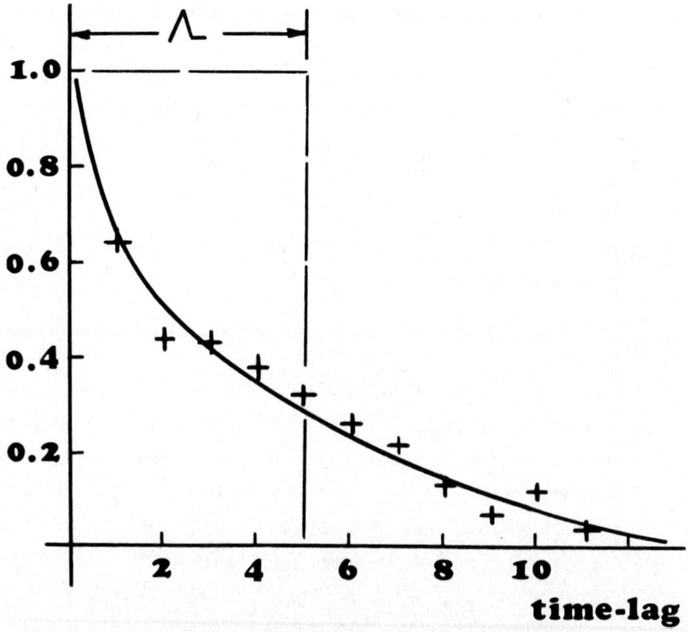

$$\mu(t) = 0.027t - 0.62 \; . \tag{6}$$

A (normalized) covariance analysis of the original velocity data relevent to Figure 1 was then done in order to get a relative measure of the integral time-scale, Λ . This result is depicted in Figure 2, which "conservatively" implies an integral time-scale of around 5-sec. This suggests that the obtained values of μ_1 and μ_0 in Equ. (6) are extremely accurate (cf. Equ. (2a)).

4. COMPARISON WITH REGRESSION SCHEMES

The method proposed herein is in general not at all the same as the well-known method of linear regression; indeed, it is more general and versatile that that scheme, and in fact reduces to linear regression only when T is "large" when compared to the "correlation length" of the data. In the linear regression approach a straight line is numerically fitted to a given set of data, $y(t)$, as

$$y(t) = \mu_0 + \mu_1 t + \varepsilon(t)$$

where $\varepsilon(t)$ is the so-called "random error" or "residual", and the errors for all t necessarily have to be both normally distributed (gaussian) and uncorrelated ($\langle\varepsilon(t)\varepsilon(t')\rangle \equiv 0$). The coefficients μ_0 and μ_1 are (usually) determined by the method of "least squares", and the complete method then denoted the method of <u>least-squares regression</u>. When the random errors are <u>not</u> uncorrelated ($\langle\varepsilon(t)\varepsilon(t')\rangle \neq 0$) this method can be generalized to the method of <u>generalized least-squares regression</u>, provided the form of $\langle\varepsilon(t)\varepsilon(t')\rangle = C(t' - t)$ is known or can be validly approximated. The "gaussian" assumption, however, is still an integral part of the generalized least-squares approach. For the method developed herein the "gaussian" assumption is never invoked nor is it ever required; furthermore, the functional form of $C(t' - t)$ does not have to be known nor assumed. All that is required is that $C(t' - t)$ have a sufficient number of derivatives for $\tau = t' - t = 0$ and that $C(\tau) \to 0$ "rapidly enough" as $\tau \to \infty$ (Treviño (1982)).

5. CONCLUDING REMARKS

A mean-value approximation for nonstationary random data has been formulated by employing the results of an earlier publication together with a "shifted-average" scheme outlined herein. The mean-value is approximated by assuming that in a sufficiently "small" time domain (0,T) the mean-value can be approximated <u>linearly</u>, i.e. as $\mu(t) = \mu_1 t + \mu_0$, where μ_1 and μ_0 are constants. The parameters μ_1 and μ_0 are then numerically computed from available data, and it is shown that the obtained values are statistically "good" estimates of the exact values of μ_1 and μ_0 . It is also shown that the method produces more general results than either <u>least-squares regression</u> or <u>generalized least-squares regression</u>, since this method can be equally applied to correlated, non-gaussian data.

FOOTNOTE

[1]Present address: Mechanical Engineering-Engineering Mechanics Department, Michigan Technological University, Houghton, MI 49931, USA.

REFERENCES

PENNELL, W.T. and LEMONE, M.A. (1974). An Experimental Study of Turbulence Structure in the Fair Weather Trade Wind Boundary Layer. J. Atmos. Sci. 37, 1121-1128.

TREVINO, G. (1982). A Method for Detrending Correlated Random Data. In Applied Time Series Analysis (Proceedings of the International Conference held at Houston, Texas, August 1981). Eds: O.D. Anderson and M.R. Perryman, North-Holland, Amsterdam and New York, 465-473.

TIME SERIES ANALYSIS: Theory and Practice 7
O.D. Anderson (editor)

STOCHASTIC MODELLING OF WIND SPEED AND DIRECTION

B. McWilliams and D. Sprevak
Department of Engineering Mathematics, The Queen's University of Belfast, Northern Ireland.

Using the distributions of orthogonal statistically independent components of wind speed, the joint distribution of wind speed and direction and the marginal distributions of wind speed and wind direction are deduced, and validated with records of observed data. A time series model for wind speed and direction is subsequently derived and, using the general properties of the spectrum of wind speeds, a methodology for estimating shorter term wind fluctuations is presented. This procedure is used to investigate the effect of different turbulence structures on the estimation of obtainable wind power from a Wind Energy Conversion System.

1. *INTRODUCTION*

With the concern expressed in recent years about dwindling energy resources, the search for feasible sources of renewable energy has increased. One such source, which has received much renewed attention, is the possibility of harnessing the wind's energy on a large scale. The efficient implementation of such a scheme obviously demands much detailed study and would therefore require substantial records of wind data. In many studies of wind power availability it is preferable to use a time series model of the wind's behaviour rather than existing wind velocity records as this eliminates the need for storage and handling of large data files. It also means that any number of years of information can be generated, thereby not restricting the analysis to the number of years of recorded data.

The distribution of wind speeds has been described in the past by the Weibull distribution, a two parameter distribution; and, in some instances, the Rayleigh distribution, the one parameter member of the Weibull family, has been found to be a suitable model; but a time series model for wind speed is not readily obtained using either of these distributions. Smith (1971) has provided some theoretical justification for the use of the Rayleigh distribution, in that it can be derived from the bivariate normal distribution when there is no prevailing wind direction, but the use of the Weibull distribution is an empirical solution to the modelling problem.

In this paper a more fundamental approach is adopted whereby orthogonal, independent components of wind velocity are modelled from which the joint distribution of wind speed and direction is derived. This approach also leads to a time series model which generates values of both wind speed and wind direction. Using spectra of wind speeds, this analysis, which is based on hourly averaged observations of wind speed and direction, is extended to give estimates of wind behaviour at shorter time intervals.

2. THE JOINT DISTRIBUTION OF WIND SPEED AND DIRECTION

Davenport (1968) and Riera, Viollaz and Reimundin (1977) have shown that the distribution of a component of wind velocity for any given direction is normally distributed, so it follows that the joint distribution of two wind velocity components will have a bivariate normal distribution. Probability contours of available data from Aldergrove (N. Ireland) were found to be approximately circular in shape, McWilliams (1983). From these data it seems sensible to assume that components of wind velocity have equal variances and negligible correlation. These assumptions reduce the number of parameters required to specify the bivariate normal distribution from five, in the general case, to three. Moreover, by considering the components of wind velocity along the prevailing wind direction and the direction at right angles, the number of parameters is again reduced by one as the component of wind velocity along the perpendicular direction to the prevailing wind direction should have zero mean.
In the case where a prevailing wind direction exists, the simplest model is then derived by assuming that the component of wind speed along the prevailing direction is normally distributed with non-zero mean and a given variance while the component of wind speed along a direction at right angles is independent and normally distributed with zero mean and with the same variance. Let

V_y = wind velocity along the predominant wind direction. It is assumed normally distributed, mean μ and variance σ^2.

V_x = wind velocity along the perpendicular to the predominant wind direction. It is assumed normally distributed, mean zero and variance σ^2. V_x and V_y refere to velocities relative to a fixed point on the ground. The joint density function of V_x and V_y is

$$f_{V_xV_y}(v_x,v_y) = \frac{1}{2\pi\sigma^2} \exp - \frac{v_x^2 + (v_y - \mu)^2}{2\sigma^2} , \quad -\infty < v_x, v_y < \infty,$$

and the corresponding joint distribution of wind speed (V) and direction (Θ) is given by

$$f_{V\Theta}(v,\theta) = \frac{v}{2\pi\sigma^2} \exp\left(-\frac{\mu^2}{2\sigma^2}\right) \exp\left(-\frac{v^2 - 2\mu v \sin\theta}{2\sigma^2}\right) , \tag{1}$$

$$0 < \theta < 2\pi, \quad 0 < v < \infty.$$

Integration of equation (1) over V gives the marginal distribution of Θ

$$f_{\Theta}(\theta) = \frac{1}{\sqrt{2\pi}} \psi(\gamma) + \gamma \sin\theta . \psi(\gamma \cos\theta) . \Phi(-\gamma \sin\theta), \quad 0 < \theta < 2\pi, \tag{2}$$

where $\Phi(x) = \int_{x}^{\infty} \psi(y)\,dy,$

$$\psi(x) = \frac{1}{\sqrt{2\pi}} \exp\left(-\frac{x^2}{2}\right)$$

and $\gamma = \mu/\sigma$.

Integration of equation (1) over Θ gives the marginal distribution of V

$$f_V(v) = \left[\frac{1}{\sigma^2} v \exp\left(-\frac{v^2}{2\sigma^2}\right)\right] \exp\left(-\frac{\mu^2}{2\sigma^2}\right) I_o\left(\frac{\mu}{\sigma^2} v\right), \quad 0 < v < \infty, \tag{3}$$

where $I_o(x)$ is the Bessel function of the first kind and zero order. It is interesting to note that when $\mu = 0$ (i.e. no prevailing wind direction) the distribution of wind direction as given by equation (2) reduces to $1/2\pi$, a uniform distribution, and the distribution of wind speed reduces to the terms in square brackets in equation (3) which is the density function of the Rayleigh distribution.

McWilliams, Newmann and Sprevak (1979) and McWilliams and Sprevak (1980) have tested the assumptions and performance of this model using data from several sites in Ireland. Comparisons with the Weibull and Rayleigh models were made and all three models were found to provide similar descriptions of the distribution of wind speed. However, the present model is the only one of the three which also provides the marginal distribution of wind direction.

3. *ESTIMATION OF THE PARAMETERS OF THE DISTRIBUTION*

The parameters of this model were estimated by resolving the wind velocity data into the two specified orthogonal directions and the sample mean and variance for each direction were computed. This estimation procedure is cumbersome as it requires observations on the joint frequency distribution of wind speed and direction. An alternative simple estimation procedure using the method of moments, which only requires the marginal distributions of wind speed and wind direction was developed. McWilliams and Sprevak (1980) have validated this simpler estimation procedure with data from several sites.

4. *TIME SERIES MODEL FOR WIND VELOCITY*

Letting $V_y(t)$ and $V_x(t)$ denote the average hourly components of wind velocity, at time t, then $V_y(t)$ has a normal distribution with mean $\mu(t)$ and variance $\sigma^2(t)$ and $V_x(t)$ has a normal distribution with zero mean and

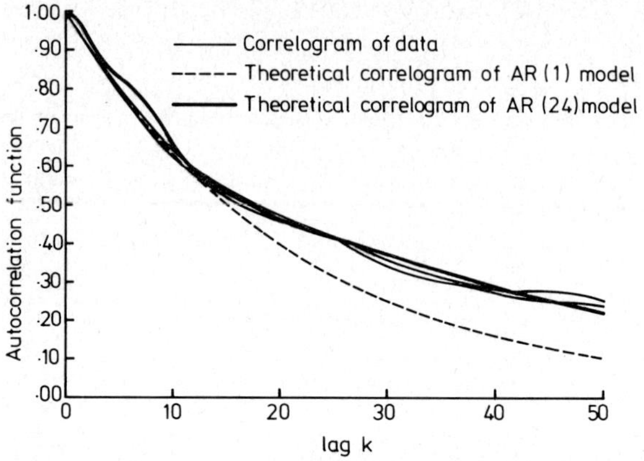

FIGURE 1 Autocorrelation functions of data for X and Y directions compared with the theoretical curves of the AR(1) and AR(24) models.

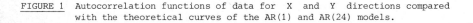

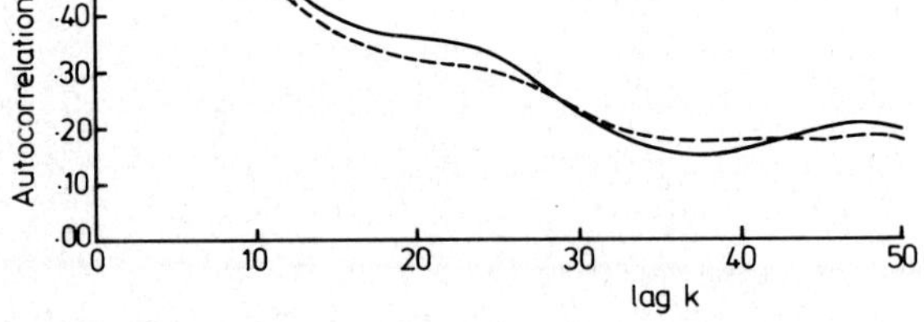

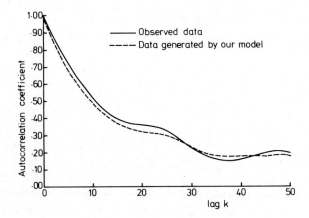

FIGURE 2 Autocorrelation functions of observed and simulated wind speeds.

variance $\sigma^2(t)$. To render the data stationary it is assumed that the main sources of deterministic evolutive behaviour are cyclical components which affect the temporal variability of $\mu(t)$ and $\sigma^2(t)$. The annual evolutive behaviour of $\mu(t)$ and $\sigma(t)$ is described in terms of harmonic components and the two series can then be rendered stationary by the transformations

$$V_x^s(t) = V_x(t)/\hat{\sigma}(t)$$

$$V_y^s(t) = (V_y(t) - \hat{\mu}(t))/\hat{\sigma}(t)$$

where $\hat{\mu}(t)$ and $\hat{\sigma}(t)$ are the harmonic series estimates of $\mu(t)$ and $\sigma(t)$ respectively.

The stationary series $V_x^s(t)$ and $V_y^s(t)$ can be modelled by standard techniques of time series analysis, in particular the Box-Jenkins (1970) procedure of identification, estimation and verification may be used. The sample auto-correlation function, which is required for the model identification stage, is given in Figure 1 for both series and shows that the two components of wind velocity have the same stochastic behaviour, thus reducing the problem to the modelling of a single stationary Gaussian series. Further evidence for this result is given by Eidsvik (1981) who shows that components of wind velocity in different directions have the same spectral shape.

The sample partial autocorrelation functions for both directions were also found to be very similar in that they both cut off after lag one which suggests an autoregressive process of order 1, AR(1), as a suitable model. However, the theoretical autocorrelation function of this process, as shown in Figure 1, decays too quickly after about lag 20. Several other low-order models were tried but none was found to be an improvement on the AR(1) model. A possible important term could be an autoregressive component at lag 24 corresponding to the daily cyclical property exhibited by wind. The model is then given by

$$V_i^s(t) = \alpha_1 V_i^s(t - 1) + \alpha_{24} V_i^s(t - 24) + \varepsilon(t),$$

where i represents x or y, α_1 and α_{24} are the parameters of the model to be estimated by minimizing the sum of the squares of the residuals and $\varepsilon(t)$ is a random perturbation, normally distributed with mean zero and variance σ_ε^2. The values of α_1 and α_{24} (found using one year of hourly observations) are 0.952 and 0.012 respectively. The functional form of the theoretical auto-correlation function for this model can also be obtained analytically, McWilliams and Sprevak (1982a). A plot of this function is included in Figure 1 and is a remarkable improvement on the AR(1) model.

Although this model provides an accurate description of the component of wind velocity it was found that when used to generate wind speeds, the amplitude of the daily cycle produced was negligible compared to that of the observed data. It was subsequently discovered that, for the data used in this analysis, $\sigma(t)$

has quite a pronounced diurnal trend for three months of the year. For these three months $\sigma(t)$ was then treated as having a deterministic cyclic component with period 24 hours, the other nine months were treated as before. Further data were generated incorporating this modification and Figure 2 shows that the model now includes a daily cycle to a certain degree, although not to the same extent as the observed data. In an effort to improve the modelling of the daily trend, the possibility of lagged correlation between the components of wind velocity was considered. The cross-correlation function of the residuals of the two components of wind velocity was calculated but no evidence of lagged correlation was found. It was concluded that, as the wind direction $\theta(t)$ shows no daily trend, resolving the wind veocity into the two components results in the loss of some information on the daily cycle, McWilliams (1983).

5. *EXTENSION OF HOURLY DATA*

The performance of a Wind Energy Conversion System (WECS) depends on the local wind structure and any evaluation of the WECS therefore requires data on the local wind behaviour. In particular, the short term wind fluctuations will affect the output from the WECS and, in the case of a horizontal-axis turbine, may have an important bearing on the design of the control mechanism which keeps the WECS facing into the wind. However, information on the short term wind fluctuations is limited, as at most sites only an average value of wind speed and direction is recorded for each hour, and Lamming, Ibbetson and Milford (1980) have shown that the use of hourly averaged wind data underestimates obtainable wind power. Eidsvik (1981) and der Kinderen, van Meel and Smulders (1977) have presented spectra of wind speeds, over a wide range of frequencies, for different sites and these spectra have been found to contain some common features: they are characterized by two peaks, one corresponding to a period of several hours, the other corresponding to a period of approximately one minute, and between these two peaks a broad spectral gap which can be assumed constant between the periods of five minutes and two hours. If hourly averaged data are available then the modelling procedure outlined can be extended, using these properties of the spectrum of wind speeds, to give an estimate of wind behaviour at shorter time intervals. These spectra have been calculated from wind speeds whereas the model developed in this paper is for wind speed components, but Panofsky and van der Hoven (1955) have shown that the spectra of wind speeds and wind speed components are the same for the range of high frequencies that are greater than the spectral gap.

For periods of two hours or greater, the spectral density function is provided by the hourly obervations. The spectral gap can be modelled by white noise, which has a constant spectrum, the spectral density being a measure of the

variance of the process. The peak in the high frequency end of the spectrum indicates the presence of an autocorrelation structure which cannot be specified exactly in a model in the absence of data. An autoregressive process or order 2, AR(2), can be used to model this section of the spectrum, but the two parameters of the model and the variance of the noise term need to be estimated with experimental data. However, in order to maintain the continuity and the stationarity of the spectrum and to obtain a peak at a specified frequency, these parameters need to satisfy certain constraints which put restrictions on the range of values which they can take, McWilliams and Sprevak (1982b).
Different turbulence structures (short term wind fluctuations) can then be obtained by varying the parameters of the AR(2) process, subject to the appropriate constraints.
This procedure for extending hourly wind data was used in a simulation study to show that a reliable estimate of obtainable wind power requires observations on the short term wind fluctuations or turbulence structure of the local wind.
The energy pattern factor, which is defined as the ratio of the mean of the cubed wind speeds $\overline{V^3}$ to the cube of the mean wind speed $(\bar{V})^3$, is a measure of the error in estimating obtainable wind power using averaged values of wind speed. However, the value of $\overline{V^3}$ is reduced due to the mechanical and electrical losses of the wind turbine and the fact that the turbine is not suitable for extracting power from very low and very high wind speeds.
The effect of a given turbulence structure on the energy pattern factor was investigated. The results of this study are given in Table 1, where ϕ_1 and ϕ_2 are the parameters of the AR(2) process which specify the short term wind fluctuations, and show that the error in estimating obtainable wind power from hourly data can vary from about 6% to 40%. Using limited amounts of actual data on these short term fluctuations, Lamming, Ibbetson and Milford (1980) have found the energy pattern factor to vary from 1.06 to 1.15.

Parameters		Energy Pattern Factor
ϕ_1	ϕ_2	
0.182	-0.1	1.06
0.333	-0.2	1.06
0.462	-0.3	1.07
0.571	-0.4	1.06
0.667	-0.5	1.07
0.750	-0.6	1.08
0.824	-0.7	1.09
0.889	-0.8	1.13
0.947	-0.9	1.24
0.974	-0.95	1.40

TABLE 1 Values of Parameters Specifying the AR(2) Process Along with the Corresponding Value of the Energy Pattern Factor

6. CONCLUSIONS

A complete modelling procedure for wind characteristics was presented. Using the distribution of wind velocity components, the joint distribution of wind speed and direction was modelled and was found to describe well the summarizing marginal distributions. The simple approach lends itself to the derivation of a stochastic model which was shown to closely follow the temporal variability of wind data. A methodology for extending hourly data to estimate short term wind fluctuations was also discussed and was used to investigate the effect of differing turbulence structures on the output of a Wind Energy Conversion System.

REFERENCES

BOX, G.E.P. and JENKINS, G.M. (1970). Time Series Analysis Forecasting and Control. Holden-Day, San Francisco.

DAVENPORT, A.G. (1968). The Dependence of Wind Loads on Meteorological Parameters. University of Toronto Press.

der KINDEREN, W.J.G.J., van MEEL, J.J.E.A. and SMULDERS, P.T. (1977). Effects of wind fluctuations on windmill behaviour. Wind Engineering, Vol. 1, No. 2, 126-140.

EIDSVIK, K.J. (1981). Estimation of time series models for horizontal wind on the Norwegian coast. Wind Engineering, Vol. 5, No. 2, 66-72.

LAMMING, S.D., IBBETSON, A. and MILFORD, J.R. (1980). Some aspects of small-scale wind structure and its effects on a vertical axis wind turbine. Wind Engineering, Vol. 4, No. 3, 125-133.

McWILLIAMS, B., NEWMANN, M.M. and SPREVAK, D. (1979). The probability distribution of wind velocity and direction. Wind Engineering, Vo. 3, No. 4, 269-272.

McWILLIAMS, B and SPREVAK, D. (1980). The estimation of the parameters of the distribution of wind speed and direction. Wind Engineering, Vol. 4, No. 4, 227-238.

McWILLIAMS, B. and SPREVAK, D. (1982a). The simulation of hourly wind speed and direction. Mathematics and Computers in Simulation 24, 54-59.

McWILLIAMS, B. and SPREVAK, D. (1982b). A simulation study of the effects of short term wind fluctuations on the estimation of obtainable wind power. In Modelling and Simulation (Proceedings of the International Conference held in Paris, July 1982) Ed: G. Mesnard, Vol. 11, 61-64.

McWILLIAMS, B. (1983). Ph.D. Thesis, The Queen's University of Belfast.

PANOFSKY, H.A. and van der HOVEN, I. (1955). Spectra and cross-spectra of velocity components in the mesometeorological range. Quarterly Journal of the Royal Meteorological Society 81, 603-606.

RIERA, J.D., VIOLLAZ, A.J. and REIMUNDIN, J.C. (1977). Some recent results on probabilistic models of extreme wind speeds. Journal of Industrial Aerodynamics 2, 271-287.

SMITH, O.E. (1971). An application of distributions derived from the bivariate normal density function. In Probability and Statistics in the Atmospheric Sciences (Proceedings of the International Symposium held in Honolulu, Hawaii) 162-168.

TIME SERIES ANALYSIS: Theory and Practice 7
O.D. Anderson (editor)

PREDICTION OF GENERATED WIND TURBINE POWER WITH HIGH FREQUENCY WIND SPEED SERIES

Jiann-Jong Lou
Robertson, Fowler & Assoc., P.C., 211 East 46th St., New York, N.Y. 10017, USA

Ross B. Corotis
Department of Civil Engineering, The Johns Hopkins University, Baltimore, Maryland 21218, USA

Time series of wind speed data and generated wind turbine power are analyzed. In an attempt to correlate wind speed with turbine power, data were sampled every two seconds from a United States Department of Energy demonstration 200 KW wind turbine installation. Wind speeds were recorded from three heights on a meteorological tower and from a turbine nacelle anemometer. Turbine power was recorded directly from the wind-driven generator. Spatial and temporal averaging was performed, and time-lagged spatial cross correlations, cross spectral density functions and coherence functions were computed. A "frozen field" mesoscale meteorological assumption was used to translate meteorological tower data to the turbine. Nonstationarity in the mean and standard deviation were investigated. These analyses form the basis for data collection procedures for initial site evaluation and for full-scale machine performance predictions. Data were 1-second averages sampled once every two seconds.

INTRODUCTION

Stochastic analyses have been performed on a data tape supplied by Battelle Pacific Northwest Laboratories from a Clayton, New Mexico site where a DOE/NASA MOD-0A 200kW wind turbine (wind energy conversion system, WECS) is in operation. The Clayton test site is located in a region where high wind potential exists (Pennell, et al., 1980). A local channeling effect, caused by a small valley oriented approximately north-south, results in a prevailing southeasterly wind.

The MOD-OA wind turbine has a 37.5m (125 ft) diameter, two-bladed rotor which drives a 200kw capacity synchronous generator through a step-up gearbox. The rotor is positioned downwind on a 30m (100 ft) steel-truss tower. The rotor is designed to operate at a constant speed of 40 rpm and it drives a 60 Hz three-phase alternator at 1800 rpm. Constant rotor speed is maintained by controlling the blade pitch angle with a feedback control system. The rotor, alternator, transmission, and associated equipment are mounted on a bedplate, which is yawed to align the rotor with the wind. Detailed description of MOD-OA wind turbine can be found elsewhere (e.g., Thomas and Richards, 1977).

The data system used for recording MOD-OA performance consists of an on-site microprocessor and four mechanical anemometers. Data were sampled once every two seconds for three hours and stored on a magnetic tape. The parameters recorded for determining performance are: alternator power output, meteorological tower wind speed and direction at 9.1m (30 ft), 30.0m (100 ft) and 45.7m (150 ft), nacelle wind speed and direction and yaw error of the rotor. The nacelle anemometer is located approximately 4.5m (15 ft) upwind of the plane of the rotor and 2.7m (9 ft) above the rotor centerline. The meteorological tower is located 152m (500 ft) from the wind turbine, and was upwind of the wind turbine when the data reported here were collected.

SPECTRAL ANALYSIS

The spectral analysis of the input wind and output turbine power is capable of addressing the question of the response of a wind turbine to a turbulent wind. It can also help determine the appropriate averaging time for which winds should be averaged so that the wind speed could be correlated with turbine power output.

The power spectral density function was computed directly from the wind speed time series by the fast Fourier transform algorithm, (Akins and Peterka, 1975; Bendat and Piersol, 1971, 1980). The power spectral density has been normalized with the variance of the time series.

The wind speed spectra (which should be considered as longitudinal velocity spectra) are remarkably smooth and follow the -5/3 power law (Hinze, 1975). This implies the high-frequency behavior is consistent with local isotropy and, in the inertial subrange, the spectra fall as $n^{-5/3}$. Since the sampling interval is two seconds, the Nyquist frequency is 0.25 Hz. Due to the relatively long duration of the records, the lowest frequency that can be reliably determined is about 0.0001. Hz. Fluctuations exist in the incident wind over a wide range of frequencies. This variation is continuous for several orders of magnitude in frequency, and indicates a broad-band process.

The computed spectra were compared with two empirical expressions for the spectral density of longitudinal turbulence (Davenport, 1961; Kaimal, 1973), The expression proposed by Davenport is widely used in predicting wind loading of structures. The expression proposed by Kaimal is height dependent and can be adjusted to include the effects of atmospheric stability. These two approaches are considered for a mean wind speed of 10m/s and an alongwind turbulence intensity of 15% in near neutral stability (statistics consistent with the Clayton wind record). It was concluded that the expression by Davenport adequately describes the data from the wind turbine test facility.

Atmospheric turbulence is characterized by energy containing periodicities with frequencies of the order of 0.002 - 0.1 Hz. The response of a WECS is also characterized by a wide variety of frequencies over a broad spectral bandwidth (0.01 - 1.0 Hz). This range of characteristic frequencies overlaps the range of atmospheric boundary layer turbulence frequencies with significant spectral energy, indicating response of the WECS to the fluctuating portion of the wind. The general shape of the wind speed spectra (in reduced normalized form) is consistent with those found in previous meteorological research for the frequency range examined. For instance, the power spectrum of horizontal wind speed in the frequency range from 0.0007 to 900 cycles per hour was investigated by Van der Hoven (1957). He noted atmospheric motions which have periods between 6 and 24 hours, associated with synoptic-scale weather systems and with diurnal heating and cooling effects. There is little activity which would produce periods between 4 hours and a few minutes but mesoscale turbulence produces gusts which have periods mainly in the range 0.5 to 2 minutes. The averaging process effectively filters out contributions from periods shorter than the averaging period. Thus, daily or monthly mean wind speeds supress information on the energy associated with diurnal or synoptic variations in wind speeds. With the gap in the spectrum between periods of a few minutes and about 4 hours, it is found that there is very little difference in the statistics of data taken over a month or more whether the averaging time is 10 minutes, 1 hour, or 2 hours.

CROSS CORRELATION ANALYSIS

Cross correlation functions were computed between wind speeds at different heights on the meteorological tower, between wind speed at the meteorological tower and wind speed at the turbine nacelle, and between wind speed at various locations (including spatially averaged wind speed) and generated turbine power.

The distance between the meteorological tower and turbine nacelle is 152m (500 ft) and the height of the turbine nacelle is approximately 33m. The purpose in comparing wind speeds at the meteorological tower and turbine nacelle is to check if wind measurements are made close enough to the wind turbine to be representative of the "same" wind that drives the rotor. These computed cross correlations functions increase with averaging time, as expected, and exhibit positive time delays, indicating the meteorological tower was upwind of the wind turbine when the data reported here were recorded. The mean wind speed measured at an elevation 30.0M on the meteorological tower is 11.4 m/s. It will take the wind approximately 12 seconds on the average to travel the distance of 152m from the meteorological tower to be registered by the turbine nacelle anemometer, which is verified by the actual cross correlation functions. Lag times vary according to each individual anemometer height on the meteorological tower. For example, there is a lag of about 6 seconds between the wind speed at elevation 9.1m and at nacelle. The 6 seconds decrease in lag time, as compared with the previous case, is because of the time delay between wind speed at elevations 9.1m and 30.0m.

A general guideline exists for the estimation of maximum separation distances between stations (wind measuring locations and/or WECS sites). For flat terrain, spatial variations in the wind structure near a measurement location can be modeled by using a "frozen field" assumption (Hinze, 1975; Lumley, 1970). The wind velocity at a point some distance ℓ downstream of a measurement location can be estimated by lagging the data at the measurement location by a time interval $\ell/\bar{u}$, where $\bar{u}$ is the mean wind speed. This concept is important in the study of correlation between wind speed and wind turbine power when spatial-lag correction is to be made. By applying dimensional reasoning, the maximum distance, L, for which the assumption is valid depends on the time scale of turbulence, t, and the wind speed. For eddies the size of the rotor disk, the time scale is given by

$$t = D/u_* \qquad (1)$$

where D is the rotor diameter and u_* is the friction velocity. If the range of eddy sizes at a particular location is larger than the diameter of a wind turbine, then the effects of the turbulence may be experienced over the entire turbine at once. There will also be some eddies which are smaller than the diameter of the turbine and these may cause local effects on the loading of the blades of the turbine, but such effects may not be noticeable in the integrated output of the turbine. For a neutral atmosphere, u_* can be obtained by rearranging the logarithmic profile of horizontal wind speed, i.e.,

$$u_* = \bar{u}\,k/\ell n(Z/Z_o) \qquad (2)$$

where $\bar{u}$ is the mean wind speed at height, Z, k is the von Karman constant, Z_o is the surface roughness length, and the physical meaning of u_* can be viewed as

$$u_* = \sqrt{\tau_{surface}/\rho} \qquad (3)$$

with $\tau_{surface}$ the surface shear stress and ρ the density of air. Note that the logarithmic profile is based on an assumption of constant shear stress in the atmospheric surface layer. If one takes the relevant wind speed as the wind at hub height ($z \simeq d/2$), then the maximum separation distance is:

$$L = (D/k)\,\ell n\,(D/2Z_o) \qquad (4)$$

For a rotor diameter of $\simeq$40m (MOD-OA) and a roughness length of 10cm, the maximum separation distance is around 600m. Local topographic features would certainly affect the choice of maximum separation distances between measurement stations.

Correlations of turbine power output with wind speed at different elevations on the meteorological tower were computed for various temporal averages. Maximum correlation values (1-second average case) are around 0.4 and are considered questionable for "significant" correlation. This is not surprising since instantaneous readings (a mechanical anemometer has a response time on the order of one second, which should be short enough to represent "instantaneous" wind speed) were used in the correlation calculation. A wind machine responds to the broad-band excitation caused by the atmospheric turbulence. The fluctuations in the machine output were made up of contributions of a wide range of frequencies. Because the variance of turbine output is the quantity which would affect the correlation value, time averaging must be employed before correlation is made. If all of the fluctuations associated with the output of a turbine were centered in a narrow range of frequencies, an averaging time could be selected which would allow an accurate variance of output power to be calculated. However, for all of the records which have been examined as a part of this study, the variation in the turbine output is continuous for several orders of magnitude in frequencies and no single averaging time would be appropriate. An increase of the correlations with averaging time was clearly evident. However, this increase begins to level off for averaging times of 60 seconds. The increase is caused by low-frequency components in the fluctuations of wind speed and turbine output. Due to the spectral gap, there is little difference in the correlation values when the averaging time is 60 seconds or more.

Since a wind turbine responds to wind over its entire swept area, an attempt was made to improve the cross-correlation by spatially averaging the wind speed at the three locations on the meteorological tower before correlating with the turbine power. Maximum correlation values do increase, but only slightly in comparison to the single wind case. Taking account the lag time for wind to reach the turbine from the meteorological tower also provides minor improvement.

Cross-correlation functions between the wind speed at the turbine nacelle and the machine power output were independent of averaging time. The maximum cross-correlation generally reaches a value around 0.7 and occurs at a lag around zero. All these phenomena imply that wind measurements were made so close to the wind turbine that the registered wind speed is in the disturbed flow field. In other words, interference effects due to rotors, nacelle and tower were included in the cross-correlation computation. This finding indicates that the turbine nacelle is not an ideal location for wind measurements to correlate with turbine power. In order to eliminate any interference effects, wind measurements must be made far enough away from the turbine, but close enough to be representative of the "same" wind that drives the rotor when spatial variations of wind structure nearby a measurement location are considered. This representativeness can be justified by cross-correlation functions between wind speeds at an appropriate meteorological tower and turbine site.

COHERENCE ANALYSIS

Coherence functions were computed for wind speed (considered as a system input and generated turbine power (considered as the system response). Wind speed at the meteorological tower and turbine power display a relatively weak coherence of 0.36 to 0.46 over the lower frequency range below 0.004 Hz. Above this range, moreover, the coherence function diminishes. The relatively weak coherence at lower frequencies probably reflects contributions to the turbine power from inputs other than the measured wind speed. The loss of coherence at the higher frequencies probably results from the decaying nature of the input power spectrum, which causes the output power spectrum to fall off in a similar manner. Note that instantaneous data were used to calculate the coherence function. One of the most important applications of the coherence is the determination of the appropriate averaging time for the wind speed so that the average wind speed can be correlated with turbine power output. The idea is to determine that frequency above which the coherence vanishes or becomes negligibly small. From the coherence functions it appears that a frequency around 10^{-2} Hz seems to be a reasonable upper bound choice for the stated purpose. The appropriate wind averaging time is thus given by $1/10^{-2}$ = 100 seconds since the averaging process effectively filters out contributions from periods shorter than the averaging period. This indicates that, in order to correlate wind speed with turbine power, data should be averaged over a period of at least a minute. Similar results were reached from the cross-correlation analysis.

The coherence between the input wind speed at the nacelle and turbine power output is quite good ($\simeq$0.7) over a rather broad range of frequencies. This is a good indication that cross-correlation between the wind speed at the nacelle and turbine power is relatively high and independent of the averaging time. However, the result may be misleading since the flow field around the turbine nacelle is disturbed by the wind machine and the supporting tower.

CONCLUSIONS

For wind energy conversion, interest in wind data acquisition is in predicting wind turbine output. Therefore, the appropriate wind characteristics are those that correlate most highly with turbine output. This information may then be used for site evaluation. Once a turbine has been installed, the performance of the turbine can be related to the wind characteristics. If studies such as turbulent forces on the blades are of interest, then data appropriate to that purpose must also be considered (Corotis, 1982).

The research summarized here has demonstrated that the application of various sampling rates and averaging time processes to a time series may be thought of as a filter, or a window, through which only a portion of the total energy in the spectrum of wind fluctuations can be observed.

The wind turbine responds to both temporally and spatially averaged wind. Spatial and temporal averaging are applied to determine the optimal digitization in terms of correlation with generated turbine power. It was found that wind speed averages on the order of 15 seconds to two minutes at the meteorological tower correlate highly with turbine power. The cross-correlation values increase with increasing averaging time, but gradually level off for longer averaging time. A spatial-lag correction to the cross-correlation between wind speed at the meteorological tower and turbine power a few hundred meters away is not necessary. Spatial averaging of wind speeds at multiple heights on the meteorological tower does not provide improved correlation with turbine power. This implies that a single point measurement of wind speed is a reasonable measurement of the wind input to the turbine. Wind speed at the nacelle correlates highly with turbine power, and the process is independent of averaging time. This phenomenon indicates that the wind field registered at the nacelle is disturbed by the presence of the wind turbine. Undisturbed wind measurements thus must be made some distance away from the turbine. The distance can roughly be determined by dimensional analysis and is best assessed by correlating wind speed at the meteorological tower with that at the turbine nacelle.

The spectral density approach allows the separation of the effects caused by different ranges of frequencies. It has the disadvantage of providing information at only one value of mean wind speed and incident turbulence intensity. Trend analysis was performed, and it was concluded that the wind fluctuating process can be regarded as stationary, at least in the wide sense, in the absence of transitory nonstationarity phenomena such as thunderstorms. The nonstationarity behavior of the wind flow field will have negligible effect on performance design, although it may be important for structural strength.

The cross-coherence of the wind with turbine power was calculated, and the frequency above which the coherence becomes negligibly small was determined. The result is used to ascertain the appropriate averaging time so that the wind speed will be highly correlated with turbine performance. Results indicate appropriate averaging times on the order of a minute.

BIBLIOGRAPHY

AKINS, R. E. and PETERKA, J. A. (1975), "Computation of Power Spectral Densities and Correlations Using Digital FFT Techniques," Colorado State University, College of Engineering, Technical Report CER75-76REA-JAP-13.

BENDAT, J. S. and PIERSOL, A. G. (1971), Random Data: Analysis and Measurement Procedures, Wiley and Sons, New York.

BENDAT, J. S. and PIERSOL, A. G. (1980), Engineering Applications of Correlations and Spectral Analysis, Wiley and Sons, New York.

COROTIS, R. B. (1982), "Statistical Analysis of Site Wind High-Frequency Wind Speed Characteristics and Wind Turbine Power," Final Report Submitted to Battelle-Pacific Northwest Laboratories, June, 1982.

DAVENPORT, A. G. (1961), "The Spectrum of Horizontal Gustiness Near the Ground in High Winds," Quarterly Journal of the Royal Meteorological Society, Vol. 87, 194-211.

HINZE, J. O. (1975), Turbulence, McGraw-Hill, New York.

KAIMAL, J. C. (1973), "Turbulence Spectra, Length Scales, and Structure Parameters in the Stable Surface Layer," Boundary Layer Meteorology, Vol. 4., 289-309.

LUMLEY, J. L. (1970), Stochastic Tools in Turbulence, Academic Press, New York.

PENNELL, W. T., BARCHET, W. R., ELLIOTT, D. L., WENDELL, L. L., and HISTER, T. R., (1980), " Meteorological Aspects of Wind Energy: Assessing the Resource and Selecting the Sites," Journal of Industrial Aerodynamics, Vol. 5, 223-246.

THOMAS, R. L. and RICHARDS, T. R. (1977), "ERDA/NASA 100-Kilowatt Mod-0 Wind Turbine Operations and Peformance," NASA TM-73825, September.

VAN DER HOVEN, I. (1957), "Power Spectrum of Horizontal Wind Speed in the Frequency Range from 0.0007 to 900 Cycles per Hour," Journal of Meteorology, Vol. 14, 160-164.

TIME SERIES ANALYSIS: Theory and Practice 7
O.D. Anderson (editor)

A COMPARISON OF TWO AUTOMATIC SYSTEMS FOR BUILDING VECTOR TIME SERIES MODELS IN AIR POLLUTION RESEARCH

S. G. Kapoor
Mechanical and Industrial Engineering, University of Illinois, Urbana, Illinois, USA

W. R. Terry
Industrial Engineering, University of Toledo, Toledo, Ohio, USA

This paper evaluates two automatic approaches, developed by Pandit (1980) and Akaike (1976) respectively, for analyzing vector time series. These approaches are compared with respect to the form of the family of models assumed and the process for selecting an adequate model. Results of utilizing each of them to develop a vector time series representation of nitrogen dioxide and sulfur dioxide data are discussed.

1. INTRODUCTION

A review of the air pollution literature and discussions with professionals in that field suggests that time series analysis methods are not being widely used by those studying air pollution. This is somewhat disconcerting since over a decade has passed since time series analysis methods were first used to model an air pollution time series. This paper is based on the belief that one of the major reasons for the apparently sparse use of time series analysis in air pollution could be that many of those in that field lack the statistical expertize necessary for identifying an appropriate structure for a time series model. However the availability of software systems, which automate the process of building time series models, has the potential for making time series analysis methods available to an increased number of persons in the air pollution field. In order for this potential to produce desirable results it will be necessary for the prospective user to determine whether a given system is likely to produce good results.

At present the computer based systems, which have been developed for automatically building linear time series models, can be classified according to whether they use an autoregressive moving average (ARMA) representation or a Markovian state space (MSV) representation of the stochastic generating process. The purpose of this paper is to critically evaluate the suitability of a representative system from each category for modeling air quality time series. In particular the system developed by Pandit (1980) will be utilized for developing ARMA representations, while the work of Akaike (1976) will be utilized to develop a state vector representation.

Several factors influenced the selection of these systems. First, both are automatic in that they do not require the analyst to exercise judgement in identifying the form of the model. Second, both use different statistical procedures for selecting an appropriate model for representing a time series. Third, the software required for utilizing them is generally available to university researchers.

Both of these systems will be used to automatically develop both univariate and multivariate stochastic generating models for nitrogen dioxide (NO_2) and sulfur dioxide (SO_2). Three factors influenced the selection of these pollutants. First, the Environmental Protection Agency has recognized that these pollutants can adversely affect public health. Second, data for both series were readily available to the researchers. Third, a study by Merz, et al (1972) found that the sample autocorrelation for NO_2 and SO_2 showed "markedly unstable behaviors." In this study linear systems analysis techniques will be used to examine the stability and oscillatory characteristics of the models fitted by each system. Since Akaike (1974) has shown that the ARMA and MSV representations are equivalent forms it was expected that the fitted models would exhibit similar stability and oscillatory characteristics.

Several automated methods were omitted due to the reasons cited below. The heuristically based 'Auto Box' developed by Reilly (1981) was not included since this system is not typically available in a university environment. A heuristically based 'Automatic Box-Jenkins' system developed by Hill and Woodworth (1980) was omitted because it is not widely available or widely known in the U.S. The ARIMAID method of Kang, Bedworth and Rollier (1982) was not considered since its model selection procedure is based on Akaike's information criterion and since it is not capable of handling vector time series.

One purpose of this paper was to determine whether or not the Pandit and the Akaike systems were capable of handling ill-behaved time series. In view of this purpose the NO_2 and SO_2 series were selected because previous researchers have concluded that they had unstable autocorrelation functions. A second purpose was to determine if the models produced by the Pandit and the Akaike systems had similar stability and oscillatory characteristics.

2. AUTOMATIC MODEL BUILDING SYSTEMS

The purpose of this section is to contrast the Pandit and the Akaike systems for automatically building vector time series models. These methods differ in two important aspects. The first is the basic form of the family of models which is available for representing a stochastic generating process. The second is the process used for selecting an appropriate model.

The Pandit and Wu method uses the ARMA time domain representation advocated by Box and Jenkins while the Akaike method uses the MSV representation first employed by system engineers. The ARMA representation is given by

$$\underset{\sim}{x}_t - \underset{\sim}{\phi}_1 \underset{\sim}{x}_{t-1} - \underset{\sim}{\phi}_2 \underset{\sim}{x}_{t-2} - \dots - \underset{\sim}{\phi}_n \underset{\sim}{x}_{t-n} = \underset{\sim}{a}_t - \underset{\sim}{\theta}_1 \underset{\sim}{a}_{t-1} - \underset{\sim}{\theta}_2 \underset{\sim}{a}_{t-2} - \dots - \underset{\sim}{\theta}_{n-1} \underset{\sim}{a}_{t-n+1} \tag{1}$$

where $\underset{\sim}{a}_t$ is a p x 1 vector discrete time white noise process with $E[\underset{\sim}{a}_t] = \underset{\sim}{0}$ and $E[\underset{\sim}{a}_t \underset{\sim}{a}_{t-k}] = \delta_k \underset{\sim}{\sigma}_a^2$; $\underset{\sim}{x}_t$ is a p x 1 vector which represents the system response to the discrete time white noise input; $\underset{\sim}{\phi}_1, \underset{\sim}{\phi}_2, \dots, \underset{\sim}{\phi}_n$ are discrete autoregressive parameter matrices; $\underset{\sim}{\theta}_1, \underset{\sim}{\theta}_2, \dots, \underset{\sim}{\theta}_n$ are discrete moving average parameter matrices and δ_k is the Kronecker delta function.

The Markovian state vector (MSV) representation of Akaike is based on the notion of a state variable which is defined to be a vector which contains the present and all of the past information which is useful for predicting the future state of the system. The general form of the MSV model is given by

$$v_{n+1} = Av_n + Bz_{n+1}$$

where

v_n = value of state vector at time n,
z_n = value of white noise input vector at time n,
A = transition matrix,
B = input matrix.

The ARMA and MSV representations differ primarily in the manner in which they store information about the past, which is useful for forecasting the future. The ARMA representation assumes that the dimensionality of the vector used to describe the system is limited to the number of variables being observed, but it permits the number of historical values of this vector to be made as large as necessary for retaining past information which could be useful for forecasting the future. In contrast the MSV representation assumes that only the most recent value of the state vector can be included in the model, but that the dimensionality of the state vector can be made as large as necessary for capturing the kinematic characteristics of the system which are useful for forecasting its future behavior.

The Pandit and the Akaike methods also utilize different processes for determining the most appropriate model structure. The Pandit method consists of fitting successively higher order ARMA (n, n-1) models until the following F-test indicates that the reduction in the sum of squares, which results from fitting the higher order model, is not statistically significant

$$F = \frac{(1-\Lambda^{1/2})/ps}{\Lambda^{1/2}/(km-2\lambda)} \tag{2}$$

with

$$k = N - r - \frac{p-s+1}{2}$$

$$m = \left(\frac{p^2s^2-4}{p^2+s^2-5}\right)^{1/2}$$

$$\lambda = \frac{ps-2}{4}$$

$$\Lambda = \frac{|A_0|}{|A_1|}$$

where

A_0 = sum of squares and cross products matrix of model with more parameters
A_1 = sum of squares and cross products matrix of model with less parameters
p = number of series,
N = number of (vector) observations,
r = p x number of matrix parameters in model with more parameters,
s = p x difference in the numbers of matrix parameters.

F, as defined above, has an F-distribution with ps and $(km-2\lambda)$ degrees of freedom.

In contrast Akaike utilizes a two phased process for determining the most appropriate dimensionality for the state vector. The first phase consists of finding the most appropriate order autoregressive model for representing the data series of interest which is assumed to be a vector valued process of dimension r. This is accomplished by fitting successively higher order autoregressive models to the observed data series denoted by y(t)

$$\underline{y}(t) = A_1\underline{y}(t-1) + A_2\underline{y}(t-2) + \dots + A_m\underline{y}(t-m) + u(t)$$

where u(t) is a vector white noise sequence. This process is stopped when the following information criterion is minimized

$$AIC = N \log (|\hat{\underline{C}}|) + 2mr^2$$

where N represents number of observations in the data series, m is the order of the autoregressive model, r represents the dimensionality of the vector valued process which has been observed, $|\hat{\underline{C}}|$ is the r dimensional covariance matrix of the random variable $\underline{y}(t)$.

The second phase of the process for determining the most appropriate dimensionality of the state vector consists of performing a sequential canonical correlation analysis between the set of present and past values of the process:

$$\underline{y}^p = \{y(t)', y(t-1)', \dots , y(t-m)'\}$$

and the set of present and future values:

$$\underline{y}^f = \{y(t)', y(t+1)', \dots , y(t+n)'\} .$$

The value of n, which represents the number of steps into the future, is successively increased until the following information criterion is minimized

$$DIC(q) = AIC(q) - AIC(t)$$

where q represents the number of non-zero canonical correlation coefficients, and t is equal to nr. AIC(q) is defined as:

$$AIC(q) = N \log \prod_{i=1}^{q} (1-c_i{}^2) + 2F(q)$$

where c_i represents the i-th largest sample correlation coefficient and F(q) represents the number of free parameters in the canonical correlation model.

3. MODELING RESULTS

Both the Pandit approach and the Akaike approach were used to develop both univariate and multivariate time series models for NO_2 and SO_2 data collected in Chicago by the Illinois Environmental Protection Agency. The NO_2 data represents the average of the readings obtained from four air quality monitoring sites, while the SO_2 data represents the average obtained from six monitoring sites. Both the NO_2 and the SO_2 data were obtained according to a six day sampling schedule specified by the U.S. Environmental Protection Agency. Plots of the SO_2 and NO_2 data which span the period from January 6, 1975 to December 27, 1977 are shown in Figures 1 and 2 respectively.

The Pandit system when applied to the NO_2 data resulted in an ARMA (4,3) model of the form

$$\underset{}{(1-}\underset{(.41)}{1.46B} - \underset{(.01)}{.09B^2} + \underset{(.47)}{.79B^3} - \underset{(.09)}{.21B^4})\, N_t = (1-\underset{(.42)}{1.32B} - \underset{(.62)}{.18B^2} + \underset{(.27)}{.62B^3})\, a_{t1}$$

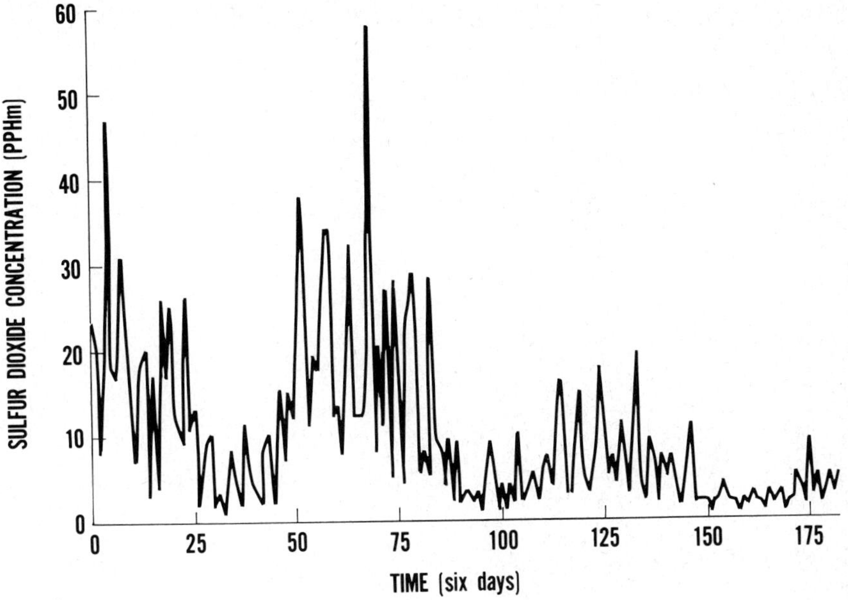

Figure 1—Plot of Sulfur dioxide concentration

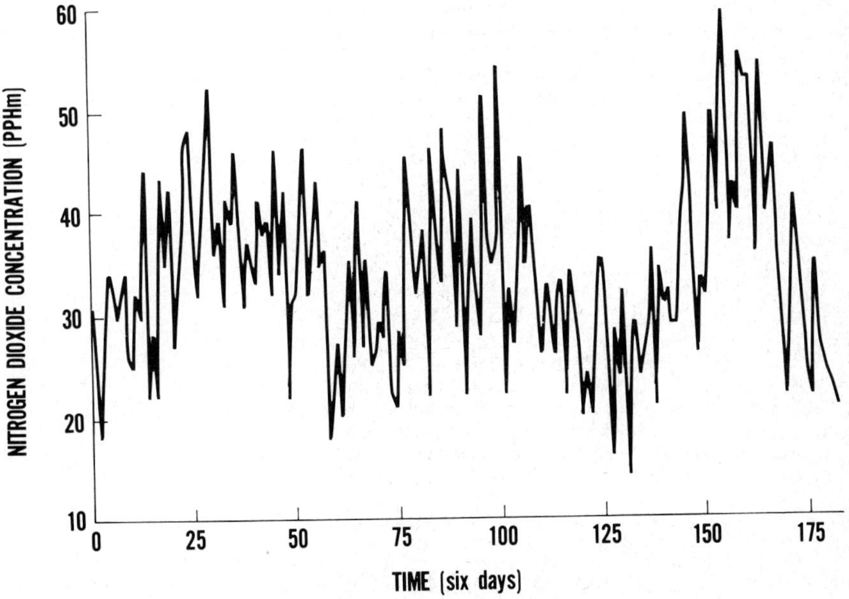

Figure 2—Plot of nitrogen dioxide concentration

where N_t represents the value of NO_2 at time t and a_t represents the random shock for time t and the standard errors are shown in brackets below the parameter estimates.

Application of the Akaike system to the NO_2 data resulted in the following MSV representation:

$$\begin{bmatrix} N_t \\ N_{t+1/t} \end{bmatrix} = \begin{bmatrix} \underset{(na)}{0} & \underset{(na)}{1} \\ \underset{(.109)}{-.034} & \underset{(.149)}{.954} \end{bmatrix} \begin{bmatrix} N_{t-1} \\ N_{t/t-1} \end{bmatrix} + \begin{bmatrix} \underset{(na)}{1} \\ \underset{(.074)}{.247} \end{bmatrix} \begin{bmatrix} E_t \end{bmatrix}$$

where N_t represents the observed value of NO_2 at time t, $N_{t+1/t}$ represents the one step ahead forecast of NO_2 at time t and E_t represents the random shock for time t.

Application of the Pandit system to the SO_2 data resulted in the following ARMA (5,4) model:

$$(1-\underset{(.74)}{.975}B + \underset{(.10)}{.259}B^2 + \underset{(.94)}{.359}B^3 - \underset{(.41)}{.642}B^4 + \underset{(.02)}{.119}B^5)S_t = (1-\underset{(.07)}{.805}B + \underset{(.24)}{.263}B^2$$

$$+ \underset{(.37)}{.327}B^3 - \underset{(.06)}{.284}B^4)a_{t2}$$

where S_t represents the value of SO_2 at time t and a_{t2} represents the random shock for time t .

Application of the Akaike system to the SO_2 data revealed the need for a state vector with four dimensions. The first element was the observed value of SO_2 at time t, while the last three elements represented the one, two and three step ahead forecasts of SO_2 at time t. However, the system did not produce an adequate model due to the fact that process, for estimating the parameters of the transition and input matrices, using the covariance matrix failed to converge.

When both NO_2 and SO_2 were considered simultaneously the Pandit system resulted in a vector ARMAV (4,3) model of the form

$$\begin{bmatrix} S_t \\ N_t \end{bmatrix} = \begin{bmatrix} -.876 & -.088 \\ .178 & .175 \end{bmatrix} \begin{bmatrix} S_{t-1} \\ N_{t-1} \end{bmatrix} + \begin{bmatrix} .214 & -.233 \\ .193 & -.365 \end{bmatrix} \begin{bmatrix} S_{t-2} \\ N_{t-2} \end{bmatrix}$$

$$+ \begin{bmatrix} +.218 & +.313 \\ -.184 & -.657 \end{bmatrix} \begin{bmatrix} S_{t-3} \\ N_{t-3} \end{bmatrix} + \begin{bmatrix} -.389 & .069 \\ -.156 & .017 \end{bmatrix} \begin{bmatrix} S_{t-4} \\ N_{t-4} \end{bmatrix}$$

$$+ \begin{bmatrix} .729 & .099 \\ -.080 & -.450 \end{bmatrix} \begin{bmatrix} a_{1t-1} \\ a_{2t-1} \end{bmatrix} + \begin{bmatrix} -.291 & .340 \\ -.096 & .069 \end{bmatrix} \begin{bmatrix} a_{1t-2} \\ a_{2t-2} \end{bmatrix}$$

$$+ \begin{bmatrix} -.111 & -.270 \\ .180 & .633 \end{bmatrix} \begin{bmatrix} a_{1t-3} \\ a_{2t-3} \end{bmatrix} + \begin{bmatrix} a_{1t} \\ a_{2t} \end{bmatrix}$$

where a_{1t} and a_{2t} represent respectively the random shocks to SO_2 and NO_2 time series. All coefficients in the above ARMAV (4,3) model are significant at the 0.05 level.

When Akaike's system was used to analyze both the SO_2 and NO_2 data the following Markovian state vector representation resulted:

$$\begin{bmatrix} S_t \\ N_t \\ S_{t+1/t} \end{bmatrix} = \begin{bmatrix} \underset{(na)}{0} & \underset{(na)}{0} & \underset{(na)}{1} \\ \underset{(.126)}{-.039} & \underset{(.088)}{.350} & \underset{(.227)}{-.267} \\ \underset{(.095)}{-.068} & \underset{(.059)}{-.056} & \underset{(.138)}{1.048} \end{bmatrix} \begin{bmatrix} S_{t-1} \\ N_{t-1} \\ S_{t/t-1} \end{bmatrix} + \begin{bmatrix} \underset{(na)}{1} & \underset{(na)}{0} \\ \underset{(na)}{0} & \underset{(na)}{1} \\ \underset{(.076)}{.163} & \underset{(.076)}{-.026} \end{bmatrix} \begin{bmatrix} ES_t \\ EN_t \end{bmatrix}$$

where $S_{t+1/t}$ represents the one step ahead forecast of SO_2 at time t, and ES_t and EN_t represent respectively the random shocks to the SO_2 and NO_2 series at time t. In the above the standard errors are shown below the corresponding estimates.

4. ANALYSIS OF FITTED MODELS

The eigenvalues for the ARMA and the MSV models fitted to the NO_2 and SO_2 provide a means for analyzing the stability characteristics of the fitted models. The eigenvalues, λ's, are the characteristic roots of the ARMAV and MSV models are obtained by solving their characteristic equations. The general form of the characteristic equation is given by

$$A - \lambda I = 0.$$

The matrix A for the ARMAV model is given by

$$A = \begin{bmatrix} \tilde{\phi}_1 & \tilde{\phi}_2 & \tilde{\phi}_3 & \tilde{\phi}_4 \\ I & 0 & 0 & 0 \\ 0 & I & 0 & 0 \\ 0 & 0 & I & 0 \end{bmatrix}$$

where all elements in this matrix are themselves 2 by 2 matrices. Thus the A for the ARMAV is an 8 by 8 matrix which has 8 eigenvalues. For the MSV model the A matrix is given by

$$A = \begin{bmatrix} \tilde{\phi}_1 & \tilde{\phi}_2 \\ I & 0 \end{bmatrix}$$

Since each of the elements of this matrix is a 2 by 2 submatrix the A matrix for the MSV is a 4 by 4 matrix which has 4 eigenvalues.

The results of the eigenvalue calculations for the fitted models are given by Tables I, II and III. The eigenvalue results in these tables show that the ARMA models contain complex conjugate pairs of eigenvalues which indicate that the data series have cyclic effects. These periods are seemingly obvious from a visual inspection of the SO_2 and NO_2 data. On the other hand the eigenvalues of the MSV models are real, which indicate that the data series do not contain any cyclic effects.

A comparison of the eigenvalue structures for the ARMA and the MSV models shown in Tables I, II and III revealed that the models produced by the Pandit procedure were more complex than those produced by the Akaike procedure.

TABLE I

EIGENVALUES FOR UNIVARIATE MODEL FOR NO_2

ARMA	MSV
.926 ± .163i	.988
.338	-.034
-.722	

TABLE II

EIGENVALUES FOR UNIVARIATE MODEL FOR SO_2

ARMA	MSV
.365 ± .838i	failed to converge
.993	
.020	
-.569	
-.765	

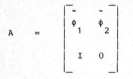

TABLE III

EIGENVALUES FOR BIVARIATE SO_2 and NO_2

ARMA	MSV
.351 ± .798i	.945
-.594 ± .621i	.468
.930 ± .013i	.070
-.659	-.085
-.013	

This might suggest that the Akaike procedure weights parsimony more heavily than does the Pandit procedure. However, an overemphasis on parsimony might result in a model which does not adequately represent cyclic phenomena. On the other hand an overemphasis on complexity might create difficulties in explaining and predicting the underlying physical phenomena.

5. CONCLUSIONS

This paper used the Pandit and Akaike approaches to fit both univariate and bivariate time series models to NO_2 and SO_2 data. Eigenvalues were calculated for each of these models for the purpose of providing a means for comparing the dynamic properties of the fitted ARMA and MSV models. These comparisons indicated that major structural differences exist between the two alternative model representations. These differences were unexpected since Akaike has shown that the ARMA and MSV models are theoretically equivalent representations. This could suggest that the procedures used by Pandit and Akaike to identify a model structure are not equivalent. At present this is an open question which is being investigated further.

This paper suggests that automatic model building procedures could breakdown. This potential for breakdown suggests that model adequacy tests should not be ignored. Further work using simulation will be done to gain a better understanding of the situations which could cause either the Akaike or the Pandit approach to breakdown.

6. REFERENCES

AKAIKE, H. (1974). Markovian Representation of Stochastic Processes and its Application to the Analysis of Autoregressive Moving Average Processes, Annals of the Institute of Statistical Mathematics, Vol. 26, 363-387.

AKAIKE, H. (1976). Canonical Correlation of Time Series and the Use of an Information Criterion. In System Identification: Advances and Case Studies, Eds: R. K. Mehra and D. G. Lainiotis, Academic Press, New York, 27-96.

HANNAN, E. J. (1976). The Identification and Parametrization of ARMAX and Statespace Forms, Econometrica, Vol. 44, 713-722.

HILL, G. W. and WOODWORTH, D. (1980). Automatic Box-Jenkins Forecasting, Journal of the Operational Research Society, Vol. 31, 413-422.

KANG, C. A., BEDWORTH, D. B. and ROLLIER, D. A. (1982). Automatic Identification of Autoregressive Integrated Moving Average Time Series, IIE Transactions, Vol. 14, 156-166.

MERZ, P. H., PAINTER, L. J. and RYASON, P. R. (1972). Aerometric Data Analyzis and Forecast and an Atmospheric Smog Diagram, Atmospheric Environment, Vol. 6, 319-342.

PANDIT, S. M. and WU, S. M. (1977). Modeling and Analysis of Closed-Loop Systems from Operating Data, Technometrics, Vol. 19, 477-485.

PANDIT, S. M. (1980). Data Dependent Systems and Exponential Smoothing. In Analyzing Time Series, Ed: O.D. Anderson, North-Holland, Amsterdam, 217-238.

REILLY, D. P. (1981). Experiences with an Automatic Box-Jenkins Modelling Algorithm. In Time Series Analysis, Eds: O. D. Anderson and M. R. Perryman, North-Holland, Amsterdam, 493-508.

TIME SERIES ANALYSIS: Theory and Practice 7
O.D. Anderson (editor)

STATE VECTOR TIME SERIES ANALYSIS OF QUEUEING NETWORK SIMULATION MODELS

W. R. Terry, Department of Industrial Engineering, The University of Toledo, Toledo, Ohio, USA

K. S. Kumar, Department of Electrical Engineering, The University of Toledo, Toledo, Ohio, USA

Most queueing network problems are not analytically tractable and simulation is frequently used to study their steady state behavior. Unfortunately simulation output reports do not provide the information necessary for analyzing the transient behavior of such networks. This paper presents a method for using state vector time series analysis to analyze the results of such simulation models to obtain a quantitative model of the transient behavior of a queueing network.

1. INTRODUCTION

Many queueing network (QN) problems are too complex to be solved by analytical methods. In such cases simulation is the only method which can be used to obtain a solution. Unfortunately simulation output reports do not provide the information necessary for analyzing a network's transient behavior.

The objective of this paper is to develop a systematic procedure for analyzing both the transient and the equilibrium behavior of a QN. This procedure consists of four phases. The first calls for developing a discrete event simulation model of the QN. The second treats this model as a sampled data system in which the size of each queue in the network is observed at equally spaced intervals of time. In the third phase the work of Akaike (1974 a & b, 1976) is utilized to develop a Markovian state vector representation of the sampled data system. The fourth phase involves finding the eigenvalues and eigenvectors for this Markovian state vector representation and then using these quantities to analyze the transient behavior of the system.

The remainder of this paper is organized as follows. The second section discusses alternative approaches for constructing simulation models which satisfy the time series requirement that data be available at equally spaced intervals of time. The third section briefly summarizes Akaike's method for developing a Markovian representation of a vector valued stochastic process. The fourth section describes a method for analyzing the system of stochastic difference equations which result from Akaike's Markovian representation. The fifth section presents three numerical examples which illustrates how the above methodology might be used in analyzing queueing network problems which arise in the design of computer systems.

2. STRATEGIC CONSIDERATIONS IN SIMULATION MODELING

The time series method, proposed in the next section, for analyzing simulation output assumes that data describing the state of the simulated system is available at equally spaced intervals of time. This section will discuss alternative approaches which may be used to obtain such data. It will also discuss the factors which should be considered in determining an appropriate length for such an interval.

In designing a system for obtaining such data several objectives are potentially important. One is to minimize computation cost. A second is to minimize the cost of developing the requisite software. A third is to obtain sampled data on the state of the simulated system sufficiently often so that all of the important components of the system's dynmaic behavior might be analyzed. The basic architecture of the simulation model can have a great effect on the first two objectives. One of the key architectural features is the method which the model uses to advance simulated time.

There are two fundamentally differrent approaches which can be used to advance time. One is the unit time advance method in which time is advanced in equally spaced intervals. The other approach is known as the next event time advance method. In this case it is necessary to maintain a chronological list of the various events which are supposed to occur during a simulation run. The simulation model is executed by removing the earliest event from this list, advancing time to the time of occurrence for that event, updating the status of the model to reflect the occurrence of that event and then repeating the above process until all items on the event list have been exhausted.

As a general rule the next event time advance method will require less time than the unit time advance when the state of the system being simulated can change abruptly at irregularly spaced intervals of time. However, this rule will not necessarily hold when the requirement that data on the state of the system be available at equally spaced intervals of time is imposed. In this case it will be necessary to develop software for periodically sampling the simulation output. On the other hand the unit time advance method eliminates the need for the development of this software. However, it is potentially less efficient in executing the simulation model in that it can require the model to be evaluated at times in which the state of the system has not changed.

A key consideration in using either of these methods is determining the optimal size of the interval for sampling the simulation models output. If the sampling interval is too small, then computation costs will be unnecessarily high and the information obtained unnecessarily redundant. This redundancy may result when the size of the sampling interval is much smaller than the average length of time between changes in the sizes of the various queues in the network. When this occurs the residuals from the fitted state vector model will tend to be highly autocorrelated. The presence of such autocorrelation thus indicates that the size of the sampling interval should be increased.

In the present study the size of the sampling interval was selected on an ad hoc basis. However, the problem of determining the size of the sampling interval is similar to the problem of determining the size of the increment when using numerical methods to solve a differential equation. The relevance of such methods to the problem of determining the size of the sampling interval is being explored and will be discussed further in a future paper.

3. MARKOVIAN STATE VECTOR MODELING OF SAMPLED DATA SYSTEMS

After developing an adequate simulation model of the queueing network, the next step is to determine the most appropriate system of stochastic difference equations for describing the dynamic behavior of the queueing network. The procedure for developing this system utilizes data obtained by observing the status of a simulation model of the queueing network at equally spaced intervals of time. This data is then analyzed according to a procedure developed by Akaike (1974 a & b, 1976) for determining the most appropriate Markovian state vector model for representing the dynamic behavior of a sampled data system.

Akaike's approach utilizes the concept of a state vector which is defined as a vector which contains all of the information about the past and present behavior of a system which is useful in predicting its future behavior. The general form of Akaike's Markovian state vector representation of a vector valued stochastic process X_T of dimension r is given by

$$Z_T = FZ_{T-1} + GE_t. \tag{1}$$

In this representation Z_t is an s dimensional state vector where $s \geq r$. The first r elements of Z_T are the values of X_T while the remaining r-s elements correspond to prior values of any of the elements of X_T which contain information useful for predicting the future values of X_T. E_t is a sequence of independent and identically distributed random vectors with mean vector equal to zero and covariance matrix equal to M. In the above representation E_t represents an exogoneous input to the system. F is an s by s transition matrix which determines how the value of the state variable at time t-1 influences the value at time t. G is an s by r input matrix which determines how the current value of the random input vector affects the current value of the state variable.

The theoretical basis for identifying the structure of the model in equation (1) was established by Akaike (1974a). This paper suggested that a canonical correlation analysis between the set of the present and past observations and the set of the present and future observations could be utilized to obtain a reasonable estimate of the structure for the state vector. However, in order to implement this approach it is necessary to have a procedure of determining the maximum number of past and future observations to include in the canonical correlation analysis. The theoretical basis for solving this problem is provided by Akaike (1974b). This paper recommended that the quantity

$$\begin{aligned}\text{AIC} = &- 2 \log_e \text{(maximum likelihood)} \\ &+ 2 \text{ (number of independently adjusted parameters)}\end{aligned}$$

be used as a criterion for evaluating the fit of a statistical model.

In a subsequent paper Akaike (1976) presented a systematic procedure for determining the most appropriate dimension for the state vector Z_T and for obtaining optimal estimates of the elements of the transition matrix F and the input matrix G. This procedure consists of six steps which can be summarized as follows: (1) A sequence of AR models of increasing order are fit and the order of the model with minimum AIC is used as the number of lags into the past to include in the canonical correlation analysis. (2) Canonical correlations of the past with an increasing number of steps into the future are calculated and the difference in the AIC for successive steps is used to determine the number of steps into the future to include in the state vector. (3) Preliminary estimates of the parameters of the transition and input matrices are obtained from the canonical correlation analysis. (4) The preliminary estimates of the matrix parameters are then used to obtain an infinite order AR representation which in turn is used to obtain a sample estimate of the residual covariance matrix. (5) The sample estimate of the residual covariance matrix is used to replace its theoretical counterpart in a log likelihood expression for the parameters of the transition and input matrices. (6) The non-linear equations which result from maximizing this log likelihood expression are solved by the Newton-Raphson method.

4. ANALYSIS OF FIRST ORDER MARKOVIAN STATE VECTOR MODEL

This section summarizes results from discrete linear systems analysis which could be useful for analyzing the behavior of a Markovian state vector model. In particular this section will discuss how the eigenvalues of the transition matrix F might be used to obtain a simpler, but equivalent representation of F.

It will also discuss how eigenvalues might be useful to analyze the stability and the oscillatory behavior of a QN.

The free response of the queueing network of the QN can be analyzed by assuming that the exogoneous input E_t in equation (1) is equal to zero. This assumption results in the following homogoneous system of difference equations

$$Z_t = FZ_{t-1} \tag{2}$$

The above representation may be simplified by transforming the transition matrix F into a diagonal matrix. If the eigenvalues of F are distinct, then such a transformation is given by

$$A = M^{-1}FM \tag{3}$$

where M is an s by s matrix the columns of which represent the eigenvectors of F and A is a diagonal matrix with the eigenvalues of F on the diagonal.

If the eigenvalues of F are not distinct, then pertubation techniques can be used to produce a transition matrix with distinct roots and the original matrix can be regarded as the limit of a collection of systems with distinct roots.

An equilibrium point of a system, denoted by $\overline{Z}$, is a point which has the

property that once the state vector reaches $\overline{Z}$ it remains there for all future time. An equilibrium point for a homogoneous system must satisfy the condition

$$\overline{Z} = F\overline{Z} \tag{4}$$

which is equivalent to stating that $\overline{Z}$ is an eigenvector of F with corresponding eigenvalue equal to one. Therefore, if an eigenvalue of A is equal to one, then any corresponding eigenvector will be an equilibrium point. However, if no eigenvalue of F is equal to one, then the origin, which is always an equilibrium point will be the only equilibrium point of the system.

A system which for any given initial condition tends to an equilibrium point as time increases is referred to as an asymptotically stable system. The requirment of asymptotic stability is equivalent to the requirement that all of the eigenvalues of F lie inside the unit circle of the complex plane, i.e. all of the eigenvalues of F must have absolute values less than one. If at least one eigenvalue has an absolute value greater than one, then the corresponding equilibrium point is said to be unstable since this leads to a solution that increases geometrically to infinity as time increases. Finally if no eigenvalues are outside the unit circle, but one or more is exactly on the boundary, then the corresponding equilibrium point is said to be marginally stable since the state vector neither tends to zero nor to infinity.

Each eigenvalue also defines both a characteristic growth rate and a characteristic frequency of oscillation. These characteristics can be deduced by considering the placement of the eigenvalues in the complex plane. This can be seen by expressing the eigenvalue in polar coordinates

$$\lambda = re^{i\theta} = r(\cos\theta + i\sin\theta). \tag{5}$$

In this case the characteristic response due to this eigenvalue is

$$\lambda^k = r^k e^{ik\theta} = r^k(\cos k\theta + i\sin k\theta). \tag{6}$$

Equation (6) implies the following:

(1) If λ is real and positive, then the response pattern is a geometric sequence r^k which increases for $r > 1$ and decreases for $r < 1$.

(2) If λ is real and negative, then the response pattern is an alternating geometric sequence.

(3) If λ is complex, then the real response to λ and its complex conjugate is of the form $r^k(A\sin k_\theta + B\cos k_\theta)$ which indicates that the system oscillates. The oscillation may be damped or explosive depending on whether r is less than or greater than one.

The largest eigenvalue determines the dominant mode of behavior of the system. This together with the above relationships provides a means for assessing the dynamic behavior of the queueing network. In particular the first relationship implies that if $\lambda > 1$, then at least one of the queues in the network has a tendency to increase indefinitely, whereas there is no such tendency if $0 \leq \lambda < 1$. The second and third relationships together indicated that if λ is real and negative or if it is complex, then at least one of the queues in the system behaves in a periodic fashion.

In practice it will not be enough to know that one of the queues in the network has a tendency to increase indefinitely. To redesign the system to eliminate this problem it will be necessary to identify which nodes have such a tendency. This can be accomplished by first diagonalizing the transition matrix. The diagonalized representation of F, denoted by Λ, can be obtained by performing the following basis transformation:

$$\Lambda = m^{-1} F m$$

where m is a matrix, the columns of which contain the eigenvectors corresponding to the various eigenvalues. It can be shown that

$$\Lambda = \begin{bmatrix} \lambda_1 & 0 & \cdots & 0 \\ \vdots & \lambda_2 & & \vdots \\ \vdots & & \ddots & \vdots \\ 0 & \cdots & & \lambda_\eta \end{bmatrix}$$

The process of diagonalizing the F matrix thus transforms it into a form in which there is no feedback between the various elements in the system. This makes it possible to study each of the observed system variables independently of the other. If $\lambda_i > 1$, then it follows that the queue corresponding to row i in the observation vector will tend to increase indefinitely.

5. NUMERICAL EXAMPLES

This section presents three numerical examples which illustrate how the above method for fitting a Markovian state vector model can be utilized to analyze QN problems. It also illustrates how the methods of discrete linear systems analysis can be used to analyze the stability and the oscillatory behavior of a QN.

The first QN selected was a single processor time sharing computer system with 256K memory. This system consists of two queues: a memory queue and a CPU queue. The scheduling policies for these queues are:

(1) Jobs are loaded into memory on a first fit policy, i.e. when a job arrives it is loaded into memory if the space required is less than that available, otherwise it joins the memory queue and waits until the required space is available. When space becomes available and there are two or more jobs which could fit into memory, then any of these may be selected at random.

(2) The CPU queue is operated on a round robin basis in which all jobs that are loaded into memory will automatically be entered into the CPU queue. Each job in the CPU queue will be serviced for a fixed time slice. When a job's time slice is exhausted then execution of the job is temporarily suspended and the job is placed at the rear of the CPU queue. The jobs in the CPU queue are serviced on a rotating basis.

The jobs arriving for service can be characterized in terms of the following parameters: job arrival rate, memory requirement and execution time.

The job parameters and their probability distributions are as shown in Table I.

TABLE I

Parameter	Distribution	
Job arrival rate	Exponential	30 sec mean
Job execution time	Exponential	20 sec mean
Job size	Uniform	16 K to 64 K

The CPU time slice was set at 1 second.

The above system differs from QN problems which can be solved analytically in several imporant ways. First, the CPU queue is operated on a round robin basis in which each job is processed for a fixed amount of time and then sent to the back of the queue. In contrast, QN problems which can be solved analytically assume that each node in the network can be regarded as a multi-channel queue in which each channel has an identical exponential service time distribution. Second, both the memory queue and the CPU queues in real world systems have finite waiting space. In contrast, analytically tractable QN problems assume that every node in the network has unlimited waiting space. Third, the goal is to investigate the transient behavior of the system whereas only a steady state solution can be obtained analytically.

A discrete event simulation model of the above system was constructed using GPSS. The GPSS model was used to simulate the system's behavior for 10 hours and the status of the following system variables was observed every 10 minutes: memory queue, CPU queue, and storage used. Akaike's approach was utilized to determine the appropriate order for the state vector and to estimate the optimal values of the parameters for the corresponding model. This resulted in the following difference equation model:

$$\begin{bmatrix} MQ_t \\ CPUQ_t \\ STORU_t \end{bmatrix} = \begin{bmatrix} .000031^{**} & .195432^{*} & -.004415^{*} \\ .631272^{**} & .200650 & -.003001 \\ 23.2268^{**} & 9.26369 & -0.165030 \end{bmatrix} \begin{bmatrix} MQ_{t-1} \\ CPUQ_{t-1} \\ STORU_{t-1} \end{bmatrix} + \begin{bmatrix} 1 & 0 & 0 \\ 0 & 1 & 0 \\ 0 & 0 & 1 \end{bmatrix} \begin{bmatrix} EMQ_t \\ ECPUQ_t \\ ESTORU_t \end{bmatrix} \quad (7)$$

where * denotes those parameters which are significant at the 10 percent level of significance and ** denotes those that are significant at the 5 percent level.

The above model provides a complete description of the dynamic behavior of the system. This behavior has two components. One is that which is caused by the structural component of the system, while the other is caused by a random process which impinges on the system. The structural component is represented by the F matrix, while the random component is represented by the G matrix. A system designer is concerned primarily with the structural component of behavior. Such behavior can be determined by analyzing the eigenvalues for the F matrix.

The eigenvalues of the transition matrix in (7) are $\lambda_1 = 0.229$; $\lambda_2 = -0.120$ and $\lambda_3 = -0.073$. Since none of the eigenvalues are complex it follows that the system is non-cyclic and since the absolute values of all of the eigenvalues are less than one it follows that the system is stable.

This indicates that none of the queues in the system have a tendency to build up indefinitely.

An examination of the statistically significant terms in the system of equations represented by (7) indicated that three relationships are important determinants of system behavior. First, the size of the memory queue is positively influenced by the size of the memory and CPU queues in the preceding time period and negatively influenced by the storage used in that period. Second, the CPU queue is positively influenced by the size of the memory queue in the preceding time period. Third, the amount of storage used is positively influenced by the size of the memory queue in the preceding time period. In summary the above relationships indicate that the memory queue in the preceding time period is an important determinant of all characteristics of the present state of the system.

The second example selected was a multiprogramming computer system. This system also contains queues for memory and CPU respectively. The scheduling policies for these queues are:

(1) Jobs are loaded into memory on a first fit policy as was the case in the preceeding example.

(2) The CPU queue is operated on a first come, first sereve priority basis. Once a job captures the CPU it is permitted to use it continuously without interruption until it encounters an input/output request. At this time it relinquishes the CPU and goes to the back of the CPU queue.

The characteristics of the jobs arriving for service and their input/output characteristics are summarized by the following table.

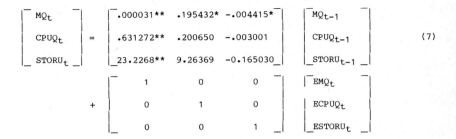

TABLE II

Parameter	Distribution	
Job arrival rate	Exponential	30 sec mean
Job execution time	Exponential	20 sec mean
Job size	Uniform	16 K to 64 K
Input/Output request	Uniform	2 to 10 sec
Input/Output service time	Uniform	1 to 5 sec

As in the previous example Akaike's approach was utilized to determine the appropriate order and parameter values for a state vector model of the output of a GPSS simulation model of the multiprogramming system. This analysis revealed that the transition matrix F contained no parameters which were significantly different from zero and that the input matrix G was an identity matrix. Such a model suggests that the behavior of such a system is not forecastable.

This result agrees with the results of analytical queueing theory. In particular the multiprogramming computer system is a QN for which the results of Jackson (1957) can be employed to decompose the network into two queues with exponential interarrival and service times which can be analyzed independently of one another. The memory less property of the exponential distribution sug-gests that the historical values of such queues do not contain any information which is useful in forecasting their future value.

The third example was selected for the purpose of investigating the effect of priority rules on the behavior of a time sharing computer system. In this case four priority classes of jobs were assumed for the memory queue and each arriving has equal chance of falling into one of the four priority classes. Jobs within each priority class are processed on a first come first serve basis with jobs in a higher class given preference over jobs in a lower class in the memory queue. However, higher priority jobs were not permitted to preempt lower quality jobs for the CPU. The CPU queue is operated on a round-robin basis.

The characteristics of jobs arriving for service are summarized by the following table.

TABLE III

Parameter	Distribution	
Job arrival rate	Exponential	30 sec mean
Job execution time	Exponential	25 sec mean
Job size	Uniform	16 K to 64 K

The following Markovian state vector model resulted from utilizing Akaike's approach to a GPSS simulation model of the above system.

$$\begin{bmatrix} MQ_t \\ CPUQ_t \\ STORU_t \end{bmatrix} = \begin{bmatrix} 0.496^{**} & 0.781^{**} & -0.018^{**} \\ 0.348^{**} & 0.096 & 0.003 \\ 8.685^{**} & 6.070 & 0.051 \end{bmatrix} \begin{bmatrix} MQ_{t-1} \\ CPUQ_{t-1} \\ STORU_{t-1} \end{bmatrix} + \begin{bmatrix} 1 & 0 & 0 \\ 0 & 1 & 0 \\ 0 & 0 & 1 \end{bmatrix} \begin{bmatrix} EMQ_t \\ ECPUQ_t \\ ESTORU_t \end{bmatrix} \quad (8)$$

where * and ** represent respectively parameters which are significant at the 10 and 5 percent levels. A comparison of the transition matrices in equations 7 and 8 reveals that they are similar both with respect to the dimensionality and the locations of the statistically significant parameters. The eigenvalues of the transition matrix in (8) are $\lambda_1 = .675$, $\lambda_2 = -.202$ and $\lambda_3 = .170$. Since none of these eigenvalues are complex it follows that the system is non cyclic. Furthermore since all of the eigenvalues are less than one in absolute value it follows that the system is stable and that none of the queues in the system have a tendency to build up indefinitely. Since similar results were obtained for a non priority system there is no reason to suspect that imposing priority rules has much of an effect on system stability. However, caution should be exercised in attempting to generalize this conclusion to other queueing network topologies and other priority systems.

6. CONCLUSION

This paper has proposed and illustrated the use of a methodology for analyzing the dynamic behavior of QN problems. The results from the numerical example indicate that using the results of the method described above could provide the system designers with a better grasp of how the various system components interact with one another throughout time. In addition, the eigenvalues of the state vector transition matrix could be used to analyze the stability and oscillatory characteristics of the system. Since analytical methods are not available for studying the dynamic behavior of queueing networks and since current statistical methods for analyzing simulation output do not provide quantitative information on dynamic behavior, it would appear that the method proposed might prove useful to engineers designing queueing networks.

7. REFERENCES

AKAIKE, H. (1974_a). Markovian Representation of Stochastic Processes and its Application to the Analysis of Autoregressive Moving Average Processes, Annals Institute of Statistical Mathematics, 363-387.

AKAIKE, H., (1974_b). A New Look at the Statistical Model Identification, IEEE Transactions on Automatic Control, AC-19, 716-722.

AKAIKE, H. (1976). Canonical Correlation Analysis of Time Series and the Use of an Information Criterion, in Advances and Case Studies in System Identification. Eds: R. K. Mehra and D. G. Lainiotis, Academic Press, New York.

JACKSON, J. R. (1957). Networks of Waiting Lines, Operations Research, 5, 518-521.

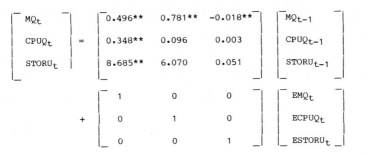

TIME SERIES ANALYSIS: Theory and Practice 7
O.D. Anderson (editor)

THE RELATIONSHIP BETWEEN PRICE AND VOLUME OF CONTRACTS ON THE WHEAT FUTURES MARKET

Thomas R. Gulledge, Jr. and James E. Willis
Department of Quantitative Business Analysis, 3190 CEBA
Louisiana State University, Baton Rouge, Louisiana 70803, USA

In recent years there has been much interest in the relationship between price and volume of contracts traded on speculative markets. This research extends earlier research by defining a specific relationship between price and volume in a single market. The resulting model is used to generate a medium term forecast over a set of test data.

1. Introduction

In recent years there has been much interest in the relationship between price and volume of contracts traded on speculative markets. This research extends earlier work by considering volume of contracts traded as an explanatory variable for the futures price of wheat on the Chicago Board of Trade. This paper follows the lead of Rogalski (1978) and examines volume as a single independent variable. In particular, a transfer function is specified, the parameters are estimated, and forecasts are generated.

2. Historical Perspective

The relationship between price and volume of contracts has been examined by many researchers. Although there is sufficient theory to suggest such a relationship (Epps and Epps, 1976; Cornell, 1981; Tauchen and Pitts, 1983), the empirical work has been conflicting. Some authors present very strong and convincing evidence of a relationship while other authors report a complete lack of relationship. When investigating stock market transactions, no apparent connection could be found between price and volume (Granger and Morgenstern, 1963; Godfrey, et. al., 1964). Labys and Granger also noted the lack of relationship between commodity prices and volume (1970, p. 114-116). Ying (1966) reported a relationship between price and volume for stock market prices. Other researchers have searched for a "statistical causality" between price and volume of contracts. For example, Rogalski (1978) uses the Haugh (1976) test of causality and concludes that there is empirical evidence to support a positive relationship between price and volume, however, the causality is not unidirectional. The supporting rationale for this particular relationship is summarized by Rogalski.

> The rationale is that given an initial equilibrium position and assuming a net increase (decrease) in demand for a security, the market will eventually clear equilibrium price. During the process of adjustment, transactions are constantly occurring as a function of demand. Under these conditions one would expect a rise in the volume of transactions with both a rise or fall in the price level.

The structure of the stock market is different from that of the commodity market, but in this research the price/volume relationship is examined once again because of the conflicting results of previous research.

The scope of this research is two-fold. First, the price/volume relationship is examined in a specific market, wheat futures on the Chicago Board of Trade.

Second, the relationship is modeled by defining an explicit functional form that relates price to volume. The purpose of the research is to develop a model which will allow one to predict the direction in which the futures price of wheat will move based on the current and past price and volume data. This information would be important to a trader's decision to hold or sell futures during a multi-month horizon.

3. The Data

The monthly midrange of the closing price (FP) of the nearest wheat future on the Chicago Board of Trade is selected as the price variable for the analysis. By concentrating on the nearest future, the delivery date is not a consideration. Total monthly volume (V) of contracts (millions of bushels) traded on the Chicago Board of Trade is the selected input variable. Both series contain 192 observations for the years 1964-1979. The recorded data was taken from the Statistical Annual of the Chicago Board of Trade.

A first examination of the series indicates that they are non-stationary in both mean and variance. Stationarity is attained by transforming the data. In this case, first differences of the natural logarithms are used. Sequence plots of the logarithms of these series are presented in Figures 1 and 2.

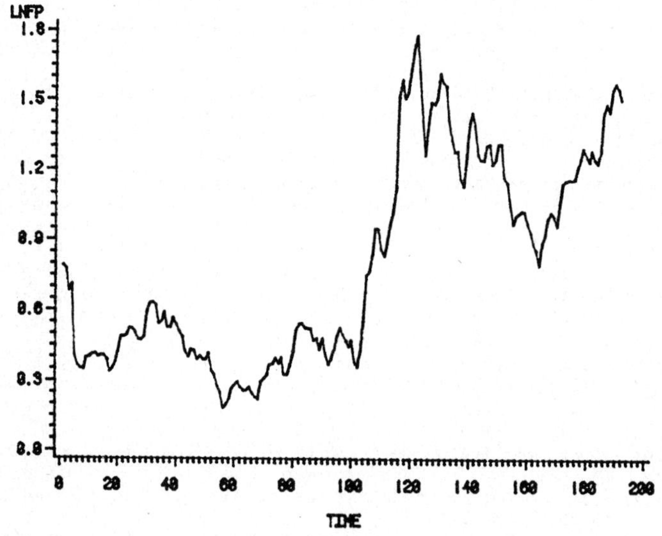

Figure 1. The natural logarithm of futures price.

One characteristic of the data requires special attention. There is a large increase in the level of the price series beginning at observation 103. This date coincides with the announcement of a large grain sale agreement between the United States and the Soviet Union (Russian Grain Transactions, 1973). This event was considered as an intervention, and an additional variable (X) was added to the analysis. The intervention variable was given a value of one for those observations greater than 102 and a value of zero elsewhere. This indicates that there was a change in the level of the series after the grain

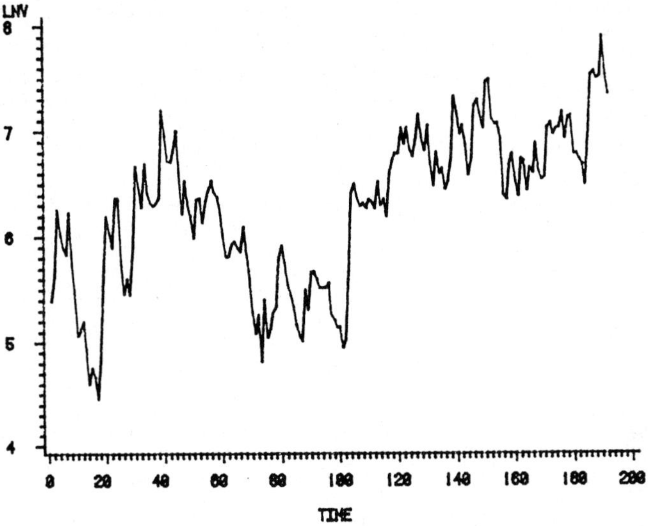

Figure 2. The natural logarithm of volume.

sale agreement. The initial change in level could be attributed to the grain transaction, but the change in mean over the remainder of the sample period requires additional explanation. This period (1972-79) was in general a period of grain shortages. This, coupled with inflation, is probably the reason the price did not return to pre-intervention levels.

4. Model Identification

The sample cross-correlation function between price and volume is presented in Figure 3. The following univariate model for volume was used for the prewhitening prior to constructing Figure 3:

$$\underset{(.072)}{(1 + .161B^2)}\underset{(.072)}{(1 + .234B^6)}(1 - B)\ \ln V_t = a_t. \tag{4.1}$$

Also, the dashed lines in Figure 3 represent 95% confidence intervals under the null hypothesis $\rho_{xy} = 0$.

After analysis of the sample cross-correlation function, several models were tentatively identified. The various models proved to be roughly comparable in terms of diagnostic tests and forecast function form; therefore the one with the simplest form was selected. This parsimoneous model had r=b=0 and s=1 for the transfer function operator orders. See Box and Jenkins (1976) for details of this identification procedure. The example of Dooley and Reilly (1982) was used as a reference for identifying the structure on the intervention component. The model is stated as

$$(1 - B)\ln FP_t = (\omega_{10} - \omega_{11}B)(1 - B)\ln V_t + \frac{(1 - B)}{(1 - \delta_{21}B)}\omega_{20}X_t + \eta_t \tag{4.2}$$

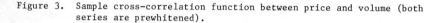

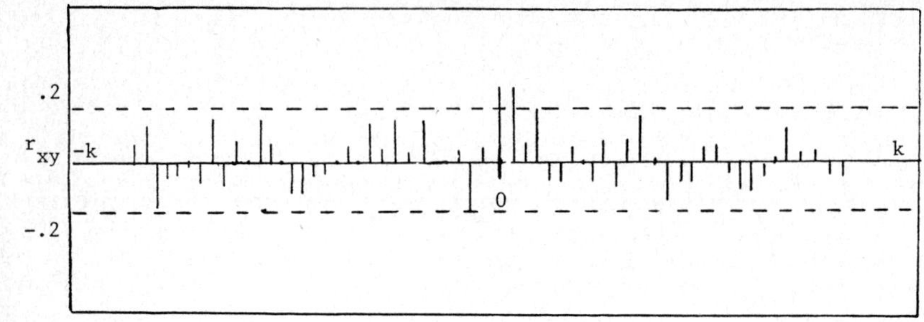

Figure 3. Sample cross-correlation function between price and volume (both series are prewhitened).

where

$$\eta_t = \frac{(1 - \theta_{14}B^{14})}{(1 - \phi_1 B)} \varepsilon_t. \tag{4.3}$$

No economic meaning could be assigned to the lag 14 noise model parameter, but it was retained in the model since there was no evidence of "abnormal" observations separated by 14 months. The estimated parameters along with the goodness-of-fit statistics are presented in Table 1.

Parameter	Estimated Value	Standard Error	$\|t\|$
ω_{10}	.0487	.0165	2.95
ω_{11}	-.0332	.0164	-2.02
ω_{20}	.0967	.0430	2.25
δ_{21}	.9175	.0570	16.10
ϕ_1	.2704	.0715	3.78
θ_{14}	-.2089	.0739	-2.82

Lag	χ^2	df	p-value
6	4.93	4	.294
12	15.61	10	.111
18	21.89	16	.147
24	24.72	22	.311

Table 1. Parameter estimates and goodness-of-fit statistics for the transfer function between price and volume.

An analysis of the residual series does not suggest a better model specification. This was confirmed by estimating several over-parameterized models. The end result was always a return to the model presented in Table 1. As final diagnostic checks it was noted that the residuals are not cross-correlated with the prewhitened input series, and the residuals are approximately normally distributed.

5. Forecasting

To evaluate the predictive ability of the model, forecasts are made by month for the year 1980. Since the model is estimated using the years 1964 through 1979, the forecasts for 1980 are a legitimate test of predictive capability. It is not critical that a commodity price forecasting model be able to predict every up and down in the market. If a trader changed position on every reversal of price direction, transactions cost would be prohibitive. For our purposes, it is the trend in the price movement that is important. The assumption is that the forecast origin is time period 192 (the last month of 1979), and the current and past values of price and volume are known with certainty. From this origin a series of forecast values are generated for comparison with the actual prices for 1980.

The forecast values, actual values, and 95% confidence intervals are presented in Table 2. A plot of the forecast is presented in Figure 4. The values are left as logarithms to preserve the symmetric interval. Since it is the direction of predicted price change that determines successful trading, it does not matter if the forecast is in terms of logs or levels.

			95% Confidence Limits	
Date	Forecast	Actual	Lower	Upper
January	1.497	1.475	1.370	1.623
February	1.490	1.521	1.283	1.697
March	1.485	1.437	1.216	1.754
April	1.487	1.393	1.168	1.807
May	1.481	1.430	1.117	1.844
June	1.487	1.420	1.084	1.891
July	1.493	1.502	1.054	1.932
August	1.514	1.512	1.043	1.986
September	1.518	1.602	1.016	2.020
October	1.512	1.642	0.981	2.043
November	1.532	1.644	0.974	2.091
December	1.531	1.588	0.947	2.116

Table 2. Actual and predicted values from the transfer function model that is presented in Table 1. Predicted values from equation (4.1) are used for the input series.

In terms of this test, the important characteristic of the forecast is the general upward trend. At least, in this case, a simple buy and hold policy would have resulted in successful trading since there is an upward trend in the actual values over the forecast horizon.

6. Conclusion

This research examines the presence of a relationship between the futures price of wheat and volume of contracts traded as suggested in the literature, and goes further to develop a forecasting model based on the relationship found.

The transfer function model specified in equations (4.2) and (4.3) is based on a full round of tentative model identification, parameter estimation, and testing for model adequacy. The ultimate test of any model is its predictive ability. For commodity prices, it is the ability to predict a general direction of change over a number of time periods that is important. The model successfully predicted the general direction of price change over the test periods. These results are certainly not conclusive, but they are indeed encouraging.

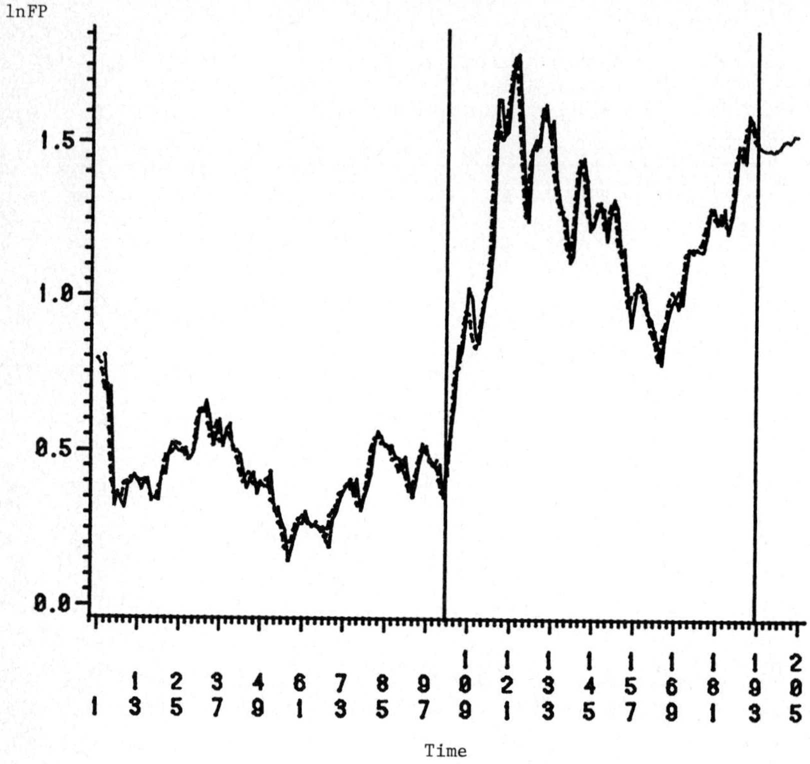

Figure 4. Actual and predicted values from the transfer function model that is presented in Table 1. (Solid line is predicted).

References

BOX, G. E. P. and JENKINS, G. M. (1976). Time Series Analysis: Forecasting and Control. Holden-Day, San Francisco.

CHICAGO BOARD OF TRADE (1964-1980). Statistical Annual of the Chicago Board of Trade. Board of Trade of the City of Chicago, Chicago.

CORNELL, BRADFORD (1981). The relationship between volume and price variability in futures makets. The Journal of Futures Markets 1, 303-16.

DOOLEY, K. and REILLY, D. P. (1982). Transfer Function modelling with a multi-step intervention variable. (Paper presented at the ORSA/TIMS Joint National Meeting held at San Diego, California, October, 1982).

EPPS, T. W. and EPPS, M. L. (1976). The stochastic dependence of security price changes and transaction volumes: implications for the mixture-of-distributions hypothesis. Econometrica 44, 305-321.

GODFREY, M. D., GRANGER, C. W. J. and MORGENSTERN, O. (1964). The random walk hypothesis of stock market behavior. Kyklos 17, 1-30.

GRANGER, C. W. J. and MORGENSTERN, O. (1963). Spectral analysis of New York stock market prices. Kyklos 16, 1-25.

HAUGH, L. D. (1976). Checking the independence of two covariance-stationary time series: a univariate residual cross-correlation approach. Journal of the American Statistical Association 71, 378-85.

LABYS, W. C. and GRANGER, C. W. J. (1970). Speculation, Hedging, and Commodity Price Forecasts. D. C. Heath, Lexington, Massachusetts.

ROGALSKI, R. J. (1978). The dependence of prices and volume. The Review of Economics and Statistics 60, 268-274.

RUSSIAN GRAIN TRANSACTIONS (1973). Hearings before the permanent subcommittee on investigations of the committee on government operations. U.S. Government Printing Office, Washington, D. C.

TAUCHEN, G. E. and PITTS, M. (1983). The price variability-volume relationship on speculative markets. Econometrica 51, 485-505.

YING, C. C. (1966). Stock market prices and volume of sales. Econometrica 34, 676-685.

TIME SERIES ANALYSIS: Theory and Practice 7
O.D. Anderson (editor)

MERGERS, STOCK PRICES, AND INDUSTRIAL PRODUCTION: AN EMPIRICAL TEST OF THE NELSON HYPOTHESIS

John B. Guerard, Jr.
Department of Finance, Lehigh University, Bethlehem, PA 18015, USA

The object of this study is to employ time series methodology to examine the hypothesized positive correlations among mergers, stock prices, and industrial production. Univariate and bivariate models are constructed using quarterly data from 1895 to 1950. A postsample period from 1951 to 1954 is reserved to test the models. The bivariate model offers significant forecasting improvement.

1. INTRODUCTION

Using quarterly data from 1895 to 1904, Nelson (1959) found a correlation coefficient of .613 between mergers and stock prices. The positive correlation between mergers and stock prices should exist because businessmen are more willing to merge their businesses when the stock prices they receive are increasing. The acquiring firms' managers are more able to pay the higher prices for the acquired firms' shares as the price-earnings multiple of the combined entity rises. Financial theory was developed by Larson and Gonedes (1969) to explain the conglomerate merger movement of the 1960s in terms of the price-earnings multiples. Larson and Gonedes hypothesized that although the price-earnings multiples of the acquiring firms exceeded those of the acquired firms, the market should drive the price-earnings multiples of the combined entity to the weighted average of the constituents' pre-combination earnings multiples. The incremental value of the combined entity would be zero if the combined entity's price-earnings multiple equaled the pre-merger weighted average multiple. The lack of merger profits is evidence of the Perfectly Competitive Acquisitions Market (PCAM) which holds that the price paid for the acquired firms is such that the acquiring firms will not profit [Mandelker (1974)]. Empirical evidence supports the larger price-earnings multiples existing for the acquiring firms than for the acquired firms in the pre-merger period [Harris, Stewart, and Carleton (1982)]. However, the acquiring firms' prices (and price-earnings multiples) in the post-merger period do not reflect merger profits and are consistant with the PCAM. A recent study by Beckenstein (1979) of large mergers occurring during the 1948-1975 period found a positive coefficient on the stock price index variable in the merger equation; however, Beckenstein dismissed the association of stock prices and mergers because of the numerically insignificant value.

The Nelson study also found a correlation coefficient of .259 between mergers and industrial production for the same time period. Mergers should increase as economic activity increases. This study tests the existence of any statistically significant correlation among mergers, stock prices, and industrial production using quarterly data from 1895 to 1950.

A univariate Box-Jenkins model is estimated for mergers from 1895 to 1950 to test if mergers follow a random walk. Forecasts from the univariate merger model are produced for the period from 1951 to 1954 to test for the appropriateness of the univariate model versus bivariate time series models employing stock prices

---

I appreciate the helpful comments of Jan de Gooijer, Richard Ashley, and Paul Shaman. Any errors remaining are the sole responsibility of the author.

and industrial production. The bivariate time series models are constructed to test if the mean square forecast errors from the univariate time series models may be reduced. Causality tests are performed in the manner developed by Ashley, Granger, and Schmalensee (1980) and Ashley (1981).

The postsample period from 1951 to 1954 is selected primarily because it was the era of the rebirth of the conglomerate merger movement of the United States. A brief history of the three major merger movements is presented to trace the development of the conglomerate merger movement and show the general pattern of mergers occurring with rising stock prices. The reader is referred to Butters, Lintner, and Cary (1951) and Nelson (1959) for more complete historic analysis of the merger movements.

2. MERGER HISTORY OF THE UNITED STATES

The first major merger movement began in 1879, with the creation of the Standard Oil Trust, and ended with the depression of 1904. During the merger movement, giant corporations were formed by the combination of numerous smaller firms. The smaller companies represented nearly all the manufacturing or refining capacity of their industries. The forty largest firms in the oil-refining industry, comprising over ninety percent of the country's refining capacity and oil pipelines for its transportation, combined to form Standard Oil. In the two decades following the rise of Standard Oil, similar horizontal mergers created single dominant firms in several industries. These dominant firms included the Cottonseed Oil Trust (1884), the Linseed Oil Trust (1885), the National Lead Trust (1887), the Distillers and Cattle Feeders (1887), and the Sugar Refineries Company (1887).

The trust form of organization was outlawed by court decisions. But merger activities continued to create "near" monopolies as the single corporation or holding company organization became dominant. The Diamond Match Company (1889), the American Tobacco Company (1890), the United States Rubber Company (1892), the General Electric Company (1892), and the United States Leather Company (1893) were created by the development of the modern corporation or holding company.

The height of the merger movement was reached in 1901 when 785 plants combined to form America's first billion-dollar firm, the United States Steel Corporation. The series of mergers creating US Steel allowed it to control 65 percent of the domestic blast furnace and finished steel output. This growth in concentration was typical of the first merger movement. The early mergers saw 78 of 92 large consolidations gain control of 50 percent of their total industry output, and 26 secure 80 percent or more.

The first major merger movement occurred during a period of rapid economic growth. The economic rationale for the large merger movement was the development of the modern corporation, with its limited liability, and the modern capital markets, which facilitated the consolidations through the absorption of the large security issues necessary to purchase firms. Nelson found the mergers were highly correlated to the period's stock prices and industry production. However, mergers were more sensitive to stock prices. The expansion of security issues allowed financiers the financial power necessary to induce independent firms to enter large consolidations. The rationale for the first merger movement was not one of trying to preserve profits despite slackening demand and greater competitive pressures. Nor was the merger movement the result of the development of the national railroad system, which reduced geographic isolation and transportation costs.

The first merger movement ended in 1904 with a depression, with whose onset coincided the Northern Securities case. Here it was held, for the first time, that anti-trust laws could be used to attack mergers leading to market dominance.

A second major merger movement stirred the country from 1916 to the depression of 1929. This merger movement was only briefly interrupted by the First World War and the recession of 1921 and 1922. The approximately 12,000 mergers of the period coincided with the stock market boom of the 1920s. Although mergers greatly affected the electric and gas utility industry, market structure was not as severely concentrated by the second movement as it was by the first merger movement. Stigler (1950) concluded that mergers during this period created oligopolies, such as Bethlehem Steel and Continental Can. Mergers, primarily vertical and conglomerate in nature as opposed to the essentially horizontal mergers of the first movement, did affect competition adversely. The conglomerate product-line extensions of the 1920s were enhanced by the high-cross elasticities of demand for the merging companies' products [Lintner (1971)]. Antitrust laws, though not seriously enforced, prevented mergers from creating a single dominant firm. Merger activity diminished with the depression of 1929 and continued to decline until the 1940s.

The third merger movement began in 1940; mergers reached a significant proportion of firms in 1946 and 1947. The merger action from 1940 to 1947, although involving 7.5 percent of all manufacturing and mining corporations and controlling five percent of the total assets of the firms in those industries, was quite small compared to the merger activities of the 1920s. The mergers of the 1940s included only one merger between companies with assets exceeding 50 million dollars and none between firms with assets surpassing 100 million dollars. The corresponding figures for the mergers of the 1920s were 14 and eight, respectively. Eleven firms acquired larger firms during the mergers of the 1920s than the largest firm acquired during the 1940s merger.

The mergers of the 1940s affected competition far less than did the two previous merger movements, with the exception of the food and textile industries. The acquisitions by the large firms during the 1940s rarely amounted to more than seven percent of the acquiring firms' 1939 assets or to as much as a quarter of the acquiring firm's growth rate from 1940 to 1947.

Approximately five billion dollars of assets were held by acquired or merged firms over the 1940-1947 period. Smaller firms were generally acquired by larger firms. Companies with assets exceeding 100 million dollars acquired, on average, firms with assets of less than two million dollars. The larger firms tended to engage in a greater number of acquisitions than smaller firms. The acquisitions by the larger, acquiring firms tended to involve more firms than did those acquired by smaller, acquiring firms. Mergers added relatively less to the existing size of the larger acquiring firms. The relatively smaller asset growth of the larger acquiring firms is in accordance with the third merger movement's generally small effects on competition and concentration. One factor contributing to the maintenance of competition was the initiative for the mergers coming from the owners of the smaller firms. Financiers and investment bankers did not play a prominent part in the early third merger movement.

The fourth merger movement, from 1951 to 1954, the forecast period of this study, was becoming a movement of conglomerate mergers. One of the 9 mergers occurring in 1951 involving acquired firms with assets exceeding 10 million dollars, 4 of these mergers were conglomerate mergers, of which 3 were product line extension combinations. The growth of the large conglomerate mergers continued throughout the forecast period. In 1954, 21 of the 37 mergers involving acquired firms with assets exceeding 10 million dollars were conglomerate in nature; 14 of the 21 conglomerate mergers were product line extensions while only 2 of the mergers were market extension combinations.

3. THE ESTIMATION OF THE UNIVARIATE TIME SERIES

The primary objectives of this study are to test the hypothesis that quarterly mergers are significantly correlated with quarterly stock prices and industrial production, and that bivariate time-series models produce lower mean square forecasting errors than does the univariate merger model.

The time series methodology of Box and Jenkins (1970), applied to quarterly series of mergers, the Dow Jones Industrial Average (DJIA), and industrial production, produces the autocorrelation and partial autocorrelation functions shown in Table 2. The series employed in this analysis are: (1) the original data merger series from the Nelson study reflecting firm disappearances during the 1895-1954 period; (2) the quarterly index of the Dow Jones average which was estimated with the arithematic mean of the three monthly indices for a given quarter; and (3) a quarterly industrial production series developed from the Nelson study. The use of the DJIA does not present a substantial change from the Beckenstein study that used the Standard and Poor's Index because of the high correlation between the two stock prices indices [Lorie and Hamilton (1973)]. The asympototic standard errors of the autocorrelations and partial autocorrelations for the functions at k lags are $1\sqrt{n}$ or .067.

TABLE 1. Range-Mean Characteristics of Sub-Period Mergers, Stock Prices, and Industrial Production.

	Mergers		DIJA		Industrial Production	
Sub-Period	Range	Mean	Range	Mean	Range	Mean
1895.1-1899.4	410.0	76.6	27.2	40.1	5.7	15.4
1900.1-1904.4	143.0	60.9	21.8	46.2	6.1	20.8
1905.1-1909.4	57.0	23.1	29.4	63.0	8.1	26.2
1910.1-1914.4	59.0	20.4	14.2	62.4	5.6	30.5
1915.1-1919.4	52.0	28.4	53.4	87.5	12.7	38.0
1920.1-1924.4	156.0	81.6	33.0	90.2	19.0	40.6
1925.1-1929.4	291.0	222.7	233.7	199.4	13.0	50.1
1930.1-1934.4	230.0	82.6	206.1	123.9	25.0	39.6
1935.1-1939.4	23.0	28.1	83.6	135.5	23.0	53.8
1940.1-1944.4	82.0	45.3	48.0	128.4	67.0	102.4
1945.1-1949.4	116.0	74.8	49.5	179.5	40.0	99.6
1950.1-1954.4	87.0	70.4	172.8	270.9	36.0	122.8

The merger, stock price, and industrial production series employ the logarithmic transformation because the plot of the range versus the mean of each sub-set of 20 observations reveals a relatively random scatter about a straight line [Jenkins (1979)]. The respective range-mean estimates are presented in Table 1.

The autocorrelation and partial autocorrelation estimates for the merger, stock price, and industrial production series are presented in Table 2.

The quarterly merger (M) series is estimated to follow an ARIMA (0,1,4) process because the partial autocorrelation function decays after the three quarter lag. The random walk hypothesis of mergers is rejected because of the significance of the moving average operator terms; the merger series does not follow the (0,1,0) random walk process. The ARIMA (0,1,4) merger model generates quarterly merger forecasts for the period from 1951 to 1954 that have a lower mean square forecasting error than the ARIMA (0,1,0) model. The ARIMA (0,1,4) process produces a residual variance that is 14.6 percent less than the random walk model; The ARIMA (0,1,4) merger model may be represented as:

$$\ln \bar{V}_{M_t} = .007 + (1-.600B + .086B^2 + .277B^3 - .104B^4)\, A_t$$

[s.e.] [.028] [.064] [.076] [.066] [.068]

The sample period residual variance of the (0,1,4) merger process from 1895 to 1950 is .769. The Box-Pierce statistic for the (0,1,4) process is 38.78 with 31 degrees of freedom, which is less than the critical chi-square value (at the 1 percent level) and we cannot reject our null hypothesis that the model residuals are random. Neither does the plot of residuals cause concern for rejecting residual normality.

TABLE 2. Autocorrelation Function Estimates

Series	Series Transformation	Lags 1	2	3	4	5	6	7	8	9	10	11	12
Mergers	$\ln V_M$	.71	.68	.68	.54	.51	.47	.36	.33	.25	.18	.18	.18
	$\ln \bar{V}_M$	-.45	-.03	.19	-.16	.02	.07	-.11	.11	-.06	-.04	-.02	.06
	$\ln \bar{V}^2_M$	-.64	.08	.19	-.18	.06	.06	-.14	.15	-.09	.01	-.02	.06
DJIA	$\ln V_{SP}$	.97	.94	.90	.86	.82	.78	.74	.71	.68	.65	.62	.60
	$\ln \bar{V}_{SP}$	.26	.18	.08	-.06	-.09	.01	-.14	-.08	-.11	-.02	-.09	-.03
	$\ln \bar{V}^2_{SP}$	-.44	.02	.02	-.08	-.08	.15	-.12	.05	-.09	.11	-.08	.02
Industrial Production	$\ln V_{IP}$	.98	.95	.93	.91	.89	.87	.85	.83	.81	.79	.77	.75
	$\ln \bar{V}_{IP}$	.31	.05	.07	-.18	-.14	-.03	-.06	-.04	.11	-.05	-.07	-.05
	$\ln \bar{V}^2_{IP}$	-.31	-.21	.20	-.20	-.06	.10	-.03	-.09	.21	-.10	-.03	-.04

Partial Autocorrelation Function

Series	Order of Difference	Lags 1	2	3	4	5	6	7	8	9	10	11	12
Mergers	$\ln V_M$	.71	.36	.26	-.12	-.03	.01	-.10	-.02	-.08	-.05	.06	.15
	$\ln \bar{V}_M$	-.45	-.30	.05	-.05	-.06	.00	-.07	.04	-.03	-.07	-.15	-.01
	$\ln \bar{V}^2_M$	-.64	-.58	-.27	-.19	-.18	-.07	-.17	-.05	-.04	-.01	-.17	-.13
DJIA	$\ln V_{SP}$	.97	-.11	.11	-.07	.04	.05	-.09	.08	-.02	.03	.02	.03
	$\ln \bar{V}_{SP}$	.26	.12	.01	-.11	-.07	.07	-.13	-.04	-.07	.06	-.08	-.02
	$\ln \bar{V}^2_{SP}$	-.44	-.22	-.08	-.13	-.22	-.01	-.10	-.07	-.19	-.03	-.08	-.11
Industrial Production	$\ln V_{IP}$	.98	-.09	.03	-.01	.02	-.01	.00	.01	-.03	.00	.02	-.02
	$\ln \bar{V}_{IP}$	.31	-.05	.08	-.25	-.01	.00	-.02	-.04	.11	-.14	-.00	-.08
	$\ln \bar{V}^2_{IP}$	-.31	-.34	.01	-.24	-.20	-.15	-.11	-.23	.03	-.11	-.03	-.24

The relatively long time frame of the merger series might lead one to question the stability of the model. The model is estimated for two sub-sets of the estimation period and the coefficients are relatively stable. The model estimate for the 1895.1-1929.2 period is:

$$\ln \bar{V}_{M_t} = .007 + (1 - .6094B + .096B^2 + .294B^3 - .124B^4)\, A_t$$

[.043] [.081] [.096] [.084] [.086]

The chi-square value is 25.83 with 25 degrees of freedom and the sample residual variance is 1.151. The model estimate for the 1928.4-1950.4 period is:

$$\ln \bar{V}_{M_t} = \underset{[.023]}{-.008} + (1 - \underset{[.100]}{.356B} - \underset{[.107]}{.064B^2} - \underset{[.102]}{.059B^3} + \underset{[.099]}{.250B^4})\, A_t$$

The chi-square value is 19.89 with 19 degrees of freedom and the sample residual variance is .108.

The univariate time series stock price (SP) model from 1895 to 1950 is represented as an ARIMA (0,1,2) process:

$$\ln \bar{V}_{SP_t} = \underset{[.009]}{.010} + (1 + \underset{[.065]}{.228B} - \underset{[.066]}{.190B^2})\, A_t$$

$$\chi^2 = 40.10 \text{ with 33 degrees of freedom, Residual variance} = .008.$$

The univariate time series industrial production (IP) model from 1895 to 1950 is represented as an ARIMA (0,1,4) process.

$$\ln \bar{V}_{IP_t} = \underset{[.066]}{.502} + (1 + \underset{[.064]}{.615B} + \underset{[.076]}{.137B^2} + \underset{[.066]}{.221B^3} - \underset{[.067]}{.160B^4})\, A_t$$

$$\chi^2 = 44.04 \text{ with 33 degrees of freedom, Residual variance} = 7.57.$$

The estimates of these series are used as pre-whitened inputs to the bivariate modelling process. The pre-whitened stock price and industrial production series are tested for significant cross-correlation with the merger series. Bivariate merger models may produce lower sample period mean square errors than the .769 error produced by the univariate model and, more importantly, reduce post-sample forecasting errors if significant correlation is found among mergers and the pre-whitened inputs.

4. THE ESTIMATION OF THE BIVARIATE TIME SERIES MODEL

The pre-whitened stock price and industrial production series are tested for significant cross-correlation with the merger series. The estimates of the cross-correlation functions must exceed twice the reciprocal of the square root of the number of observations to be statistically significant at the 5 percent level. In this study we are only concerned with unidirectional causality; feedback is not considered. The cross-correlation estimates from 1895 to 1950 are found in Table 3. The merger series is significantly correlated with the pre-whitened stock prices series contemporaneously and with a four quarter lag. There is no statistically significant correlation between mergers and pre-whitened industrial production. A bivariate time series model is built with the stock price and merger series because of the significant correlation of a bivariate time series with the merger and industrial production series.

TABLE 3. Cross-Correlation Function Estimates

Series	Series Transformation	Lags 0	1	2	3	4	5	6
Mergers and DJIA	$\ln\bar{V}_{M_t}$, $\ln\bar{V}_{SP_{t-k}}$	.112	-.002	.070	.025	-.163	.089	.052
		7	8	9	10	11	12	
		-.008	-.012	-.014	-.147	.147	-.089	

TABLE 3. Cross-Correlation Function Estimates, Continued

Series	Series Transformation	Lags 0	1	2	3	4	5	6
Mergers and Industrial Production	$\ln\bar{V}_{M_t}, \ln\bar{V}_{IP_{t-k}}$	-.054	.040	-.063	.077	-.051	.029	-.041
		7	8	9	10	11	12	
		.079	-.087	-.087	-.072	.085	-.119	

The bivariate model is estimated from 1895 to 1950 to be of the form:

$$\ln\bar{V}_{M_t} = \underset{[.023]}{-.003} + (1 - \underset{[.066]}{.642B} - \underset{[.077]}{.014B^2} + \underset{[.066]}{.207B^3} - \underset{[.067]}{.175B^4})\, A_t$$

$$+ \frac{\underset{[.534]}{2.197}}{(1 + \underset{[.163]}{.341B} + \underset{[.156]}{.133B^2} - \underset{[.167]}{.468B^3})} \ln\bar{V}_{SP_t}$$

The Box-Pierce statistic is 27.27 with 31 degrees of freedom using the bivariate model and the sample period residual variance is .698, approximately 10.2 percent below the variance produced by the univariate merger model. The critical test of the appropriateness of the bivariate model is its mean square forecasting error from 1951 to 1954 relative to the error produced by the univariate model.

The relative stability of the bivariate model is shown with the estimation of the model over the 1895.1-1929.2 and 1928.4-1950.4 subperiods. The respective estimations are:

Period | Model

1895.1-1929.2
$$\ln\bar{V}_{M_t} = \underset{[.034]}{.014} + (1 - \underset{[.085]}{.651B} + \underset{[.100]}{.004B^2} + \underset{[.100]}{.222B^3} - \underset{[.085]}{.176B^4})\, A_t$$

$$+ \frac{\underset{[.807]}{2.702}}{(1 + \underset{[.185]}{.426B} + \underset{[.189]}{.212B^2} + \underset{[.205]}{.458B^3})} \ln\bar{V}_{SP_t}$$

$$\chi^2_{19} = 14.19, \text{ Sample Residual Variance } = 1.050.$$

1928.4-1950.4
$$\ln\bar{V}_{M_t} = \underset{[.019]}{-.015} + (1 - \underset{[.101]}{.326B} - \underset{[.106]}{.142B^2} - \underset{[.108]}{.174B^3} + \underset{[.103]}{.256B^4})\, A_t$$

$$+ \frac{\underset{[.210]}{.755}}{(1 + \underset{[.148]}{.302B} - \underset{[.144]}{.181B^2} + \underset{[.162]}{.571B^3})} \ln\bar{V}_{SP_t}$$

$$\chi^2_{19} = 20.88, \text{ Sample Residual Variance } = .101.$$

The Ashley et al. (1980) tests are employed to test if the bivariate model one-step-ahead mean square forecasting error (MSFE) is less than that for the univariate model. The bivariate model MSFE is 239.8, some 35.7 percent less than the univariate model one-step-ahead MSFE which is 373.4. Following the Ashley (1981) test of causality, we define:

$DIF_t = e_t^{univariate} - e_t^{bivariate}$, $SUM_t = e_t^{univariate} + e_t^{bivariate}$

The regression that tests for causality is:

$$DIF_t = \beta_1 + \beta_2 \; [SUM_t - E(SUM_t)] \; + \sigma_t.$$

One tests the null hypothesis that β_1 and β_2 are equal to zero and there is no causality. The alternative hypothesis is that β_1 or β_2 exceeds zero. The estimated regression equation using the postsample merger data is:

$$DIF_t = \underset{[4.90]}{5.713} + \underset{[2.88]}{.080} \; [SUM_t - E(SUM)], \quad R^2 = .271.$$

One can reject the null hypothesis that β_2 is equal to zero; thus, there appears to be statistically sufficient evidence to support the hypothesis that stock prices cause mergers. Newbold (1982) has urged caution in the use of the Ashley tests because of the difficulty in dividing the data into model fitting and forecasting segments. There seems to be little agreement to the solution of the data division problem which is beyond the scope of this study.

It appears that the Nelson hypothesis of the positive correlation between mergers and stock prices is upheld using quarterly data from 1895 to 1950. Furthermore, the construction of a bivariate time series model generates a lower mean square forecasting error than the univariate merger model from 1951 to 1954. Moreover, causality testing allows one to reject the hypothesis that increasing stock prices do not causally affect mergers.

5. CONCLUSION

The inclusion of the stock prices series in a bivariate time series model reduced the standard error of forecasting when compared to the univariate model. Stock prices aid in the forecasting of mergers in the period of the rebirth of the conglomerate merger movement; moreover, the forecasting improvement is statistically significant.

TABLE 4. Merger Model Residual Variances and Mean Square Forecasting Errors

Model	Sample Residual Variance	One-Step Ahead Postsample MSFE
ARIMA (0,1,0)	.884	427.9
ARIMA (0,1,4)	.769	373.4
Bivariate Model	.698	239.8

REFERENCES

ASHLEY, R.A. (1981). Inflation and the distribution of price changes across markets: a causal analysis. Economic Inquiry 19, 650-660.

ASHLEY, R.A., GRANGER, C.W.J., and SCHMALENSEE, R.L. (1980). Advertising and aggregate comsumption: an analysis of causality. Econometrica 48, 1149-1168.

BECKENSTEIN, A.R. (1979). Merger activity and merger theories: an empirical investigation. Antitrust Bulletin 24, 105-128.

BOX, G.E.P. and JENKINS, G.M. (1970). Time Series Analysis: Forecasting and Control. Holden-Day, San Francisco.

BUTTERS, J.K., LINTNER, J., and CARY, W.L. (1951). Corporate Mergers. Division of Research, Graduate School of Business, Harvard University, Boston.

GRANGER, C.W.J. and NEWBOLD, P. (1977). Forecasting Economic Time Series. Academic Press, New York.

GORT, M. (1969). An economic disturbance theory of mergers. Quarterly Journal Economics 93, 724-742.

HARRIS, R.S., STEWART, J.F., and CARLETON, W.T. (1982). Financial characteristics of acquired firms. In Mergers and Acquisitions: Current Problems in Perspective. Eds: M. Keenan and L. White, Lexington Books. Lexington, MA, 223-242.

JENKINS, G.M. (1979). Practical experiences with modelling and forecasting time series. In Forecasting (Proceedings of the National Conference, Cambridge University, England, July 1976). Ed: O. D. Anderson, North-Holland, Amsterdam & New York, 43-166.

LARSON, K.D. and GONEDES, N.J. (1969). Business combinations: an exchange ratio determination model. Accounting Review 44, 720-728.

LINTNER, J. (1971). Expectations, mergers, and equilibrium in purely competitive markets. American Economic Review 61, 101-112.

LORIE, J.H. and HAMILTON, M.T. (1973). The Stock Market: Theories and Evidence. Richard D. Irwin, Inc., Homewood, Illinois.

MANDELKER, G. (1974). Risk and return: the case of merging firms. Journal of Financial Economics 1, 303-335.

NELSON, R. (1959). Merger Movements in American Industry, 1895-1956. Princeton University Press, Princeton.

NEWBOLD, P. (1982). Causality testing in economics. In Time Series Analysis: Theory and Practice 1 (Proceedings of the International Conference held at Valencia, Spain, June 1981). Ed: O. D. Anderson, North-Holland, Amsterdam & New York, 701-716.

STIGLER, G.J. (1950). Monopoly and oligopoly by merger. American Economic Review 40, 23-34.

TIME SERIES ANALYSIS: Theory and Practice 7
O.D. Anderson (editor)

AUTOREGRESSIVE MODELLING OF ACCOUNTING EARNINGS AND SECURITY PRICES*

Jeffrey L. Callen
Jerusalem School of Business Administration, Hebrew University of Jerusalem, and Faculty of Business, McMaster University, Hamilton, Ontario, Canada, L8S 4M4

M.W. Luke Chan and Clarence C.Y. Kwan
Faculty of Business, McMaster University, Hamilton, Ontario, Canada, L8S 4M4

In this study the relationship between reported accounting earnings and equity prices is examined using the notion of Granger causality. By relating Granger causality to the concept of market efficiency in the semi-strong form, this study provides a natural test for market efficiency without relying on the conventional residual analysis. The causal relationship is tested using Hsiao's methodology on a cross-sectional temporal data base of quarterly earnings and prices. The results indicate that prices cause earnings but not the reverse, and that equity markets are efficient in the semi-strong form.

I. INTRODUCTION

The purpose of this paper is to establish the relationship between reported accounting earnings and market prices of equity securities using the notion of Granger [1969] causality. In the ordinary usage of the term causality, earnings clearly cause prices. This is obvious because, from past earnings and other sources of information, investors can form expectations about future earnings (cashflows) and assess the associated risks. Thus, share prices can be determined by discounting expected future earnings (cashflows) at some risk-adjusted discount rates.[1] While it is straightforward to conceptualize earnings causing price, we know of no theory which advocates the reverse in the ordinary sense of causality.

However, what causality usually means is not the same as Granger causality.

---

*The authors would like to thank Cheng Hsiao, O.D. Anderson and the referees for comments.

[1]If the income smoothing literature is correct (e.g., Ronen, Sadan and Snow [1977]), the relationship running from earnings to prices is even more direct. According to this literature, management will manipulate accounting earnings to enable shareholders to make better predictions about equilibrium prices.

As explained below, the concept of Granger causality is more closely related to market efficiency than it is to the ordinary notion of causality.[2] If markets are efficient in the semi-strong form, one would expect prices to cause earnings but not the reverse in the Granger sense.[3] Only in inefficient markets, would earnings cause prices. Thus, we have a natural test for market efficiency by studying the causal relationship between earnings and prices. This actually represents an alternative approach in testing market efficiency which does not rely on the conventional residual analysis.

This paper is organized in the following manner. Section II defines Granger causality and relates the concept to market efficiency. Section III introduces the model relating prices and earnings. Section IV describes the data and the empirical results. Finally, Section V concludes the paper.

II. MARKET EFFICIENCY AND CAUSALITY

The concept of Granger causality stems from the simple idea that the future cannot cause the present or the past. To appreciate Granger causality in the context of this study, let us consider a bivariate stochastic process X_t with components $\{E_t, P_t\}$ where E_t is earnings per share and P_t price per share at time t. Let $\overline{X}_t$, $\overline{P}_t$, and $\overline{E}_t$ denote the set of past values of X_t, P_t, and E_t before time t, respectively. We also define $\sigma^2(P_t|\overline{X}_t)$ to be the minimum mean square linear prediction error of P_t given the information set X_t.

Definition: (Causality) If $\sigma^2(P_t|\overline{X}_t) < \sigma^2(P_t|\overline{X}_t - \overline{E}_t)$, (i.e., the information set which includes past earnings data yields a more accurate prediction of prices than the same information set without past earnings data -- in the mean squared error sense), then E (earnings) are said to cause P (prices). Similarly, if

[2]For a characterization of Granger causality and some reservations about this definition on philosophical grounds, see Newbold [1982].

[3]The markets are efficient in the semi-strong form if no market participant can earn excess returns from trading based on publicly available information, such as historical prices and earnings announcements. For further discussions, see Copeland and Western [1983, pp. 285-311].

$\sigma^2(E_t|\bar{X}_t) < \sigma^2(E_t|\bar{X}_t - \bar{P}_t)$, then P are said to cause E.

Definition: (Feedback) If $\sigma^2(P_t|\bar{X}_t) < \sigma^2(P_t|\bar{X}_t - \bar{E}_t)$ and $\sigma^2(E_t|\bar{X}_t) < \sigma^2(E_t|\bar{X}_t - \bar{P}_t)$ then feedback between P and E is said to occur.

The relationship between market efficiency and Granger causality is now immediate. Obviously, if markets are efficient with respect to processing publicly available earnings information, $\sigma^2(P_t|\bar{X}_t)$ should be equal to $\sigma^2(P_t|\bar{X}_t - \bar{E}_t)$ since the two information sets are the same except for the information contained in past earnings. Also, if markets are efficient, prices must have incorporated all relevant information about market expectations of the firm's future earnings (see Beaver, Lambert, and Morse [1980]). Specifically, in a rational expectations world, prices provide unbiased forecasts of future cashflows, and reported earnings are noisy, periodic realizations of these cashflows. Therefore, as long as earnings are not manipulated, current prices will convey information about <u>average</u> future earnings. Putting it somewhat differently, as long as the noise surrounding earnings is truly random, earnings will on average reflect the underlying cashflows. Thus, prices provide unbiased forecasts not only of future cashflows, but also of future earnings. One would, therefore, expect that including past prices in the information set would yield better predictions of future earnings, i.e., $\sigma^2(E_t|\bar{X}_t) < \sigma^2(E_t|\bar{X}_t - \bar{P}_t)$.

On the other hand, suppose that markets are inefficient such that current prices do not reflect all publicly available information. In that case, the earnings series would contain information not already encapsulated in current prices so that earnings would cause price, i.e., $\sigma^2(P_t|\bar{X}) < \sigma^2(P_t|\bar{X}_t - \bar{E}_t)$. Also, if the market is so inefficient that prices do not reflect much information about expectations of future earnings, then prices would not cause earnings. However, if prices reflect some but not all earnings information, then prices would still cause earnings. In such a case, because earnings cause prices, feedback would occur between earnings and prices.

Considering all this, one can test for market efficiency in the semi-strong form with respect to earnings information as follows. If prices are found to cause earnings but not the reverse, markets are efficient. If earnings cause prices but not the reverse, or if feedback is exhibited, markets cannot be considered efficient. However, if no causal relationship is observed between earnings and prices, no conclusive statement can be made with respect to market efficiency.

III. THE MODEL

It is well-known (e.g., Masani [1966]) that a regular full rank stationary stochastic process $\{E_t, P_t\}$ can be modelled, under fairly general conditions, by the autoregressive representation

$$P_t = \psi_{11}(L)P_t + \psi_{12}(L)E_t + v_t \tag{1}$$

$$E_t = \psi_{21}(L)P_t + \psi_{22}(L)E_t + u_t. \tag{2}$$

Here, L denotes the lag operator with $(L)P_t = P_{t-1}$ and $\psi_{ij}(L)$ is the lag polynomial $\sum_{k=1}^{Q} \psi_{ijk}L^k$. The $\{v_t, u_t\}$ are zero mean white noise innovations with constant variance-covariance matrix.

Causal relationships enter the model in a very natural way. If $\psi_{12}(L) \equiv 0$ (i.e., $\psi_{12k} = 0$ for all k) then it is clear from equation (1) that past earnings have no effect on predicting prices, that is earnings do not cause prices. Similarly, if $\psi_{21}(L) \equiv 0$, prices do not cause earnings. Thus, in theory, one could determine the causal relationships between earnings and price by first of all fitting equations (1) and (2) by least squares -- yielding estimates which are consistent and asymptotically normally distributed -- and then testing to see if the $\psi_{ij}(L) = 0$, $i \neq j$. This approach is problematic however, because the test of $\psi_{ij}(L) = 0$ is generally quite sensitive to the order of the lags of the $\psi_{ij}(L)$ (see Hsiao [1979a,b]). Instead, following Hsiao, we will use the data itself to determine the lag structure. Specifically, the optimal order of the lag for each $\psi_{ij}(L)$ in each of equations (1) and (2) is determined using Akaike's [1969a,b] Final Prediction Error (FPE) criterion. The FPE (of P_t) is defined to be the (asymptotic) mean squared prediction error

$$E(P_t - \hat{P}_t)^2 \tag{3}$$

where P_t is the predictor of P_t given by

$$\hat{P}_t = \hat{b} + \hat{\psi}^m_{11}(L)P_t + \hat{\psi}^n_{12}(L)E_t. \tag{4}$$

The superscripts m and n denote the order lags of $\psi_{11}(L)$ and $\psi_{12}(L)$, respectively, where m and n are bounded above by the maximum lag order investigated, say Q. $\hat{\psi}^m_{11}(L)$, $\hat{\psi}^n_{12}(L)$ and $\hat{b}$ are the least squares estimates of $\psi_{11}(L)$, $\psi_{12}(L)$, and the constant term b, respectively. Akaike estimates the Final Prediction Error by

$$FPE_P(m,n) = \left(\frac{T+m+n+1}{T-m-n-1}\right) \sum_{t=1}^{T} (P_t-\hat{P}_t)^2/T. \tag{5}$$

The first term of the product on the right hand side of equation (5) is a measure of estimation error while the second term represents a measure of modelling error. By choosing the lag structure with minimum FPE, Akaike's criterion tries to balance the bias from choosing too small a lag order against the increased variance from a higher lag order specification. More specifically, Akaike has shown that, if the dependence between ψ_{ij} and recent values of prices and earnings decreases as the length of past history increases, the FPE is comprised of two components. The first component is due to the FPE of the best linear prediction for given m and n while the second component is due to the statistical deviation of $\hat{\psi}^m_{11}(L)$ and $\hat{\psi}^n_{12}(L)$ from $\psi^m_{11}(L)$ and $\psi^n_{12}(L)$. Generally, as m and n are increased, the first component decreases whereas the second component increases for a finite length of observations of prices and earnings.

The FPE approach yields a number of distinct benefits in terms of identifying the model. First, as we have already pointed out, the data itself is used to determine the lag structure rather than presupposing some arbitrary lag order specification. Second, the FPE criterion does not constrain the lag structure of each variable to be identical i.e., in general, $m \neq n$. Third, it has been shown that the FPE criterion is equivalent to choosing the model specification on the basis of an F test with varying significance levels. (See Hsiao [1981]) In other words, rather than specifying an ad hoc significance level of 5 or 10%, the choice of whether to include a variable is determined by an explicit optimality criterion, namely, minimizing the mean square prediction error. In particular, the FPE criterion is more generous towards the inclusion of a variable than conventional tests with significance levels of 5% or 10%.

In addition to identifying the model, the FPE criterion can be used to determine causal relationships directly. Suppose we are to test if earnings cause prices. Then Hsiao [1979a] has suggested the following sequential procedure. First, the FPE criterion is used to determine the optimal order of the one dimensional autoregressive process for prices alone. Call this order q, so that the resulting FPE is $FPE_p(q,0)$. Second, fix the lag structure for prices at q and use the FPE criterion to specify equation (1). Let r be the resulting order for the lag operator $\psi_{12}(L)$ so that the FPE is $FPE_p(q,r)$. Third, holding the order of the lag operator $\psi_{12}(L)$ at r, let the order of lag operator $\psi_{11}(L)$ vary from 0 to q. Choose the order of $\psi_{11}(L)$ that gives the smallest FPE, say s (where s is not necessarily equal to q), thereby yielding $FPE_p(s,r)$. This third step is a check to see whether the lag structure for prices (which was originally of length q) is sensitive to the lag structure of the manipulated variable earnings. Finally, if $FPE_p(q,0) \leq FPE_p(s,r)$, then prices are best represented by a one dimensional autoregressive process so that earnings do not cause prices. Conversely, if $FPE_p(q,0) > FPE_p(s,r)$, then we would conclude that earnings cause prices. A similar approach would be used on equation (2) to test if prices cause earnings.

It is worth noting that this FPE procedure for testing causality directly is consistent with the Granger definition of causality. In both cases, the criterion for cauality is minimizing the mean square prediction error. Furthermore, as we have already noted, minimizing the FPE involves a tradeoff between the size and power of the test. The usual alternative approach involving standard hypothesis testing arbitrarily specifies the size (i.e., the significance level) of the test. The latter need bear no relationship to the underlying Granger criterion of minimizing the mean square prediction error.

IV. EMPIRICAL RESULTS

The model [Equations (1) and (2)] was estimated from pooled cross-sectional and time-series data. Besides attenuating the problem of multicolinearity in the time series, pooling was necessitated, in most cases, by insufficient degrees of freedom for individual firm estimation. This is not surprising considering the potential number of lagged variables in the model.[4] While additional data could have been collected for each firm, we were hard pressed to rationalize using decade old data for explaining more recent price-earnings relationships. It was felt that pooling the data was in fact the lesser evil.

The data used in the analysis are comprised of a total of 972 quarterly observations covering 27 manufacturing firms over a nine year period from the first quarter of 1969 to the fourth quarter of 1977. Each firm in the sample has a December 31 year-end. In addition, most firms used either straight-line depreciation or, in a few cases, a combination of straight-line and accelerated depreciation techniques. All firms valued inventories on a FIFO basis.

Akaike's FPE criterion defined in equation (5) was developed for a single time series and not for temporal cross-sectional data. However, assuming that the data is generated from repeated measurements of the same autoregressive vector-valued process, the original FPE can be modified quite readily.[5] The modified FPE criterion is

$$FPE_P(m,n) = \left(\frac{NT+m+n+1}{NT-m-n-1}\right)\left(\frac{\sum_{i=1}^{N}\sum_{i=1}^{T}(P_{it}-\hat{P}_{it})^2}{NT}\right) \tag{6}$$

where N denotes the number of cross-sectional units and T denotes the number of time-series observations used to obtain the least square estimates of

---

[4]The maximum lag investigated is Q = 10. Thus, if pooling is not applied, nine years of quarterly data less 20 lagged price-earnings variables (and a constant term) leaves only 16 degrees of freedom for each firm.

[5]See also Chan, Hsiao, and Keng (1982).

$$\hat{P}_{it} = \hat{b} + \hat{\psi}^{m}_{11}(L)\, P_{it} + \hat{\psi}^{n}_{12}(L)\, E_{it} \quad (i = 1,\ldots, N;\ t = 1,\ldots, T). \tag{7}$$

We took fourth quarter differences, i.e. $(1-L^4)$, in the logarithms of the earnings per share data. The rationale for taking this is provided by an extensive empirical literature (e.g., Griffin [1977], Brown and Rozeff [1979] and Foster [1977]) on the time-series properties of quarterly earnings. This literature has demonstrated quite conclusively that quarterly earnings data has a fourth-period seasonal component. Similarly, we took fourth quarter differences in the logarithms of the price data. But since there is no empirical or theoretical justification for this choice with price data, we will also present results for successive quarter differencing, i.e. $(1-L)$, in the logarithms of the price data.[6] Notice that the data transformations for both earnings and price have actually controlled the size effect in the pooling of cross-sectional and time-series data.

Table 1 lists the FPE for earnings and price per share where these variables are treated as one-dimensional autoregressive processes. P^1 denotes the price variable where successive quarter differencing was used and P^4 denotes the price variable where (seasonal) fourth quarter differencing was used. The smallest FPE (underlined) for E, P^1 and P^4 are at lags 8, 8 and 9, respectively. In Table 2, we have listed the estimated coefficients and standard errors associated with the optimum (in the minimum FPE sense) of the univariate autoregressive models.

Table 3 presents the FPE of the earnings and price variables for the bivariate processes. The term controlled variable refers to the variables on the left hand side of equations (1) and (2). These variables are controlled in the sense that their lag structure is fixed at the optimum lag determined in Table 1. That optimum lag is indicated in parentheses beside each controlled variable. The FPE

---

[6]Taking successive quarter differences in the logarithms transforms prices into quarterly returns whereas fourth quarter differencing yields annual returns. Our data was such that there were no negative earnings. In the event of losses, (gross) returns would have to be calculated directly rather than by taking log relatives.

of the controlled variable is computed by varying the order of the manipulated variable from one to ten. Q=10 is the (*a priori* specified) maximum possible lag for each variable. Column 3 in Table 2 gives the lag for the manipulated variable which yields the minimum FPE for the controlled variable. The minimum FPE is given in the last column. The estimated coefficients and their respective standard errors for these bivariate models are listed in Table 4.

Tables 1 and 3 correspond to the first two steps of the Hsiao sequential procedure for testing causality. The third step -- determining if the lag structure of the controlled variable is sensitive to the lag structure of the manipulated variable -- did not change the minimum FPE's. Therefore, Tables 1 and 3 are sufficient for determining causal relationships. First, let us consider the case where earnings is the controlled variable and P^1 the manipulated variable. Since the minimum FPE for the univariate earnings series (.0812) is greater than the minimum FPE for the bivariate process (0.808), prices cause earnings. On the other hand, when price (P^1) is the controlled variable and earnings the manipulated variable, earnings enters optimally with one lag but it is not significant in the FPE sense. That is, we obtain the same FPE for prices (P^1) using either the univariate price series or the bivariate series. Thus, earnings do not cause prices. The results are very similar when P^4 is substituted for P^1. Again, prices cause earnings since the bivarate FPE is .0804 as opposed to the univariate FPE of .0812. Also, earnings do not cause price since the minimum FPE is .0251 for both the univariate and bivariate processes. It is worth noting that despite the fact that on the basis of Akaike's FPE criterion prices cause earnings, the evidence is not overwhelming. Whether one uses P^1 or P^4 for the price variable, the FPE of the controlled earnings variable is reduced marginally by introducing the price series to the earnings series.[7] *A priori* one would expect prices to convey more information about future earnings than is indicated by the FPE criterion. Of course, this may simply reflect the fact that the market expectations about future earnings are not being realized. In other words, market participants may simply not be good forecasters of future earnings. Some preliminary evidence to this effect is provided by the Beaver, Morse and Lambert [1980] study (see their Table 4). The ability of their price-earnings forecast model to predict earnings is really not much better than a simple random walk with drift. This is so despite the fact that they predict portfolio earnings behaviour rather than earnings for individual securities. Conceptually, our work is related to the Beaver study. However, the results of the two studies are basically non-comparable for the following reasons. First, their underlying model is moving average and ours is autoregressive. Second, the test methodologies are very different from each other. Third, they utilize annual data and we have quarterly data. This point is important because the time-series of quarterly earnings is likely to provide more information about future earnings than will a time-series of annual earnings. See Lorek [1979], for example. Perhaps, one reason why prices only have a marginal impact on explaining future earnings is that so much information about future earnings is captured by the quarterly time-series structure of past earnings.

V. CONCLUSION

We have related the notion of Granger causality to market efficiency. In particular, we argued that in an efficient equity market, prices should cause earnings but not the reverse. Utilizing a cross-sectional temporal data base of quarterly earnings and prices, we were able to test for causal relationships using the Hsiao methodology. Specifically the underlying model was assumed to be bivariate autoregressive. The lag structure of the model was estimated and causal relationships determined by Akaike's Final Prediction Error criterion. We found that prices cause earnings but earnings do not cause prices, thereby confirming the (semi-strong) efficiency of equity markets. However, the extent to which prices cause earnings was marginal, suggesting that little information about future earnings is conveyed by current prices.

---

[7]Nevertheless, we can still conclude that prices cause earnings. This is similar to ordinary hypothesis testing where the critical value is 2 (at the 5% level) and the test statistic is 2.2, say, allowing us to reject the null hypothesis albeit marginally.

Table 1: FPE of a One-Dimensional Autoregressive Process Fit for E, P^1, and P^4

The Order of Lags	FPE of E	FPE of p^1	FPE of p^4
1	.1129	.0200	.0339
2	.1131	.0200	.0340
3	.1115	.0200	.0338
4	.0945	.0200	.0308
5	.0854	.0201	.0295
6	.0857	.0202	.0294
7	.0843	.0202	.0288
8	.0812	.0196	.0275
9	.0814	.0196	.0251
10	.0814	.0197	.0252

Table 2. Optimum (Minimum FPE) Univariate Autoregressive Models for E, P^1 and P^4 (standard errors in parenthesis)

Independent Variables	Dependent Variables E_t		P^1_t		P^4_t
Constant	0.1215 (0.0160)	Constant	-0.0033 (0.0059)	Constant	0.0054 (0.0067)
E_{t-1}	0.4348 (0.0420)	P^1_{t-1}	-0.1296 (0.0397)	P^4_{t-1}	0.7792 (0.0386)
E_{t-2}	0.1114 (0.0454)	P^1_{t-2}	-0.0117 (0.0401)	P^4_{t-2}	0.0647 (0.0474)
E_{t-3}	0.0537 (0.0459)	P^1_{t-3}	0.0876 (0.0404)	P^4_{t-3}	0.1238 (0.0478)
E_{t-4}	-0.6609 (0.0448)	P^1_{t-4}	0.0442 (0.0407)	P^4_{t-4}	-0.6261 (0.0486)
E_{t-5}	0.3256 (0.0506)	P^1_{t-5}	-0.0082 (0.0407)	P^4_{t-5}	0.4064 (0.0504)
E_{t-6}	0.1148 (0.0527)	P^1_{t-6}	-0.0124 (0.0399)	P^4_{t-6}	0.0133 (0.0478)
E_{t-7}	-0.0212 (0.0513)	P^1_{t-7}	-0.0328 (0.0388)	P^4_{t-7}	-0.0443 (0.0475)
E_{t-8}	-0.2214 (0.0448)	P^1_{t-8}	-0.1724 (0.0384)	P^4_{t-8}	-0.4303 (0.0465)
				P^4_{t-9}	0.2927 (0.0360)
DW = 1.9459 Log-likelihood = -88.1535		DW = 2.0070 Log-likelihood = 333.882		DW = 1.9647 Log-likelihood = 261.484	

Table 3. The Optimum Lags of the Manipulated Variable and the FPE of the Controlled Variable

Controlled Variable*	Manipulated Variable	The Optimum Lag of Manipulated Variable	FPE
E(8)	P^1	4	.0808
P^1(8)	E	1	.0196
E(8)	P^4	2	.0804
P^4(9)	E	1	.0251

*The numbers in parentheses indicate the order of the autoregressive operator in the controlled variable.

Table 4. Optimum (Minimum FPE) Bivariate Autoregressive Models for Earnings and Prices (standard error in parenthesis)

Independent Variables	Dependent Variables E_t		E_t		P^1_t		P^4_t
Constant	0.1164	Constant	0.1178	Constant	0.0069	Constant	0.0085
	(0.0160)		(0.0159)		(0.0063)		(0.0072)
E_{t-1}	0.4282	E_{t-1}	0.4244	P^1_{t-1}	-0.1250	P^4_{t-1}	0.7823
	(0.0422)		(0.0420)		(0.0397)		(0.0386)
E_{t-2}	0.1084	E_{t-2}	0.1056	P^1_{t-2}	-0.0113	P^4_{t-2}	0.0626
	(0.0456)		(0.0454)		(0.0400)		(0.0474)
E_{t-3}	0.0541	E_{t-3}	0.0571	P^1_{t-3}	0.0905	P^4_{t-3}	0.1254
	(0.0461)		(0.0456)		(0.0404)		(0.0478)
E_{t-4}	-0.6640	E_{t-4}	-0.6644	P^1_{t-4}	0.0507	P^4_{t-4}	-0.6230
	(0.0452)		(0.0445)		(0.0408)		(0.0487)
E_{t-5}	0.3311	E_{t-5}	0.3211	P^1_{t-5}	-0.0049	P^4_{t-5}	0.4066
	(0.0504)		(0.0505)		(0.0408)		(0.0504)
E_{t-6}	0.1179	E_{t-6}	0.1232	P^1_{t-6}	-0.0096	P^4_{t-6}	0.0115
	(0.0525)		(0.0524)		(0.0399)		(0.0478)
E_{t-7}	-0.0149	E_{t-7}	-0.0150	P^1_{t-7}	-0.0310	P^4_{t-7}	0.0444
	(0.0511)		(0.0510)		(0.0388)		(0.0475)
E_{t-8}	-0.2236	E_{t-8}	-0.2215	P^1_{t-8}	-0.1716	P^4_{t-8}	-0.4277
	(0.0447)		(0.0446)		(0.0384)		(0.0465)
P^1_{t-1}	-0.0363	E_{t-1}	0.0185	E_{t-1}	-0.0250	P^4_{t-9}	0.2704
	(0.0810)		(0.0631)		(0.0448)		(0.0360)
P^1_{t-2}	0.1237	E_{t-2}	0.1198			E_{t-1}	-0.0227
	(0.0814)		(0.0628)				(0.0190)

P^{1}_{t-3}	0.1830 (0.0819)			
P^{1}_{t-4}	0.1738 (0.0830)			
	DW = 1.9578 Log-likelihood = -82.714	DW = 1.9520 Log-likelihood = -83.145	DW = 2.0092 Log-likelihood = 335.007	DW = 1.9663 Log-likelihood = 262.205

REFERENCES

AKAIKE, H. (1969a). Statistical Predictor Identification, *Annals of Institute of Statistical Mathematics* 21, 203-217.

AKAIKE, H. (1969b). Fitting Autoregressions for Prediction, *Annals of Institute of Statistical Mathematics* 21, 243-247.

BEAVER, W., LAMBERT, R., and MORSE, D. (1980). The Information Content of Security Prices, *Journal of Accounting and Economics* 2, 3-28.

BROWN, L., and ROZEFF, M. (1979). Univariate Time-Series Models of Quarterly Accounting Earnings per Share: A Proposed Model, *Journal of Accounting Research* 17, 179-189.

CHAN, M.W.L., HSIAO, C., and KENG, C.W.K. (1982). Defense Expenditure and Economic Growth in Developing Countries: A Temporal Cross-sectional Analysis, *Applied Time Series Analysis* edited by O.D. Anderson and M.R. Perryman (North-Holland, Amsterdam), 53-64.

COPELAND, T.E., and WESTON, J.F. (1983). *Financial Theory and Corporate Policy*. Addison-Wesley Publishing Company, Don Mills, Ontario, Canada.

FOSTER, G. (1977). Quarterly Accounting Data: Time-Series Properties and Predictive Ability Results, *Accounting Review* 52, 1-21.

GRANGER, C.W.J. (1969). Investigating Causal Relations by Econometric Models and Cross-spectral Methods, *Econometrica* 37, 424-438.

GRIFFIN, P.A. (1977). The Time-Series Behaviour of Quarterly Earnings: Preliminary Evidence, *Journal of Accounting Research* 15, 71-83.

HSIAO, C. (1979a). Autoregressive Modeling of Canadian Money and Income Data, *Journal of American Statistical Association* 74, 553-560.

HSIAO, C. (1979b). Causality Tests in Econometrics, *Journal of Economic Dynamics and Control* 1, 321-346.

HSIAO, C. (1981). Autoregressive Modeling and Money-Income Causality Detection, *Journal of Monetary Economics* 7, 85-106.

LOREK, K.S. (1979). Predicting Annual Net Earnings with Quarterly Earnings Time-Series Models, *Journal of Accounting Research* 17, 190-204.

MASANI, P. (1966). Recent Trends in Multivariate Prediction Theory, *Multivariate Analysis, I* (Academic Press, New York), 351-382.

NEWBOLD, P. (1982). Causality Testing in Economics, Time Series Analysis: Theory and Practice 1 edited by O.D. Anderson (North-Holland, Amsterdam), 701-716.

RONEN, J., SADAN, S., and SNOW, C. (1977). Income Smoothing: A Review, Accounting Journal, 11-26.

TIME SERIES ANALYSIS: Theory and Practice 7
O.D. Anderson (editor)

CAUSALITY TESTING BASED ON EX ANTE FORECASTS

Jack Y. Narayan
Department of Mathematics, State University of New York at Oswego,
Oswego, New York 13126 U.S.A.

Celal Aksu
Department of Accounting, School of Management, Syracuse University,
Syracuse, New York 13210 U.S.A.

Granger proposes that the definition of causality involve prediction and stresses that the inference about causal direction should be based on post-sample forecasts. The purpose of this article is to use a test proposed by Granger to study the causal relationships that may exist among Gross National Product, Money Supply and the Index of Net Business Formation. Multivariate and univariate state space models are used to obtain the post-sample forecasts.

1. INTRODUCTION

Empirical studies on causality testing are almost always based on in-sample tests of hypotheses of interest. Granger (1980) proposed that the definition of causality involve prediction and argued that the inference about causal direction should be based on post-sample forecasts. Thus in order to test whether x causes y one should use the sample data to construct the best possible bivariate model for the series, and also the best univariate model for y. Post-sample one-step predictions from these two models would then be compared in accuracy. Causality may be inferred if the forecasts for y from the bivariate model are significantly better than the forecasts for y from the univariate model. Ashley, Granger and Schmalensee (1980) give a description and illustration of such a procedure. Newbold (1982) supports the idea of post-sample testing and suggests that the analysis should not be limited to the bivariate case and states that forecast horizons should be extended beyond one.

Granger's concept of causality is different from that of the philosophers although Nelson and Schwert (1980) conceded that the former does correspond to restrictions on bivariate time series models that are of interest to economists.

The purpose of this paper is to use the Granger test to study the causal relationships that may exist among Gross National Product, Money Supply and the Index of Net Business Formation. In particular we will determine whether the causality from Money Supply to Gross National Product, as established from in-sample studies, is present in a post-sample situation. The state space procedure is used for estimating the univariate and multivariate models needed to carry out the test. The state space routine from SAS (1981) was utilized for the computations. The quarterly data was obtained from Business Conditions Digest and covered the period January 1948-December 1982.

2. GRANGER'S CAUSALITY

'A variable x causes another variable y with respect to a given universe or information set that includes x and y, if present y can be better predicted by using past values of x than by not doing so, all other information contained in the past of the universe being used in either case.' Granger (1969). This definition can be stated in a formal way as follows:
Let $R_n: \{x_{n-j}, y_{n-j}, z_{n-j}, \ldots, j \geq 0\}$ represent all the information available on some set of time series at time n. Suppose that at time n optimum forecasts for y_{n+1} are obtained first, by using all the information in R_n, and then by using all the information apart from the past and present values of x_{n-j}, $j \geq 0$, of the series x_t. If the first forecast, using all the information, is superior to the second, then the series x_t is said to cause y_t. The simplest case is when R_n consist of just values from the series x_t and y_t. Granger proposes, as a working definition of causality, consideration of the mean-square error of optimum linear predictors. Thus if $MSE(x,y)$ is the population mean-square of the one-step forecast error of y_{n+1} using the optimum linear forecasts based on x_{n-j}, y_{n-j}, $j \geq 0$, then x causes y if $MSE(x,y) < MSE(y)$, where $MSE(y)$ is the mean-square errors of the univariate model. If, in addition, y causes x, the two series are said to exhibit feedback.

For a finite data set, a test of significance described by Ashley et al. (1980), could be used to test if two mean-square errors are significantly different. This test is employed later in this paper to test for causality from Money Supply to Gross National Product.

Causal relationships among macroeconomic variables were investigated by many authors who used different procedures and reported contradictory results. Feige and Pearce (1979) found no evidence of causal relations between the supply of money and aggregate nominal income in strong contradiction to the conclusions reached earlier by Sims (1972). Pierce (1977) studied relationships among pairs of various monetary, financial, and macroeconomic variables and found little evidence of causation, even in cases such as growth of demand deposits and yield on treasury bills, which theory suggests are related. Also Feige and Pearce (1979) report no significant relationship between growth of the money supply and the inflation rate. Thury (1983) reported similar results in studies with macroeconomic series from Austria.

Newbold (1982) obtained causality from Money Supply to Gross National Product. His conclusion was based on an in-sample analysis in which a multivariate ARMA model was estimated for series consisting of 132 quarterly observations each. Fey and Jain (1982), in an in-sample study, supported the findings of Newbold (1982). They utilized state space procedures and reported that the bivariate model for Gross National Product and Money Supply resulted in a thirteen percent reduction in error variance for GNP. Their multivariate model in which prices was included with the other two series resulted in a seventeen percent reduction in error variance. The study was done with quarterly data for the period 1955-1970. Nelson and Schwert (1982) observed that since different investigators use different testing procedures, the explanations for these inconsistencies presumably lie in the relative size and power of the alternative test statistics.

Our study differs from those mentioned above in that the test for causality was based on post-sample analysis. We use an adaptive modeling procedure in which a new model was obtained for each post-sample forecast. To our knowledge, the power of the Ashley et al. (1980) test has not been fully investigated. The reductions in error variances obtained in our post-sample study were of the same order of magnitude as those obtained by Fey and Jain (1982) in their in-sample analysis. However these differences did not prove to be significant when they were subjected to the Ashley et al. (1980) test.

3. THE STATE SPACE MODEL

We give a brief outline of the state space procedure as described by Akaike (1976). Let $\underline{y}_T$ denote a stationary multivariate time series of dimension r. Then the state space representation of $\underline{y}_{T+1}$ is given by

$$\underline{Z}_{T+1} = \underline{F}\,\underline{Z}_T + \underline{G}\,\underline{E}_{T+1}\,; \quad \underline{y}_{T+1} = \underline{H}\,\underline{Z}_{T+1} \tag{1}$$

where $\underline{Z}_T$, called the state vector, is a vector process of dimension s whose first r components comprise $\underline{y}_T$, $\underline{F}$ is an s by s transition matrix, $\underline{G}$ is an s by r input matrix and $\underline{E}_{T+1}$ is a sequence of zero mean, independent, s-dimensional random vectors with constant covariance matrix. The matrix $\underline{H}$ has the form $[\underline{I}, \underline{0}]$ where $\underline{I}$ is an r by r identity matrix and $\underline{0}$ is an r by $(s-r)$ zero matrix. The vector $\underline{Z}_T$ contains the set of present and past information sufficient to predict future values. The representation (1) is by no means unique. For example, the change of variable $\underline{\theta}_T = \underline{M}\,\underline{Z}_T$, where $\underline{M}$ is any nonsingular matrix will also be an acceptable state space representation. However, Akaike (1976) used the concept of predictor space to show that a general multivariate linear stochastic system has a canonical representation which is unique.

The state space method requires that the models be identified in their canonical forms. An appealing aspect of the canonical representation is its relationship to the concept of parsimony of a model. The components of the state vector, in the canonical representation, are selected by performing a sequence of canonical correlation analyses between the set of present and past values, and the set of present and future values. An information criterion based on Bartlett's Chi-square is used to determine whether a specific element ought to be included in the state vector.

The amount of past information to be used in the canonical correlation analysis is decided upon by fitting a sequence of autoregressive models of increasing orders and computing the Akaike information criterion (AIC). The order of the AR model for which AIC is minimum determines the number of lags into the past that are relevant to the analysis. The optimum AR model also provides information for the estimation of the $\underline{G}$-matrix. Estimates for the $\underline{F}$-matrix are obtained during the canonical correlation analysis. These estimates are then used as initial values in a maximum likelihood procedure for obtaining final estimates.

A residual analysis is used to test for model adequacy. Specifically, if the model is correctly specified, then for large lags the residuals are uncorrelated, normally distributed random variables with mean zero and variance $1/n$, where n is the number of observations in the stationary series. Thus a model might be considered inadequate if several of the residual correlations are greater than $2/\sqrt{n}$.

Akaike (1974), showed that any autoregressive moving average (ARMA) process has a state space representation. Thus the state space model can be used to produce forecasts with the same properties as those from a corresponding ARMA model. The major advantage of the state space procedure over the multivariate ARMA approach lies in the identification phase of the modeling. In the univariate ARMA procedure, one way to recognize particular mixed models is through visual inspection of the sample autocorrelations. Newbold (1983) remarked that, in the multivariate case, this is a pretty hopeless proposition, as the corresponding patterns would now be followed by matrices of order m by m, where m is the number of time series. In the state space procedure, the identification problem is resolved by the automatic selection of the unique canonical form.

4. AN EXAMPLE

We now give an example to show the equivalence between the ARMA and state space representation for the bivariate case. Suppose we have the following ARMA model

$$\begin{bmatrix} x_{t+1} \\ y_{t+1} \end{bmatrix} = \begin{bmatrix} \phi_{11} & \phi_{12} \\ \phi_{21} & \phi_{22} \end{bmatrix} \begin{bmatrix} x_t \\ y_t \end{bmatrix} + \begin{bmatrix} \varepsilon_{t+1} \\ \eta_{t+1} \end{bmatrix} - \begin{bmatrix} \theta_{11} & \theta_{12} \\ 0 & 0 \end{bmatrix} \begin{bmatrix} \varepsilon_t \\ \eta_t \end{bmatrix} \tag{2}$$

where $(\varepsilon_t, \eta_t)'$ is a sequence of zero mean independent random vectors with constant covariance matrix. The corresponding scalar equations are

$$x_{t+1} = \phi_{11}x_t + \phi_{12}y_t + \varepsilon_{t+1} - \theta_{11}\varepsilon_t - \theta_{12}\eta_t \tag{3}$$

$$y_{t+1} = \phi_{21}x_t + \phi_{22}y_t + \eta_{t+1} \tag{4}$$

In order to get the state space formulation we must first determine the state vector which contains the relevant variables and lags for forecasting. We start with the two components x_t and y_t. Denoting conditional expectation by the symbol $|$, we have from (3)

$$x_{t+1|t} = \phi_{11}x_t + \phi_{12}y_t - \theta_{11}\varepsilon_t - \theta_{12}\eta_t$$

Since this equation has additional information over x_t and y_t, the element $x_{t+1|t}$ is added to the state vector. Conditional expectation on (4) yields

$$y_{t+1|t} = \phi_{21}x_t + \phi_{22}y_t \tag{5}$$

Since $y_{t+1|t}$ is completely determined by the first two components x_t and y_t of the state vector, it is not added to the state vector. Similarly $y_{t+i|t}$, $i \geq 0$ will also be a linear combination of the elements of the state vector so far determined and no more y's will be included in the state vector.

Shifting the index in (3) and taking conditional expectation with time origin t, we get

$$\begin{aligned} x_{t+2|t} &= \phi_{11}x_{t+1|t} + \phi_{12}y_{t+1|t} \\ &= \phi_{11}x_{t+1|t} + \phi_{12}(\phi_{21}x_t + \phi_{22}y_t) \end{aligned} \tag{6}$$

Since $x_{t+2|t}$ is a linear combination of the components of the present state vector, it is not added. Similarly $x_{t+i|t}$, $i \geq 3$ can also be expressed as a linear combination of the components of the state vector. Thus the state vector is $(x_t, y_t, x_{t+1|t})'$ where the symbol $'$ denotes the transpose of a matrix.

We now construct the state space form by starting with the identities

$$x_{t+1} = x_{t+1|t} + \varepsilon_{t+1} \tag{7}$$

$$y_{t+1} = y_{t+1|t} + \eta_{t+1}$$

The second identity together with (5) imply

$$y_{t+1} = \phi_{21}x_t + \phi_{22}y_t + \eta_{t+1} \tag{8}$$

Replacing t by t+1 in (3) and taking conditional expectation we get

$$x_{t+2|t+1} = \phi_{11}x_{t+1} + \phi_{12}y_{t+1} - \theta_{11}\varepsilon_{t+1} - \theta_{12}\eta_{t+1} \tag{9}$$

Finally equations (7), (8) and (9) together imply

$$x_{t+2|t+1} = \phi_{11}x_{t+1|t} + \phi_{12}\phi_{22}y_t + \phi_{12}\phi_{21}x_t$$
$$+ (\phi_{12} - \theta_{12})\eta_{t+1} + (\phi_{11} - \theta_{11})\varepsilon_{t+1} \tag{10}$$

The scalar equations (7), (8) and (10) in matrix form is the following state space system

$$\begin{bmatrix} x_{t+1} \\ y_{t+1} \\ x_{t+2|t+1} \end{bmatrix} = \begin{bmatrix} 0 & 0 & 1 \\ \phi_{21} & \phi_{22} & 0 \\ \phi_{12}\phi_{21} & \phi_{12}\phi_{22} & \phi_{11} \end{bmatrix} \begin{bmatrix} x_t \\ y_t \\ x_{t+1|t} \end{bmatrix}$$
$$+ \begin{bmatrix} 1 & 0 \\ 0 & 1 \\ (\phi_{11} - \theta_{11}) & (\phi_{12} - \theta_{12}) \end{bmatrix} \begin{bmatrix} \varepsilon_{t+1} \\ \eta_{t+1} \end{bmatrix} \tag{11}$$

We can now identify the terms in equation (1)

$$\underline{Z}_t = (x_t, y_t, x_{t+1|t})', \quad \underline{Y}_t = (x_t, y_t)'$$

$$\underline{H} = \begin{bmatrix} 1 & 0 & 0 \\ 0 & 1 & 0 \\ 0 & 0 & 0 \end{bmatrix}$$

The matrices $\underline{F}$ and $\underline{G}$ are respectively the 3 by 3 and the 3 by 2 matrices on the right hand side of equation (11).

Conversely, if the state space model is given as (11) we can obtain the corresponding ARMA form as follows. The system is equivalent to the scalar equations (7), (8) and (10). Substituting for the term $\phi_{22}y_t$ from (8) and for $x_{t+1|t}$ from (7) in equation (10) we get

$$x_{t+2|t+1} = \phi_{12}\phi_{21}x_t + \phi_{12}(y_{t+1} - \phi_{21}x_t - \eta_{t+1})$$
$$+ \phi_{11}(x_{t+1} - \varepsilon_{t+1}) + (\phi_{11} - \theta_{11})\varepsilon_{t+1} + (\phi_{12} - \theta_{12})\eta_{t+1}$$
$$= \phi_{12}y_{t+1} + \phi_{11}x_{t+1} - \theta_{11}\varepsilon_{t+1} - \theta_{12}\eta_{t+1} \tag{12}$$

Finally substituting for $x_{t+2|t+1}$ from (7) and replacing t by t-1 in the resulting equation, we get

$$x_{t+1} = \phi_{12}y_t + \phi_{11}x_t - \theta_{11}\varepsilon_t - \theta_{12}\eta_t + \varepsilon_{t+1} \tag{13}$$

Equations (8) and (13) are the scalar equations corresponding to the bivariate ARMA model in (2).

We remark again that the main advantage in using the state space method rather than the ARMA is to be found in the model identification step. The state space procedure offers an easier and more systematic approach.

5. PROCEDURE

The data set which consisted of 136 observations for each series was divided into two sets. The first contained 112 observations of each series and was used for initial model identification and estimation. This set is referred to as the sample data. The second set, referred to as the post-sample data, consisted of the last 24 observations of each series and was used to calculate ex ante forecast errors for the Ashley et al. (1980) test.

The sample data was used to estimate a univariate state space model for the Gross National Product series and forecasts were made for periods one through four in the post sample data. The sample data was then updated to include the 113th data point and the entire modeling process of identification, estimation and diagnostic checking was repeated to get a new model from which forecasts for periods 114 through 117 were obtained. This process of updating the sample data, remodeling and forecasting was repeated for the entire post-sample data. Thus genuine ex ante forecasts with horizons one through four were obtained.

Bivariate and multivariate models were then estimated and forecasts were obtained for each period in the post-sample data in a stepwise fashion as described above. In the bivariate models, Gross National Product and Money Supply were analyzed. In the multivariate models, the Index of Net Business Formation was included with the other two series.

An adaptive state space modeling procedure was used to accommodate the changes that take place in model structure of the univariate, bivariate, and multivariate procedures as these series evolve. Thus, a full model, that is, one including the significant as well as the insignificant parameters, had to be obtained for each time period. The insignificant parameters were then restricted to be zero in a stepwise manner to get, what is referred to as, a restricted model. Residual analysis was done to test for model adequacy. We used the log transformation in order to stabilize the variance of each series.

6. RESULTS

Since the number of estimated models is very large, it is not possible to present the equations in every case. In order to show the structures involved, we give the equations for only three of these models estimated on the entire data set of 136 points. The equations are given in their ARMA forms. Let ∇GNP, ∇MS and ∇BF represent the first differences of GNP, the Money Supply and the Index of Net Business Formation respectively. The equations for the univariate, bivariate and multivariate models are:

$$(\nabla GNP)_{t+1} = .361\ (\nabla GNP)_t + \varepsilon_{t+1} \qquad (14)$$

$$\begin{bmatrix} \nabla MS \\ \nabla GNP \end{bmatrix}_{t+1} = \begin{bmatrix} .487 & 0 \\ .306 & .266 \end{bmatrix} \begin{bmatrix} \nabla MS \\ \nabla GNP \end{bmatrix}_t + \begin{bmatrix} a \\ b \end{bmatrix}_{t+1} \qquad (15)$$

and

$$\begin{bmatrix} \nabla MS \\ \nabla BF \\ \nabla GNP \end{bmatrix}_{t+1} = \begin{bmatrix} .455 & 0 & 0 \\ .504 & .398 & 0 \\ 0 & .176 & .207 \end{bmatrix} \begin{bmatrix} \nabla MS \\ \nabla BF \\ \nabla GNP \end{bmatrix}_t + \begin{bmatrix} a \\ b \\ c \end{bmatrix}_{t+1} \qquad (16)$$

The parameters all have t-ratios larger than 2. The error variances for the GNP component from the three models are .000109772, .000101423 and .000095784 respectively. Thus the percentage reductions in error variance attributed to the bivariate and multivariate models are 7.6 and 12.7 respectively. The zero entries in equations (15) and (16) represent parameters which were either identically equal to zero or were insignificant and had to be restricted to zero in a stepwise manner. From the bivariate model it is clear that Money Supply affects GNP. The multivariate model implies that GNP is affected by the Index of Net Business Formation which in turn is affected by Money Supply.

As mentioned earlier, Fey and Jain (1982) modeled GNP and Money Supply on a different time period and reported a reduction in error variance of approximately 13%. This reduction was one of the criterion used for supporting their conclusion that there was causality from Money Supply to GNP. Newbold (1982) also reported the same relationship between Money Supply and GNP. His conclusion was based on a likelihood ratio test for causality which involves building a parsimonious ARMA (p,q) bivariate model. This test, which is described in the appendix, was applied to our bivariate model on the entire data set consisting of 136 data points and indicated that there was causality from Money Supply to GNP. However, we agree with Newbold (1982) that one should not rest causality conclusions on the results of fitting a set of models to a fixed set of data points. We remark that changes occur in the coefficients as well as in the structure of the models as the data set changes. For example there are a few cases in our study in which the univariate model turns out to be a mixed ARMA model. The same can be reported for our multivariate models.

However, the bivariate models were very stable in the sense that the structure was always the same as that presented in (15). In each of the 25 bivariate models, the parameter relating GNP to Money Supply was insignificant and was restricted to be zero. The coefficients as well as the amount of reductions in error variance achieved by these bivariate models still, of course, changed with the data set. Thus the conclusion about causality may depend on the specific time period on which the study is done.

The one-step forecast errors were uncorrelated in each of the three models. Moreover the three sets of one-step forecast errors were not cross correlated at any lag other than zero. However the multistep forecast errors showed significant autocorrelation and cross correlation at lag one. Table 1 contains a summary of our post-sample analysis. The mean of squared errors (MSE) and the mean absolute deviation (MAD) were computed for the univariate, bivariate and multivariate models for forecast horizons one through four.

TABLE 1. POST SAMPLE ERROR STATISTICS.

STATISTIC	FORECAST HORIZON	UNIVARIATE MODELS	BIVARIATE MODELS	MULTIVARIATE MODELS
MSE	1	272.89	226.175	240.51
MAD	1	12.92	11.85	12.33
MSE	2	604.78	480.74	566.61
MAD	2	19.76	17.94	18.97
MSE	3	1177.59	903.66	1052.36
MAD	3	26.37	23.34	24.00
MSE	4	1275.81	1016.24	1205.09
MAD	4	28.28	26.71	27.60

It is evident from Table 1 that the one-step forecasts from the bivariate and multivariate models are better than those from the univariate models. The percentage reductions in MSE and MAD attributed to the bivariate models are 17.1% and 8.2% respectively. The corresponding figures for the multivariate models are 11.9% and 4.6% respectively. If the reductions in MSE and/or MAD are regarded as being large enough, it can be concluded from the bivariate case that there is causality from Money Supply to Gross National Product.

An application of the Ashley et al. (1980) test, carried out in the appendix, indicates the reduction in variance of the one-step forecast errors attributed to the bivariate model was significant only at the 12% level. This is hardly overwhelming evidence to support the claim that Money Supply causes GNP. Even though sizeable reductions in variances of the multistep forecast errors were obtained with the use of the bivariate models, the forecast errors for horizons greater than one showed significant autocorrelations at lag 1. Thus, Ashley et al. (1980) test for significance was not applied to multistep forecast errors.

7. CONCLUSION

We have investigated the causal relationships which exist among the three series Gross National Product, Money Supply and the Index of Net Business Formation. The test for the causal relationships was based on the forecasting performance of univariate, bivariate and multivariate models on a post-sample period. The problem was reduced to determining whether forecasts for Gross National Product from bivariate and multivariate models were more accurate than the corresponding forecasts from a univariate model.

Naturally a criterion is needed to decide if one forecast is superior to another. The usual procedure is to compare relative sizes of the variances of in-sample forecast errors. We used Ashley et al. (1980) criterion which is concerned with the comparison of the mean-square errors of the post-sample forecasts. We found that although the one-step forecasts from the bivariate models, in which Gross National Product was modeled simultaneously with Money Supply, were approximately 17.1% better than the univariate forecasts for Gross National Product, the difference between the two types of forecasts was significant only at the 12% level. The multivariate model was outperformed by the bivariate model for all forecast horizons considered. This indicates that the Index of Net Business Formation is not useful for forecasting GNP when Money Supply is included. From the point of view of a forecaster, the notion of causality testing based on the improvement in accuracy of post-sample forecasts is appealing. However, as pointed out by Newbold (1982), there are problems associated with this type of modeling, the most difficult of which is how to divide the data satisfactorily into segments for model fitting and forecast evaluation.

APPENDIX

TESTS FOR CAUSALITY:

1) ASHLEY ET. AL (1980) TEST.

The following is an application of the Ashley et al. (1980) test for evaluating post-sample forecasting performance. Let $e1_t$ and $e2_t$ be the forecast errors at time t for the univariate and bivariate models, respectively. Then, for the entire post-sample period:

$$MSE(e1) - MSE(e2) = [s^2(e1) - s^2(e2)] + [m(e1)^2 - m(e2)^2] \qquad (1)$$

where MSE denotes the sample mean of squared errors, s^2 denotes the sample variance and m denotes the sample mean. If

$$D_t = e1_t - e2_t \quad \text{and} \quad P_t = e1_t + e2_t \tag{2}$$

then equation (1) can be rewritten as follows:

$$MSE(e1) - MSE(e2) = [\hat{COV}(D,P)] + [m(e1) - m(e2)][m(e1) + m(e2)] \tag{3}$$

where $\hat{COV}$ denotes the sample covariance over the post-sample period. In our analysis both error means are negative and hence $m(P) < 0$ holds. Let COV and μ denote the population covariance and population mean corresponding to the sample parameters $\hat{COV}$ and m. Then from (3) it can be concluded that the bivariate model outperforms the univariate model if the joint null hypothesis $COV(D,P) = 0$ and $\mu(D) = 0$ is rejected in favor of the alternative hypothesis that (i) $COV(D,P) \geq 0$ and $\mu(D) \leq 0$ and (ii) $COV(D,P) > 0$ or $\mu(D) < 0$. Consider the regression equation

$$D_t = \beta_1 + \beta_2[P_t - m(P_t)] + U_t \tag{4}$$

where U_t is an error term with mean zero and is regarded as being independent of P_t. Then the test outlined above is equivalent to testing the null hypothesis $\beta_1 = 0 = \beta_2$ against (i) $\beta_1 \leq 0$ and $\beta_2 \geq 0$ and (ii) $\beta_1 < 0$ or $\beta_2 > 0$. The regression analysis resulted in $\hat{\beta}_1 = -3.0128$ and $\hat{\beta}_2 = .0371$. Since both estimates have the correct sign, an F test of the null hypothesis that both population values are zero can be employed. The F-statistic (with 2 and 22 degrees of freedom) corresponding to the null hypothesis that both population values are zero was 1.46, significant at the 24 percent level. The Durbin-Watson statistic was 2.25. However, as Ashley et al. (1980) pointed out, this test, is in essence, four tailed; it does not take into account the signs of the estimated coefficients. They made the point that if the signs of the estimated parameters agreed with those of the alternative hypothesis, then, one can perform an F test and report a significance level equal to half that obtained from the tables. In view of this observation, the null hypothesis, that the two models have equal mean of squared errors, can be rejected at approximately 12 percent level of significance in favor of the superiority of the bivariate model.

2) NEWBOLD'S TEST

Suppose a bivariate series $\underline{x}_t$ with components $x1_t$ and $x2_t$ has a vector ARMA (p,q) representation of the form

$$\begin{bmatrix} \phi_{11}(B) & \phi_{12}(B) \\ \phi_{21}(B) & \phi_{22}(B) \end{bmatrix} \begin{bmatrix} x1_t \\ x2_t \end{bmatrix} = \begin{bmatrix} \theta_{11}(B) & \theta_{12}(B) \\ \theta_{21}(B) & \theta_{22}(B) \end{bmatrix} \begin{bmatrix} a1_t \\ a2_t \end{bmatrix} \tag{5}$$

where B is the back-shift operator on the index of time series and

$\phi_{ii}(0) = 1 = \theta_{ii}(0) \quad (i = 1,2)$

$\phi_{ij}(0) = 0 = \theta_{ij}(0) \quad i \neq j$

Then it can be shown that x2 does not cause x1 if and only if

$\phi_{12}(B) = 0 = \theta_{12}(B)$.

The hypothesis x2 does not cause x1 could be checked, through a likelihood ratio test as follows. The first step is to fit an appropriate ARMA (p,q) model. Next a restricted model, implied by the hypothesis, "x2 does not cause x1", is estimated. Under this null hypothesis, the likelihood ratio test statistic

$$LR = n \,\mathrm{Log}[|\hat{\Sigma}_r| \,/\, |\hat{\Sigma}|] \qquad (6)$$

is asymptotically distributed as a Chi-square with $p + q$ degrees of freedom, when the identified model is ARMA (p,q) . In (6) $\hat{\Sigma}$ and $\hat{\Sigma}_r$ are residual covariance matrices from respective estimation of the full and restricted models.

Using this test, we tested the hypothesis that "Money Supply does not cause Gross National Product". We used 136 data points and fitted an ARMA (1,0) model to the first difference of the logarithm of the original series. The test statistic was 10.87 which is significant by reference to the Chi-square distribution with one degree of freedom. Consequently the null hypothesis is clearly rejected.

REFERENCES

AKAIKE, H. (1974). Markovian Representation of Stochastic Process and its Applications to the Analysis of Autoregressive Moving Average Processes. Annals of Institute of Statistical Mathematics 26, 363-387.

AKAIKE, H. (1976). Canonical Correlations of Time Series and the Use of an Information Criterion. In Advances and Case Studies in System Identification. Eds: R. Mehra and D.G. Lainiotis, Academic Press, New York, 27-95.

ASHLEY, R., GRANGER, C.W.J., SCHMALENSEE, R. (1980). Advertising and Aggregate Comsumption: An Analysis of Causality. Econometrica 48, 1149-67.

FEIGE, E.L. and PEARCE, D.K. (1979). The Causal Relationship Between Money and Income: Some Caveats for Time Series Analysis. Review of Economics and Statistics 61, 521-533.

FEY, R.A. and JAIN, N.C. (1982). Identification and Testing of Optimal Lag Structures and Causality in Economic Forecasting. In Applied Time Series Analysis (Proceedings of the International Conference held at Houston, Texas, August 1981). Eds: O.D. Anderson and M.R. Perryman, North-Holland, Amsterdam and New York, 65-73.

GRANGER, C.W.J. (1969). Investigating Causal Relations by Econometric Models and Cross Spectral Methods. Econometrica 37, 424-438.

GRANGER, C.W.J. (1980). Testing for Causality: A Personal Viewpoint. Journal of Economic Dynamics and Control 2, 329-52.

NARAYAN, J.Y. and AKSU, C. (1983). Forecasting Performance of Univariate and Multivariate Time Series Models for Treasury Bill Rates. Paper presented to Third International Symposium on Forecasting, Philadelphia.

NELSON, C.R. and SCHWERT, G.W. (1982). Tests for Predictive Relationships Between Time Series Variables: A Monte Carlo Investigation. Journal of American Statistical Association 77, 11-18.

NEWBOLD, P. (1982). Causality Testing in Economics. In Time Series Analysis: Theory and Practice 1 (Proceedings of the International Conference held at Valencia, Spain, June 1981). Ed: O.D. Anderson, North-Holland, Amsterdam and New York, 701-716.

NEWBOLD, P. (1983). ARIMA Model Building and Time Series Analysis Approach to Forecasting. Journal of Forecasting 2, 23-35.

PIERCE, D.A. (1977). Relationships - and the Lack thereof - Between Economic Time Series, with Special Reference to Money and Interest Rates. Journal of American Statistical Association 72, 11-16.

SAS (1981). SAS Institute Inc., Box 800, Cary, North Carolina.

SIMS, C.A. (1972). Money, Income and Causality. American Economic Review 62, 540-552.

THURY, G. (1983). Forecasting Consumer Expenditure and Disposable Personal Income of Austria Using a Seasonal Vector ARIMA Model. In Time Series Analysis: Theory and Practice 5 (Proceedings of the International Conference, held at Nottingham University, 11-15 April 1983). Ed: O.D. Anderson, North-Holland, Amsterdam and New York, 183-192.

TIME SERIES ANALYSIS: Theory and Practice 7
O.D. Anderson (editor)

UNIFIED ECONOMETRIC MODEL BUILDING USING SIMULTANEOUS TRANSFER FUNCTION EQUATIONS

Lon-Mu Liu
Department of Quantitative Methods, University of Illinois at Chicago, BOX 4348, Chicago, Illinois 60680, USA

Gregory Hudak
Scientific Computing Associates, P.O. Box 625, DeKalb, Illinois 60115, USA

Developments and applications of time series analysis in social and economic studies have occurred rapidly during the last two decades. Recently, the focus in this area has shifted from univariate or single equation to multivariate and simultaneous equation models. Several classes of multivariate time series models have been studied including vector autoregressive-moving average and simultaneous transfer function (STF) models. The STF models are a natural extension of the classical econometric models and allow for explicit contemporaneous relationships among the endogenous variables. More importantly, they facilitate the inclusion of prior knowledge into the system and substantially alleviate the "parameter explosion" problem of multivariate time series analysis. This paper describes statistical methodology for identification, estimation, and forecasting of STF models.

1. INTRODUCTION

Since the beginning of this century, a large number of econometric models have been produced. All important economic sectors in the United States appear to have several versions of these models. The traditional econometric modeling approach has faced new challenges since Box and Jenkins (1968, 1970) published their important work on time series analysis. It is frequently found that the forecasting performance of a complicated econometric model may not be even as good as a simple univariate ARIMA model. As a result and also due to the simplicity of the approach credited to Box and Jenkins as well as the availability of computer programs, much of the attention of economic data analysis seems to have shifted from traditional econometric modeling to that of time series analysis. This is particularly true in recent years, as time series analysis has been extended from univariate to multivariate models (e.g., Zellner and Palm 1974, Jenkins and Alavi 1981, Tiao and Box 1981) and seems to be able to meet various needs of econometric modeling. Despite this change of interest, it is probably wise to combine the strengths of both traditional econometric models and Box-Jenkins time series models rather than to insist on one or the other. Attempts have been made in this endeavor, particularly by Fair (1970), Hendry (1971), Chow and Fair (1973), Zellner and Palm (1974), Wall (1976), Reinsel (1979), Zellner (1979), Newbold and Reed (1979), and Hanssens and Liu (1983). The purpose of this paper is to provide a general framework for the integration of econometric and time series model building, particularly in the iterative procedure of tentative model identification, estimation, diagnostic checking and forecasting.

Level of Knowledge in Econometric Model Building

One of the most difficult and important problems in econometric model building for time series data is the tentative identification of appropriate models. Since the properties of parameter estimates in a model are based upon the model being adequately represented, the importance of this issue is clear. Tentative model identification may be handled in a variety of ways, depending on the knowledge of the system at hand. In the terminology of Hanssens and Manegold (1980), the level of knowledge for a model builder may be classified as:

Level-0: only the information set is known;

Level-1: the distinction between endogenous and exogenous variables in the information set and the explanatory variables in each equation are known;

Level-2: the functional forms and lag structure of the the relationships in the entire system are known.

Econometricians typically assume level-2 knowledge and attempt to determine a model completely based on economic theory. Hence, the only remaining problem is the estimation of the parameters in the model. In contrast to such approach, time series analysts usually assume only level-0 knowledge and use data as the sole source to derive an appropriate model. In practice, a priori economic theory is very seldom able to completely identify the lag structure of an econometric model, while structural analysis can be difficult and hazardous if only level-0 knowledge is available. For the purposes of structural model building, it is probably more appropriate to assume the model builder has level-1 knowledge and may use available data to attain level-2 knowledge.

2. FORMULATION OF ECONOMETRIC TIME-SERIES MODELS

A number of econometric time series models have been proposed in recent years. Hendry (1971) and Chow and Fair (1973) studied classical simultaneous equation models assuming autoregressive noise processes. Reinsel (1979) extended this class of models to include vector ARMA disturbances (or noise processes). In contrast, Zellner and Palm (1974) treated any econometric time series model as a special case of the linear multiple time series model proposed by Quenouille (1957).

The model we shall present for the study of a dynamic econometric system is the simultaneous transfer function equation (STF) model, also referred to as the rational distributed lag structural form model (Wall 1976, Hanssens and Liu 1983). Although not the most general in its class, it is a very useful and sufficiently flexible for the purpose of structural analysis. This class of models combines classical econometric models with Box-Jenkins transfer functions and ARMA disturbances. In its most general form, the STF model can be written as

$$\underline{\Phi}(B)\underline{Y}_t = \underline{C} + \underline{\Gamma}(B)\underline{X}_t + \underline{N}_t\,, \qquad t=1,2,\ldots,n$$

$$N_{it} = \frac{\theta_i(B)}{\phi_i(B)} a_{it} \qquad i=1,2,\ldots,k \tag{1}$$

where $\underline{Y}_t$ is a $k \times 1$ vector containing observations of k endogenous variables at time t, $\underline{X}_t$ a $m \times 1$ vector of exogenous (or predetermined) variables at time t, $\underline{a}_t$ a series of $k \times 1$ white noise vectors, which are independently and identically distributed as multivariate normal $N(\underline{0}, \Sigma)$, and $\underline{C}$ a $k \times 1$ vector of constants for each equation. $\underline{\Phi}(B)$ and $\underline{\Gamma}(B)$ are matrix operators whose elements have the form $\omega(B)/\delta(B)$ (to be defined in (3)) for the endogenous and exogenous terms, respectively; $\theta_i(B)/\phi_i(B)$ is the ARMA noise process for the i-th equation; and B the backshift operator. The lag 0 matrix coefficient of $\underline{\Phi}(B)$, denoted by $\underline{\Phi}_0$, represents the contemporaneous endogenous relationships in the system. It is assumed that the system in (1) is identifiable (Hannan 1971, Granger and Newbold 1977, Kohn 1979), the noise series are stationary and invertible (i.e., the roots of $\phi_i(B)$ and $\theta_i(B)$ polynomials lie outside the unit circle), and the roots of $|\underline{\Phi}(B)|$ lie outside the unit circle.

The model described in (1) is a k equation simultaneous system. Each equation in the system is similar to the transfer function model described in Box and Jenkins

(1970). However, the endogenous variables in the system may be contemporaneously related and are explicitly represented by some parameters in the matrix Φ_0, hence the terminology "simultaneous transfer function equation model."

STF models can be applied in any area where simultaneous and/or transfer function modeling is appropriate. In this article, we focus on the application of STF models in econometric studies. Without loss of generality, we limit our discussion and illustrations in this paper to a simultaneous model with two endogenous variables, Y_1 and Y_2, and two exogenous variables, X_1 and X_2. It is assumed that the input variables for each equation are known and the equations for the endogenous variables are

$$
\begin{aligned}
Y_{1t} &= C_1 + \beta_{11}(B)X_{1t} + \beta_{12}(B)Y_{2t} + N_{1t} \\
Y_{2t} &= C_2 + \beta_{21}(B)X_{2t} + \beta_{22}(B)Y_{1t} + N_{2t}
\end{aligned}
\tag{2}
$$

where $\beta_{ij}(B)$'s are linear or rational polynomials in the backshift operator B. The dynamic regression coefficients $\beta_{ij}(B)$'s are also referred to as the transfer functions between the corresponding input and output variables. In general, the transfer function may be expressed as $\omega(B)/\delta(B)$ with

$$
\begin{aligned}
\omega(B) &= (\omega_0+\omega_1 B+\ldots+\omega_{s-1}B^{s-1})B^b \\
\delta(B) &= 1-\delta_1 B-\delta_2 B^2-\ldots-\delta_r B^r
\end{aligned}
\tag{3}
$$

where all roots of the $\delta(B)$ polynomial lie outside the unit circle. Two examples of the model in (2) are given below:

$$
\begin{aligned}
Y_{1t} &= (-0.6-0.4B)X_{1t} + \frac{0.7B}{1-0.6B}Y_{2t} + (1-0.6B)a_{1t} \\
Y_{2t} &= \frac{0.5B}{1-1.1B+0.3B^2}X_{2t} - 0.5Y_{1t} + \frac{1}{1-0.7B}a_{2t}
\end{aligned}
\tag{4}
$$

and

$$
\begin{aligned}
Y_{1t} &= (-0.6-0.4B)X_{1t} + \frac{0.7B}{1-0.6B}Y_{2t} + (1-0.6B)a_{1t} \\
Y_{2t} &= \frac{0.5B}{1-1.1B+0.3B^2}X_{2t} - 0.5B^2Y_{1t} + \frac{1}{1-0.7B}a_{2t} .
\end{aligned}
\tag{5}
$$

The model in (4) is an STF model with a contemporaneous relationship between Y_{1t} and Y_{2t} in the equation for Y_{2t}, while (5) is an STF model without a contemporaneous endogenous relationship in the entire system. In both (4) and (5), a_{1t} and a_{2t} may be correlated.

The model described in (1) is very general and contains several important classes of model. In the single equation case, model (1) includes univariate ARIMA, distributed lag, and transfer function models. In the case of a multiple equation, model (1) is the classical econometric model if the polynomials $\phi_i(B)$, $\theta_i(B)$ and $\delta(B)$ are all equal to 1.

3. REDUCED FORM MODELS

The simultaneous equation model in (1) may or may not contain contemporaneous relationships among endogenous variables. When no contemporaneous endogenous relationships exist (i.e., Φ_0 in an identity matrix), the STF model is in reduced form. Such a reduced form model is called a "reduced form structural model". In reduced form models, current values of endogenous variables can be expressed as functions of the lagged endogenous and current and lagged exogenous variables. The contemporaneous relationships present in structural models are frequently caused by temporal aggregation. If the observations can be taken at smaller time intervals, such contemporaneous relationships may be replaced by some lag endogenous relationships resulting in a reduced form structural model. Therefore reduced form models may often arise in structural modeling. It will be shown later that model identification for reduced form models is usually easier than for structural models with contemporaneous relationships.

Decreasing the time interval for observations is not the only means to obtain a reduced form model. A reduced form model may be derived from a structural model by multiplying $\underline{\Phi}_0^{-1}$ on each matrix operator; thus, the corresponding reduced form model for (1) is

$$\underline{\Phi}_0^{-1}\underline{\Phi}(B)\underline{Y}_t = \underline{\Phi}^{-1}\underline{C} + \underline{\Phi}_0^{-1}\Gamma(B)\underline{X}_t + \underline{\Phi}_0^{-1}\underline{N}_t \ . \tag{6}$$

The above model is referred to as the "derived reduced form" model for the structural model in (1). It is also possible to build a reduced form model simply based on available data, i.e., including all statistically significant variables in each equation of the system. A reduced form model derived in this manner is referred to as an "unrestricted reduced form model". Note that each structural form model has a corresponding reduced form model, but not vice versa. For this reason, structural interpretation of unrestricted reduced form models should be proceeded with great caution.

When there are no equality constraints on parameters across equations, a reduced form model is a system of seemingly unrelated transfer function models. In such cases, the maximum likelihood estimates under single equation estimation may not be efficient, but are still consistent. Therefore, tentative model identification can proceed equation by equation before joint estimation is attempted.

4. RELATIONSHIP BETWEEN STF AND VECTOR ARMA MODELS

Unrestricted reduced form STF models may be considered as alternative parameterizations of vector ARMA models. Any vector ARMA model may be expressed in STF models and conversely. Thus, these two classes of models are interchangeable and both have their own strengths and weaknesses. STF models allow rational polynomial operators with each variable of an equation while ARMA models permit only linear operators. However, if the MA matrices of a vector ARMA model are not diagonal, then the vector ARMA model can be a more parsimonious parameterization than reduced form STF models.

Tentative model identification for a vector ARMA process is usually more straightforward than the STF models. However, in general, an initially identified vector ARMA model has more parameters than an STF model. When the number of variables in the system is large, the "parameter explosion" phenomenon in the vector ARMA model can cause difficulty in parameter estimation in terms of computer memory and execution time. Since STF models allow unrelated variables to be excluded from an equation and also allow the division of a system into several subsystems that may be modeled separately, STF models can be used for modeling a simultaneous system with a large number of variables. Owing to the special structure inherent in such models, parameter estimation for STF models can be achieved more easily and rapidly than for other models such as those considered by Zellner and Palm (1974), Reinsel (1979), and Tiao and Box (1981). The presence of one, and only one, random noise term in each equation permits the computation of the entire vector residual sequence to be carried out equation by equation. Therefore, the estimation of STF models can be achieved in more efficient algorithms and are capable of handling more variables in a model.

5. MODEL BUILDING

The class of STF models described in (1) is extensive and may contain a large number of variables and parameters. Given a set of time series $\underline{Y}_t$ and $\underline{X}_t$ of finite length, an aim is to find a model that contains as few parameters as possible and at the same time adequately represents the dynamic and stochastic relationships in the data at hand. Extending the basic ideas in Box and Jenkins (1970) for the univariate time series, a three phase iterative approach for building STF models is proposed: (i) tentative model identification, (ii) estimation, and (iii) diagnostic checking.

In general, tentative model identification is the most laborious phase in model building and relies heavily on the model builder's skills and judgement. Methods for tentative identification of STF models will be discussed in Section 6. After a model is tentatively identified for a simultaneous system, it can then be estimated. A tentatively identified model may or may not be appropriate; therefore, it is important to check if the statistics of the residual series are consonant with those of white noise processes. If not, the model should be modified, using information revealed in diagnostic checking.

6. TENTATIVE IDENTIFICATION OF AN STF MODEL

When embarking upon an empirically oriented research project, the researcher may have different levels of a priori knowledge as discussed in Section 1. Hanssens and Liu (1983) proposed a structural model identification procedure that is applicable to the level-1 case with the aim to move the researcher to level-2 knowledge, where a wide array of powerful parameter estimation techniques are available.

When only the information set is known (i.e., level-0 case), an attempt to build a structural model directly based on data can be dangerous. It is advisable to build a reduced form model first (e.g., an unrestricted reduced form STF model, or a vector ARMA model) and move to structural form model building afterward. The task to move the user from level-0 to level-1 knowledge can be nontrivial. In addition to time series technique (e.g., stepwise autoregression and cross correlation analysis), some subject matter information or theory should also be considered.

A tentative model identification procedure for STF models described in Hanssens and Liu (1983) is slightly modified and outlined below. The procedure can be divided into two steps:

(1) estimating transfer function weights (impulse response weights) and identifying an ARMA noise model for each equation;

(2) expressing linear transfer functions as rational transfer functions.

6.1 Estimating the Transfer Function Weights

Depending on the absence or presence of a contemporaneous endogenous relationship in an equation, the method for the estimation of transfer function weights may be different. The procedure for each case is outlined below:

<u>Equations without contemporaneous endogenous relationships</u>

When an equation in an STF model does not contain contemporaneous endogenous relationships, the least squares identification procedure described in Liu and

Hanssens (1982) and Liu et al. (1983) can be applied to the equation directly. Assuming no contemporaneous relationship between Y_{1t} and Y_{2t} in the first equation of model (2), the transfer function weights for $\beta_{11}(B)$ and $\beta_{12}(B)$ can be estimated from the linear model:

$$Y_{1t} = C_1 + V_{11}(B)X_{1t} + V_{12}(B)Y_{2t} + N_{1t} \tag{7}$$

where $V_{11}(B)$ and $V_{12}(B)$ are linear transfer functions with sufficient numbers of terms, say K_{11} and K_{12}. Note that the linear operator $V_{12}(B)$ must not contain a lag zero parameter.

To obtain more efficient estimates of transfer function weights and to simplify the identification procedure, it has been found useful to include a low order AR polynomial (e.g., an AR(1) process) in the disturbance, N_{1t}. When the disturbance contains possibly a seasonality of s period, this low order AR polynomial may be of the form $\phi_1(B)\phi_2(B^s)$ (e.g., $(1-\phi_1 B)(1-\phi_2 B^s)$). Using this modification, the model in (7) may be written as

$$Y_{1t} = C_1 + V_{11}(B)X_{1t} + V_{12}(B)Y_{2t} + \frac{1}{\phi_1(B)\phi_2(B^s)} a'_{1t} . \tag{8}$$

After the transfer function weights are estimated, an estimate of the disturbance can then be computed using

$$\hat{N}_{1t} = Y_{1t} - \hat{C}_1 - \hat{V}_{11}(B)X_{1t} - \hat{V}_{12}(B)Y_{2t} . \tag{9}$$

The computations involved in (8) and (9) can be accomplished easily (see e.g. Liu and Hudak 1983, and Liu et al. 1983).

After the estimated disturbance, $\hat{N}_{1t}$, is calculated, an ARMA model appropriate for the disturbance can then be identified.

Equations with contemporaneous endogenous relationships

For an equation containing contemporaneous endogenous relationships, the random errors (a_{it}) in the equation may be contemporaneously related with some endogenous input variables in the same equation. In such cases, the least squares estimates of the transfer function weights are inconsistent. Therefore, the identification procedure described above needs to be revised. Hanssens and Liu (1983) proposed an instrumental variables (IV) approach to obtain consistent estimates for the transfer function weights. Using the model described in (2), and assuming that the second equation contains contemporaneous endogenous relationships, a modified procedure of Hanssens and Liu can be summarized as follows:

(1) obtain an approximation of Y_{1t} via lag regression over X_{1t} and X_{2t}, using a sufficiently large number of lags. Thus may be written as

$$Y_{1t} = c_1 + \pi_{11}(B)X_{1t} + \pi_{12}(B)X_{2t} + \frac{1}{\phi_1(B)} u_{1t} . \tag{10}$$

The instrument of the endogenous variable Y_{1t}, say $\hat{Y}_{1t}$, is computed using

$$\hat{Y}_{1t} = \hat{c}_1 + \hat{\pi}_{11}(B)X_{1t} + \hat{\pi}_{12}(B)X_{2t} . \tag{11}$$

Since X_{1t} and X_{2t} are exogenous, $\hat{Y}_{1t}$ is independent of a_{2t}. Consequently, we can apply least-squares estimation of transfer function weights for the second equation, using the model

$$Y_{2t} = C_2 + V_{21}(B)X_{2t} + V_{22}(B)\hat{Y}_{1t} + \frac{1}{\phi_1(B)} a'_{2t} \tag{12}$$

where $\phi_1(B)$ must be the same as that in (10). The use of the common operator $\phi_1(B)$ in (10) and (12) is equivalent to <u>common filtering</u> described in Liu and Hanssens (1982) and Hanssens and Liu (1983). When the lag coefficients for $\phi_1(B)$ are small, the $\phi_1(B)$ polynomial can be treated as 1; therefore, (12) can be reduced to

$$Y_{2t} = C_2 + V_{21}(B)X_{2t} + V_{22}(B)\hat{Y}_{1t} + a'_{2t} . \tag{13}$$

On the other hand, if u_{1t} in (10) possesses strong seasonality, then AR term of the form $\phi_1(B)\phi_2(B^s)$ should be used in (10) and (12). If a'_{2t} in (12) or (13) is rather different from a white noise process, an appropriate ARMA term should be included in the model.

(2) After the linear form transfer functions are appropriately estimated, the estimates of N_{2t} in (2) may be computed, using

$$\hat{N}_{2t} = Y_{2t} - \hat{C}_2 - \hat{V}_{21}(B)X_{2t} - \hat{V}_{22}(B)Y_{1t} . \tag{14}$$

An appropriate ARMA model for N_{2t} based on $\hat{N}_{2t}$ can then be identified.

~~This ins~~trumental variables procedure provides initial information about the lag structure of the transfer functions for each input variable. The estimates of the transfer function weights under the instrumental variables method are consistent but not necessarily efficient, since we do not use the information that a_{1t} and a_{2t} may be correlated.

In general, the instrumental variables method will provide satisfactory results if the goodness of fit for the instruments is satisfactory. When the instrument-R^2 is low, the result from this method may be rather inefficient. In such cases, Hanssens and Liu (1983) suggest using both the instrumental variables and the least squares procedures to reconcile a slightly over-parameterized model. If the system is only slightly over-parameterized, the spurious parameters probably will be identified and removed in the estimation stage. When the goodness of fit is small for all the instruments, the system is not driven by the exogenous variables. Such a system behaves almost like a closed-loop time series model and any attempt to model contemporaneous relationships may not be fruitful.

6.2 Determination of the Rational Polynomial $\omega(B)/\delta(B)$

After the transfer function weights are obtained, the form of the transfer function $\omega(B)/\delta(B)$ can then be determined. If the linear transfer function $V(B)$ has a cut-off pattern, then $\delta(B)=1$ and $\omega(B)$ comprises the significant terms in $V(B)$. If $V(B)$ has a die-out pattern, the corner method proposed by Liu and Hanssens (1982) can be used to determine the values r, s, and b in the rational polynomial $\omega(B)/\delta(B)$.

7. ESTIMATION OF AN STF MODEL

Once an STF model is tentatively identified, the parameters and the covariance matrix $\underline{\Sigma}$ can be estimated by maximizing the corresponding likelihood function. The maximum likelihood estimation for structural models is usually referred to as full-information maximum likelihood (FIML) estimation in econometrics. Extensive literature exists on properties of the FIML estimates (see, for example, Sargan 1961, Hendry 1971, Chow and Fair 1973, Palm and Zellner 1980). When there are no definitional equations, Wall (1975) derived the following likelihood function for model (1),

$$L(\underline{\eta}) \propto (|\underline{\Sigma}|^{\frac{1}{2}n} / |\underline{\Phi}_0|^n) \exp\{-\tfrac{1}{2} \sum_{t=1}^{n} \underline{a}_t' \underline{\Sigma}^{-1} \underline{a}_t\} \tag{15}$$

where $\underline{\Phi}_0$ is the lag 0 structure matrix that contains the contemporaneous endogenous terms.

When the model contains definitional equations, it is possible to eliminate these exact equations via substitution in order to obtain a system of k equations. The likelihood function in (15) then can be used exactly as it stands. This tedious substitution is performed by computer program automatically as long as the definitional equations are correctly specified.

8. DIAGNOSTIC CHECKING

Once the parameters are estimated, various diagnostic checks should be performed on the estimated residual series to determine the adequacy of the model fit and to search for directions of improvement, if necessary. Useful methods include plotting the individual residual series to spot possible outliers and to study the cross correlation matrices among $\{\underline{a}_t\}$ and the prewhitened input series to determine if it is consonant with that of a white noise vector process and check if there is any lack of fit.

9. FORECASTING FUTURE OBSERVATIONS

In addition to structural analysis, STF models can also be used for forecasting. In general, either non-stochastic or stochastic variables may be used as exogenous variables. If non-stochastic variables are used, the variables will not contribute stochastic variations to the forecasts since the information on these variables are completely predetermined. A complete STF model may contain a sequence of individual models or one joint model for all stochastic exogenous variables. These models may be univariate ARIMA, transfer function, vector ARMA or another STF model. Without loss of generality, we will assume that all exogenous variables are stochastic in the following discussion.

For the purpose of illustration, the model in (1) is rewritten in the following form:

$$[\underline{\Phi}_0 + \underline{\Phi}_1(B)]\underline{Y}_t = \underline{C} + [\underline{\Gamma}_0 + \underline{\Gamma}_1(B)]\underline{X}_t + \underline{\Delta}(B)\underline{a}_t , \qquad \underline{a}_t \sim N(\underline{0}, \underline{\Sigma}) . \tag{16}$$

In general, the model for the stochastic exogenous variables can be written as

$$[\underline{\Phi}_{x0} + \underline{\Phi}_{x1}(B)]\underline{X}_t = \underline{C}_x + \underline{\Theta}(B)\underline{a}_{xt} , \qquad \underline{a}_{xt} \sim N(\underline{0}, \underline{\Sigma}_x) . \tag{17}$$

Combining the above two models, the following model is obtained:

$$\begin{bmatrix} \underline{\Phi}_0 & -\underline{\Gamma}_0 \\ \underline{0} & \underline{\Phi}_{x0} \end{bmatrix} + \begin{bmatrix} \underline{\Phi}_1(B) & -\underline{\Gamma}_1(B) \\ \underline{0} & \underline{\Phi}_{x1}(B) \end{bmatrix} \begin{bmatrix} \underline{Y}_t \\ \underline{X}_t \end{bmatrix} = \begin{bmatrix} \underline{C} \\ \underline{C}_x \end{bmatrix} + \begin{bmatrix} \underline{\Delta}(B) & \underline{0} \\ \underline{0} & \underline{\Theta}(B) \end{bmatrix} \begin{bmatrix} \underline{a}_t \\ \underline{a}_{xt} \end{bmatrix} \tag{18}$$

or

$$(\underline{\Phi}_* + \underline{\Phi}_{**}(B))\underline{Z}_t = \underline{C}_* + \underline{\Theta}_* \underline{b}_t , \qquad \underline{b}_t \sim N(\underline{0}, \underline{\Sigma}_*) \tag{19}$$

where $\underline{Z}_t' = [Y_t', X_t']$, $b_t' = [\underline{a}_t', \underline{a}_{xt}']$, $\underline{C}_*' = [\underline{C}', \underline{C}_x']$,

$$\underline{\Phi}_* = \begin{bmatrix} \underline{\Phi}_0 & -\underline{\Gamma}_0 \\ \underline{0} & \underline{\Phi}_{x0} \end{bmatrix}, \quad \underline{\Phi}_{**}(B) = \begin{bmatrix} \underline{\Phi}_1(B) & -\underline{\Gamma}_1(B) \\ \underline{0} & \underline{\Phi}_{x1}(B) \end{bmatrix}$$

$$\underline{C}_* = \begin{bmatrix} \underline{C} \\ \underline{C}_x \end{bmatrix}, \quad \underline{\Theta}_* = \begin{bmatrix} \underline{\Lambda}(B) & \underline{0} \\ \underline{0} & \underline{\Theta}(B) \end{bmatrix} \quad \text{and} \quad \underline{\Sigma}_* = \begin{bmatrix} \underline{\Sigma} & \underline{0} \\ \underline{0} & \underline{\Sigma}_x \end{bmatrix}.$$

In the above formulation, we assume that the model for the exogenous variables in (17) does not contains lagged endogenous variables. When lagged endogenous variables appear in (17), the joint model can still be written in (18) and (19) except that the zero submatrix and $\underline{\Phi}_{x1}(B)$ matrix in $\underline{\Phi}_{**}(B)$ need to be modified accordingly.

The model in (19) can be approximately written in a linear form model as:

$$(\underline{\phi}_0-\underline{\phi}_1B-\underline{\phi}_2B^2-\ldots-\underline{\phi}_pB^p)\,\underline{Z}_t = \underline{C} + (\underline{I}-\underline{\theta}_1B-\underline{\theta}_2B^2-\ldots-\underline{\theta}_qB^q)\,\underline{b}_t \tag{20}$$

This model is similar to a vector ARMA(p,q) model except that the contemporaneous structure matrix $\underline{\phi}_0$ ($\underline{\phi}_0=\underline{\Phi}_*$) may not be an identity matrix.

Let $\hat{\underline{Z}}_t(l)$ be the minimum mean squared error forecast of $\underline{Z}_{t+l}$ made at time origin t and $\underline{e}_t(l) = \underline{Z}_{t+l} - \hat{\underline{Z}}_t(l)$ be the corresponding vector of forecast errors. The l step ahead forecast $\hat{\underline{Z}}_t(l)$ can be recursively computed from

$$\hat{\underline{Z}}_t(l) = \underline{\phi}_0^{-1}(\underline{C}+\underline{\phi}_1\hat{\underline{Z}}_t(l-1)+\ldots+\underline{\phi}_p\hat{\underline{Z}}_t(l-p)-\underline{\theta}_1E(\underline{b}_{t+l-1})-\ldots-\underline{\theta}_qE(\underline{b}_{t+l-q})) \tag{21}$$

with $\hat{\underline{Z}}_t(j) = \underline{Z}_{t+j}$ $(j \le 0)$, $E(\underline{b}_{t+j}) = \underline{0}$ $(j > 0)$, and $E(\underline{b}_{t+j}) = \underline{b}_{t+j}$ $(j \le 0)$.

Under the assumption that $\{\underline{b}_t\}$ is a white noise sequence, the error vector $\underline{e}_t(l)$ is normally distributed with zero mean and covariance matrix

$$V(\underline{e}_t(l)) = \sum_{i=0}^{l-1} \underline{\psi}_i\,\underline{\Sigma}\,\underline{\psi}_i'$$

where the $\underline{\psi}$ matrices satisfy the following relationship

$$(\underline{\phi}_0-\underline{\phi}_1B-\ldots-\underline{\phi}_pB^p)(\underline{\psi}_0+\underline{\psi}_1B+\underline{\psi}_2B^2+\ldots) = \underline{I}-\underline{\theta}_1B-\underline{\theta}_2B^2-\ldots-\underline{\theta}_qB^q \tag{22}$$

Since the model in (20) is a linear approximation of the rational model in (19), it is more accurate to use the model in (19) to compute the forecasts. The $\underline{Z}_t(j)$ and $E(\underline{b}_{t+j})$ required in the computation are the same as in (21).

10. AN EXAMPLE

In order to demonstrate the usefulness of STF models, it is now applied to some actual data. This example demonstrates the incorporation of level-1 knowledge into the initial STF modeling stage and provides the FIML estimates of a tentatively identified model. Some details of the tentative model identification stage for this model may be found in Hanssens and Liu (1983).

Quenouille (1957) analyzed a multivariate data set comprised of the annual hog supply, hog price, corn supply, corn price, and farm wages of the United States.

This set of data has been studied in a number of papers (e.g., Box and Tiao 1977, Tiao and Tsay 1982, Hanssens and Liu 1983). A fairly simple econometric model can be postulated to underlie movements in hog supply and hog price, but it is difficult to specify the dynamics of such a model a priori. The supply of hogs is expected to vary positively with the price of hogs and the supply of corn, but negatively with the price of corn. The price of hogs, in turn, is negatively affected by the hog supply and positively affected by the cost factors, corn price and farm wages. No exact a priori information is available on the dynamics of these relation, except that the current hog price can only affect next year's (and subsequent years') hog supply because of a one-year production cycle. Hence, a reasonable model relating the exogenous variables corn supply (CS), corn price (CP) and farm wages (FW) to the endogenous variables hog supply (HS) and hog price (HP) is:

$$\begin{aligned} HS_t &= C_1 + \gamma_1(B)HP_{t-1} + \beta_1(B)CS_t + \beta_2(B)CP_t + N_{1t} \\ HP_t &= C_2 + \gamma_2(B)HS_t + \beta_3(B)CP_t + \beta_4(B)FW_t + N_{2t} . \end{aligned} \tag{23}$$

The procedure described in Section 6 leads to the following structural model:

$$\begin{aligned} HS_t &= c_1 + \omega_1 HP_{t-1} + \omega_2 CS_{t-1} + (\omega_3 B^2 + \omega_4 B^3)CP_t + \frac{1-\theta_1 B^2}{1-\phi_1 B} a_{1t} \\ HP_t &= c_2 + (\omega_5 + \omega_6 B)HS_t + \omega_7 FW_t + \frac{1}{1-\phi_2 B} a_{2t} . \end{aligned} \tag{24}$$

FIML estimates of the above model were obtained using the SCA statistical System (Liu et al. 1983) and are presented below. Standard errors of the estimates are given in parentheses.

$$\begin{aligned} HS_t &= \underset{(46.768)}{359.102} + \underset{(.025)}{.308}HP_{t-1} + \underset{(.032)}{.279}CS_{t-1} + (\underset{(.022)}{-.245}B^2 \underset{(.024)}{-.157}B^3)CP_t + \frac{1+\overset{(.121)}{.395}B^2}{1-\underset{(.086)}{.726}B} a_{1t} \\ HP_t &= \underset{(180.089)}{981.019} + (\underset{(.178)}{-.658}\underset{(.182)}{-1.430}B)HS_t + \underset{(.097)}{1.394}FW_t + \frac{1}{1-\underset{(.073)}{.775}B} a_{2t} \end{aligned} \tag{25}$$

$$\hat{\Sigma} = \begin{bmatrix} 474.14 & \\ 133.10 & 3261.54 \end{bmatrix}$$

A similar model building procedure may be used to identify and estimate an unrestrictive reduced form model for CS, CP and FW. In the above model, the signs of the parameter estimates are consistent to theory.

REFERENCES

BOX, G.E.P., and G.M. JENKINS (1968). Some Recent Advances in Forecasting and Control, Part I. Applied Statistics 17: 91-109.

BOX, G.E.P., and G.M. JENKINS (1970). Time Series Analysis, Forecasting and Control. San Francisco: Holden-Day.

CHOW, G.C., and R.C. FAIR (1973). Maximum Likelihood Estimation of Linear Equation Systems with Autoregressive Residuals. Annals of Economic and Social Measurement 2: 17-28.

FAIR, R.C. (1970). The Estimation of Simultaneous Equation Models with Lagged Endogenous Variables and First Order Serially Correlated Errors. Econometrica 37: 507-16.

GRANGER, C.W.J. and P. NEWBOLD (1977). Forecasting Economic Time Series. Academic Press, New York.

HANNAN, E.J. (1971). The Identification Problem for Multiple Equation System with Moving Average Errors. Econometrica 39: 751-765.

HANSSENS, D.M., and L.-M. LIU (1983). Lag Specification in Rational Distributed Lag Structural Models. Journal of Business and Economic Statistics 1: 316-325.

HANSSENS, D.M. and J.G. MANEGOLD (1980). Descriptive Model Building on Multivariate Time-Series Data, UCLA Center for Marketing Studies, Working Paper 98.

HENDRY, D.F. (1971). Maximum Likelihood Estimation of Systems of Simultaneous Regression Equations with errors generated by a vector autoregressive process. International Economic Review 17: 362-376.

JENKINS, G.M. and A.S. ALAVI (1981). Some Aspects of Modeling and Forecasting Multivariate Time Series. Journal of Time Series Analysis 2: 1-47.

KOHN, R. (1979). Identification Results for ARMAX Structures. Econometrica 47: 1295-1304.

LIU, L.-M. and D.M. HANSSENS (1982). Identification of Multiple-Input Transfer Function Models. Communications in Statistics - Theory and Methods 11: 297-314.

LIU, L.-M. and G.B. HUDAK (1983). An Integrated Time Series Analysis Computer Program: The SCA Statistical System. Time Series Analysis: Theory and Practice 4. Ed: O.D. Anderson, North-Holland, Amsterdam and New York, 291-308.

LIU, L.-M., G.B. HUDAK, G.E.P. BOX, M.E. MULLER, and G.C. TIAO (1983). The SCA System for Univariate-Multivariate Time Series and General Statistical Analysis. Scientific Computing Associates, P.O. Box 625, DeKalb, Illinois 60115.

NEWBOLD, P. and G.V. REED (1979). The Implications for Economic Forecasting of Time Series Model Building Methods. Forecasting. Ed: O.D. Anderson, North-Holland, Amsterdam and New York, 185-214.

PALM, F. and A. ZELLNER (1980). Large Sample Estimation and Testing Procedures for Dynamic Equation Systems. Journal of Econometrics 12: 251-283.

QUENOUILLE, M.H. (1957). The Analysis of Multiple Time-Series. New York: Hafner Publishing Company.

REINSEL, G. (1979). FIML Estimation of the Dynamic Simultaneous Equations Model with ARMA Disturbances. Journal of Econometrics 9: 263-281.

SARGAN, J.D. (1961). The Maximum Likelihood Estimation of Economic Relationships with Autoregressive Residuals, Econometrica 19: 414-426.

TIAO, G.C. and G.E.P. BOX (1981). Modeling Multiple Time Series with Application. Journal of American Statistical Association 76: 802-816.

TIAO, G.C. and R.S. TSAY (1983). Multiple Time Series Modeling and Extended Sample Cross-Correlations. Journal of Business and Economic Statistics 1: 43-56.

WALL, K.D. (1975). FIML Estimation of Rational Distributed Lag Structural Form Models. NBER Working Paper 77.

WALL, K.D. (1976). FIML Estimation of Rational Distributed Lag Structural Form Models. Annals of Economic and Social Measurement 5: 53-64.

ZELLNER, A. and F. PALM (1974). Time Series Analysis and Simultaneous Equation Econometric Models. Journal of Econometrics 2: 17-54.

ZELLNER, A. (1979). Statistical Analysis of Econometric Models. Journal of American Statistical Association 74: 628-651.

TIME SERIES ANALYSIS: Theory and Practice 7
O.D. Anderson (editor)

AN EMPIRICALLY DERIVED VECTOR ARMA MODEL FOR MONETARY AND FINANCIAL VARIABLES

Kenneth J. Jones
Brandeis University, Waltham, MA 02154, USA

A set of empirically derived forecast models is described and evaluated. The models forecast some twenty seven financial, monetary and exchange rate variables as weekly averages. Several criteria of merit are described. It is concluded that the models' forecasts are superior to that of a random walk.[1]

INTRODUCTION

For some time there seems to have been a logical controversy over the relationship between monetary variables and financial prices. Theorists (Friedman, 1969) argue for a logical causal interconnectedness which would imply the forcastability of such financial variables, as interest or exchange rates. On the other side, a legion of empiricists since Cootner (1964) have failed to find predictive models. Even the large econometric macro models are doubted to do better than a random-walk. Evaluations of these (McNees, 1979) present a mixed picture, but considering the large number of variables involved and the timeliness of the forecasts, one would be hard pressed to reject a random walk hypothesis.

In order to get the empirical results to fit with theory (and for that matter common sense), we have the efficient markets hypothesis which tells us that information is absorbed "instantaneously" by the markets; hence, we cannot us it for forecasting. This seems to be an incredible device for explaining the phenomenon. It would seem that a more likely explanation would be that our prediction models and the process by which we have been building them are inadequate.

It seems clear that the methodology of choice is vector-autoregressive-moving average (Box and Tiao, 1981). Jones et al. (1982) have shown that this methodology is capable of developing forecasts considerably better than that of a random-walk; and outperforming a large econometric model. The vector ARMA modeling procedure can be carried out in a completely a-theoretic fashion; or the models can be restricted by theoretical considerations. If the process is properly carried out, one can attain a model which extracts the maximum linear predictive information from the data. However, the model may not be unique. The simultaneous equations obtained are in reduced form but still may be very informative from a theoretic point of view.

METHODS

In order to explore the interconnectedness of monetary and financial data, a set of 27 variables was chosen. Since a weekly frequency was selected, data availability was a consideration for variable selection in addition to

[1]I wish to thank Dr Daniel F.X. O'Reilly for writing the computer program which produced the clock diagrams and to Data Resources, Inc. of Lexington, Massachusetts for supplying the computer time.

theoretical meaningfullness. A more detailed discussion of the rational for variable and frequency selection appears in Jones and O'Reilly (1983). Table 1 lists the variables under analysis for this study. Where the data were available on a daily basis, weekly averages were computed.[2]

Since current hardware limitations keep models of more than 20 variables out of practical reach, five separate overlapping models were estimated; four of sixteen variables and one of seventeen. All variables were analyzed as week to week changes. In addition, a method of modeling the changing variances was employed.

The data for one model encompassed the period from July 11, 1980 through January 14, 1983, 132 weekly observations. The other models used the period July 11, 1980 through June 3, 1983 for the fitted sample. The latter comprises 152 weekly observations. The out-of-sample forecast period is made up of the period from the end of fitted sample through August 5, 1983.

Table 1 presents a summary of data from the forecasts which represent the relative forecast accuracy for each variable. The column of major interest indicates the percent of correctly anticipated week to week changes in the out of sample data, for two weeks from forecast. It is demarcated by double column lines. The reason it is of major interest is that it is beyond the reach of the Working artifact[2], and it is the first forecast which would be useful for any trading scheme.

From Table 1 it is clear that some variables are fitted better than others; and that the degree of shrinkage in forecast accuracy varies. However, the preponderance of values greater than 50% indicates that most of the series are forecastable in their first differences. The turning point data indicates that the system is also anticipating changes in direction of changes.[3]

DISCUSSION

The major question concerns whether or not the forecasts <u>ex ante</u> relative to the model are significantly different from, or better than, random-walk predictions of no change. Since the series are stationary over the past, one may assume an expected forecast accuracy of fifty-percent under the null hypothesis. Almost all of the models' forecasts exceed this. The standard error for 29 observations is roughly nine-percent; so for a one-tail five-percent bound a forecast accuracy in excess of sixty-five percent is needed. A number of variables exceed this level. The smaller <u>ex ante</u> sample of nine is too small to provide a meaningful test, but the fact that these also exceed the one half correct level is suggestive. It does not seem reckless to conclude that it is possible to extract information from the multivariate data-set which will contribute to the forecastability of financial variables and that this is not instantaneously discounted in the auction markets.

---

[2]It has been shown by Working (1960) that the averaging of five elements of a random chain will introduce a first order correlation of first differences of expected value equal to 0.235, and the new process will be MA(1) in differences. Since this level of correlation is non-significant in the present sample and since the model contains few lag one coefficients, it is assumed that the artifact does not account for any substantial proportion of the results.

[3]The expected value for anticipated turning points for a trend-follower model would be 0% and for a random model would be 50%. A trend-contrary model would, of course, get 100% of the turning points.

Table 1. MODEL FORECAST ACCURACY

Variables		Periods							
		Fitted[4]				Ex Ante[5]			
	%variance[6]	one step		two step[8]	N	one step		two step	N
		%right signs	%turn[7] points	%right signs		%right signs	%turn points	%right signs	
Money									
1. Demand	73	81	83	81	132	72	69	72	29
2. Savings	30	87	56	87	132	52	53	52	29
3. Loans	42	75	87	76	132	66	72	62	29
4. US Deposits	42	70	63	70	132	59	82	59	29
5. Foreign Deposits	35	70	87	65	132	52	63	45	29
Prices									
6. Spot Index	19	72	53	70	132	52	11	62	29
7. Spot Oil	23	65	35	57	132	66	46	61	29
8. #2 Oil (future)	49	76	63	77	152	22	20	11	9
9. NYSE composite	25	75	47	76	132	55	57	52	29
10. Prime	55	38	100	37	132	N	NC[9]	NC	29
11. Fed Funds	27	60	51	56	132	62	59	55	29
12. T Bills (90 day)	19	64	56	71	152	89	86	78	9
13. T Bills (future)	25	66	73	64	152	100	100	100	9
14. GNMA (future)	19	62	63	67	152	67	75	78	9
Foreign Exchange									
15. Swiss Franc (spot)	26	71	63	64	152	56	50	75	9
16. Deutchmark (spot)	24	69	58	67	152	67	50	63	9
17. Yen (spot)	29	69	65	64	152	67	75	13	9
18. Pound (spot)	26	68	72	68	152	56	50	50	9
19. Swiss Franc (futures)	54	76	67	81	132	66	67	58	9
20. Deutchmark (futures)	32	77	67	64	152	67	50	78	29
21. Yen (futures)	30	70	71	68	132	76	64	72	29
22. Pound (futures)	28	77	71	76	132	52	50	57	29
23. Canadian $ (futures)	22	62	42	60	152	44	25	56	9
24. Gold (spot)	40	70	61	69	132	55	56	75	29
25. Gold (futures)	32	68	55	79	152	56	50	50	9
26. Silver (futures)	27	70	64	79	152	56	50	38	9
27. Copper (futures)	21	59	59	62	152	56	50	50	9

---

[4]Interval = weekly from July 11, 1980 through January 14, 1983 (N=132) or June 3, 1983 (N=152).

[5]Interval = weekly from January 21, 1983 through July 29, 1983 (N=29) or June 10, 1983 through July 21, 1983 (N=9).

[6]% variance is percent of the variance of the first differences (week to week changes in average level) explained by the model.

[7]% turning points is the percent of changes of direction of the changes in weekly averages correctly anticipated by the model.

[8]The two step ahead forecast is for the average value two weeks from the Friday on which the forecast is computed.

[9]NC indicates no turning point or no change.

REFERENCES

BOX, G.E.P. and TIAO, G.C. (1981) Modelling Multiple Time Series with Applications. JASA 76, 802-810.

COOTNER, P.H. (ed) (1964). The Random Character of Stock Market Prices. MIT Press, Cambridge.

FRIEDMAN, M. (1969). The Optimum Quantity of Money and Other Essays. Chicago.

JONES, K.J, O'REILLY, D.F.X., HUI, B.S., and SHEEHAN, K. (1983) The Use of Vector ARMA in Macroeconomic Forecasting. In Time Sries Analysis: Theory and Practice 4 (Proceedings of International Conference held in Cincinnati, Ohio, August 1982). Ed: O.D. Anderson, North-Holland, Amsterdam & New York, 193-197.

JONES, K.J. and O'REILLY, D.F.X, (1983). High Frequency Vector ARMA Models. Third International Synposium on Forecasting, June 5-8, Philadelphia.

MCNEES, S.K. (1979). The Forecasting Record for the 1970's. New England Economic Review, Sept. - Oct., 33-53.

WORKING, H. (1960). A Note on the Correlation of First Differences of Averages in a Random Chain. Econometrica 28, 916-918.

TIME SERIES ANALYSIS: Theory and Practice 7
O.D. Anderson (editor)

DYNAMIC ADJUSTMENT OF DISAGGREGATED UNEMPLOYMENT SERIES

Houston H. Stokes
University of Illinois at Chicago Box 4348, Chicago, Illinois 60680, USA

This paper uses vector autoregressive models and spectral tests, developed by Geweke, to study the dynamic structure of disaggregated unemployment series. Well known variants of the Phillips curve model are estimated with the new procedure and their results are compared to prior interpretations of the data. Important questions asked include whether the long and the short run should be defined in terms of lag length or frequency. The paper concludes that the frequency definition is more appropriate and that further work utilizing this approach is warranted.

1. INTRODUCTION

The relationship between unemployment and measures of inflationary pressure has been studied for almost 30 years without completely resolving empirically the underlying theoretical issues. While a number of economists feel that the long run Phillips curve is vertical (for example, Meltzer (1977), Stein (1978) and Wasserfallen (1983)), others, such as Eckstein and Girola (1978) and Tobin (1972) have reported research that supports a negative slope for the Phillips curve. Since the Phillips curve research has been surveyed in Santomero and Seater (1978), it will not be discussed here; however, it is worth noting that much of the controversy focuses on whether the Phillips curve relationship exists in the long run, the short run or at all. Another important area of research is whether the relationship is different in disaggregated data. Stokes-Jones-Neuberger (1975), looking at disaggregated national and regional data, found no evidence for the structural Phillips curve theory that there was a relationship between unemployment and real wages. Evidence was found at the national level for the expectations model that postulated a relationship between unemployment and nominal wages. Stokes (1983b) extended this work to the Midwest region and found, using a VAR model, that there are differences in the lead/lag relationship between unemployment rates in the Midwest states and unemployment rates in the United States which is not surprising given the variation in industrial structure across regions. It was difficult to make statements about the long and short run from inspection of VAR model coefficients, which, in their raw form, can be used to measure Granger (1969) causality and the speed of adjustment. In the Phillips curve literature, there has been much confusion between the concept of the speed of adjustment and the terms long and short run. It is not always correct to assume that, in a transfer function model of the form

$$y_t = V(B)x_t + \psi(B)e_t \tag{1}$$

(where $V(B)$ and $\psi(B)$ are polynomials in the lag operator B ($Bx_t = x_{t-1}$)), short-run effects of x_t on y_t are manifest in significant low-order terms of $V(B)$ and long-run effects are manifest in significant high-order terms of $V(B)$.[1] What the

---

[1]In Box and Jenkins (1976, 57-58), it is shown that both the order and magnitude of the coefficient determines the spectrum. Given an AR(1) Model $(1 - \phi B)x_t = e_t$, the spectrum at frequency λ becomes $p(\lambda) = 2\sigma_e^2/(1 + \phi^2 - 2\phi \cos 2\pi\lambda)$. If ϕ is a large positive number, the autocorrelation function decays slowly and the spectrum shows domination at low frequencies. If ϕ is a large negative number, the autocorrelation function will alternate in sign as it decays and the spectrum will be dominated by high frequencies. The importance of this simple example is that it illustrates how the frequency depends on both the order and the magnitude of the estimated coefficient in the AR model. In the VAR case, as we will see later, the relationship becomes more complex and the need to decompose the VAR model into the frequency domain more important.

possibly significant low-order and high-order terms in V(B) are measuring is the speed (time lag) with which a change in x_t is transmitted to y_t. The linking of the speed of adjustment with the "short run" and the "long run", as was done in modeling the effect on interest rates of monetary changes (Stokes and Neuburger (1979) and Stokes (1983c)), is not strictly correct in the Phillips curve literature, which stresses the concept that in the "short run" workers might be fooled into working more, if the rate of price increase increases (thus lowering unemployment), but that in the "long run" workers realize that the price movements are only nominal changes, not real changes, and thus the rate of unemployment returns to the natural rate.

Within the spirit of the Phillips curve literature, the short run could be construed to be high-frequency relationships between variables and the long run low-frequency relationships between variables. In order to test this interpretation, a vector autoregressive (VAR) model between disaggregated unemployment measures and inflation pressure variables will be transformed to the frequency domain, following the suggestions of Geweke (1982a, b, c; 1983)[2], to test the natural rate and monetary models of unemployment proposed and tested using single equation OLS methods by Rea (1983). The next section will discuss the models estimated and give a brief overview of the Geweke (1982a) procedure, while the last section will discuss the findings that have been summarized in Tables 1 through 17.

2. THE STATISTICAL CONSIDERATIONS

A linear time series process can be represented as

$$A(B)Z_t = D(B)e_t \tag{2}$$

where Z_t' is a row vector of k random variables and A(B) and D(B) are each k by k matrices whose elements are finite polynomials in the lag operator B. It is assumed that the zeros of the determinantal polynomials $|A(B)|$ and $|D(B)|$ are outside the unit circle so the VARMA model in equation (2) could be written as a pure VAR model

$$G(B)Z_t = D(B)^{-1}A(B)Z_t = e_t \tag{3}$$

or a pure VMA model

$$Z_t = P(B)e_t = A(B)^{-1}D(B)e_t. \tag{4}$$

The maximum order in A(B) is assumed to be p and the maximum order in D(B) is assumed to be q. The VARMA form of the model (equation 2) has, in general, lower-order elements than the VAR or VMA forms of the model. A(B) and D(B) can be expressed as

$$A(B) = a_0 + \dots + a_p B^p \tag{5}$$

$$D(B) = d_0 + \dots + d_q B^q \tag{6}$$

where a_i is the k by k matrix of the ith order elements of matrix A(B) and d_j is the k by k matrix of the jth order elements of the matrix D(B). If it is assumed

---

[2]Geweke (1982b) has used his frequency decomposition procedure at the aggregate level to study the relationship among measures of monetary policy, labor market activity and wage inflation and found strong bidirectional feedback between labor market activity and the monetary growth rate, little effect of nominal wage inflation on the monetary growth rate and unemployment, and strong feedback between labor market activity variables and the rate of wage growth.

that the zero-order elements, a_0 and d_0, are identity matrices, the form of the model is what Granger and Newbold (1977, 223) have called model A. Using this setup, instantaneous causality is seen as significant off diagonal cross correlations between error vectors. Tiao, Box, Grupe, Hudak, Bell and Chang (1980) have discussed how equation (2) might be estimated. The VAR form of the model, equation (3), is easier to interpret than the VARMA form of the model and has the additional advantage that it can be estimated by OLS. In this study the VAR form of the model will be estimated by OLS and the results will be reported in terms of significant coefficients (see Table 2). The VAR model will also be estimated by Whittle's (1963) procedure and the results decomposed by frequency, following procedures suggested by Geweke (1982a). The findings of these two approaches will be compared and contrasted[3] for various variables in the Z_t vector.

The VAR model given in equation (3) can be written as

$$Z_t = G^*(B)Z_t + e_t, \qquad \text{var}(e_t) = \textstyle\sum^* \tag{7}$$

where $G^*(B)$ has no zero-order elements, $G^*(B) = -G(B) - I$. Following the method suggested by Geweke (1982a), the k element vector Z_t can be decomposed into the k_1 element vector X_t and the k_2 element Y_t, where $k_1 + k_2 = k$.[4] X_t and Y_t can be modeled as VAR models of the form:

$$X_t = E_1^*(B)X_t + u_{1t}, \qquad \text{var}(u_{1t}) = \textstyle\sum_1 \tag{8}$$

$$Y_t = G_1^*(B)Y_t + v_{1t}, \qquad \text{var}(v_{1t}) = T_1 \tag{9}$$

as VAR models containing only first-order or greater lags:

$$X_t = E_2^*(B)X_t + F_2^*(B)Y_t + u_{2t}, \qquad \text{var}(u_{2t}) = \textstyle\sum_2 \tag{10}$$

$$Y_t = G_2^*(B)Y_t + H_2^*(B)X_t + v_{2t}, \qquad \text{var}(v_{2t}) = T_2 \tag{11}$$

or as VAR models containing zero-order lags[5]

$$X_t = E_3^*(B)X_t + F_3(B)Y_t + u_{3t}, \qquad \text{var}(u_{3t}) = \textstyle\sum_3 \tag{12}$$

$$Y_t = G_3^*(B)Y_t + H_3(B)X_t + v_{3t}, \qquad \text{var}(v_{3t}) = T_3. \tag{13}$$

Since $|\sum_1| \geq |\sum_2| \geq |\sum_3| > 0$ and $|T_1| \geq |T_2| \geq |T_3| \geq 0$, feedback from Y to X F(Y to X), can be defined as $\ln(|\sum_1|/|\sum_2|)$ and feedback from X to Y, F(X to Y), can be defined as $\ln(|T_1|/|T_2|)$.[6]

---

[3]Additional explanation of how the instantaneous relationship between the series, seen in the off diagonal elements of the covariance matrix, can be decomposed in the time domain can be found in Granger and Newbold (1977, 223) and Stokes (1983c).

[4]In all subsequent notation the matrices superscripted with a * do not contain zero order elements.

[5]If $k_1 = k_2 = 1$, equation (10) - (13) would be transfer function models.

[6]The motivation for this definition rests on the fact that the natural log of 1 is zero and the fact, as Geweke notes, that these measures can be interpreted in terms of the proportionate increase in the variance of the one-step-ahead population errors (see Geweke 1983b, 165).

Instantaneous feedback $F_{X.Y} = \ln(|\Sigma_2|/|\Sigma_3|) = F_{Y.X}$. Geweke has proved that

$$\ln(|\Sigma_2|/|\Sigma_3|) = \ln(|T_2|/|T_3|) = \ln(|T_2|\cdot|\Sigma_2|/|\Sigma_*|), \tag{14}$$

where $\Sigma*$ was defined in equation (7). If we define $F_{X,Y}$ as meaning linear dependence, then

$$F_{X,Y} = \ln(|\Sigma_1|\cdot|T_1|/|\Sigma_*|) = F(Y \text{ to } X) + F(X \text{ to } Y) + F_{X.Y}. \tag{15}$$

In this notation, Granger (1969) causality of X to Y implies F(Y to X) = 0.

The above measures can be decomposed into the frequency domain. Given f(X to Y, λ) is defined as the feedback from X to Y at frequency λ, this section discussed the conditions under which

$$\frac{1}{2\pi}\int_{-\pi}^{\pi} f(X \text{ to } Y, \lambda)\, d\lambda = F(X \text{ to } Y) \text{ and} \tag{16}$$

$$\frac{1}{2\pi}\int_{-\pi}^{\pi} f(Y \text{ to } X, \lambda)\, d\lambda = F(Y \text{ to } X). \tag{17}$$

Consider the block recursive system, consisting of equations (10) and (13), which, by construction, has the property that $\text{cov}(u_{2t}, v_{3t}) = 0$. If the bivariate system, equation (3), can be inverted, the block recursive system of equation's (10) and (13) can be inverted to form the VMA system:

$$\begin{bmatrix} X_t \\ Y_t \end{bmatrix} = \begin{bmatrix} P(B) & Q(B) \\ R(B) & S(B) \end{bmatrix} \begin{bmatrix} u_{2t} \\ v_{3t} \end{bmatrix}. \tag{18}$$

The first k_1 equations of (18) are

$$X_t = P(B)u_{2t} + Q(B)v_{3t}. \tag{19}$$

Equation (19) expresses X_t as a function of two orthogonal innovations. One, v_{3t}, represents new information coming from Y_t in the block recursive system and one, u_{2t}, represents new information coming from X_t and from both X_t and Y_t.[7] Geweke (1982a, 308) shows how equation (19) can, by using the Cramer representation, be transformed to the frequency domain as

$$S_x(\lambda) = \tilde{P}(\lambda)\, \Sigma_2\, \tilde{P}(\lambda)' + \tilde{Q}(\lambda)\, T_3\, \tilde{Q}(\lambda)', \tag{20}$$

where $S_x(\lambda)$ is the spectral density at frequency λ and where $\tilde{P}(\lambda)$ and $\tilde{Q}(\lambda)$ are the Fourier transforms of the polynomial matrices P(B) and Q(B).[8] Equation (20) is a very interesting way of representing the process. Geweke (1982b, pp 169) shows, "when X is univariate, $|P(\lambda)|^2\Sigma_2/|S_x(\lambda)|$ is the fraction of $S_x(\lambda)$ attributed to u_{2t}. When X is not univariate, the corresponding, generalized variance ratio is $|P(\lambda)\ \Sigma_2\ P(\lambda)|/|S_x(\lambda)|$." Feedback from Y to X at frequency λ is

[7] The reason that v_{3t} represents new information from Y_t alone arises from the fact that equation (13) contains current information from X_t on the right-hand side, because by assumption the polynomial matrix $H_3(B)$ contains zero-order terms. Equation (10), in contrast, does not contain zero-order terms for Y_t on the right-hand side in the polynomial matrix $F_2^*(B)$. Hence, u_{2t} contains new information coming from X_t and from both X_t and Y_t.

[8] The prime refers to the conjugate transpose.

$$f(Y \text{ to } X, \lambda) = \ln|S_x(\lambda)|/|\tilde{P}(\lambda)\ \Sigma_2 \tilde{P}(\lambda)'| \tag{21}$$

which approaches zero if v_{3t} is not reflected in X at frequency λ, and approaches infinity if the contribution of u_{2t} at frequency λ goes to zero. Geweke has proved that

$$F(Y \text{ to } X) \geqslant (1/2\pi) \int_{-\pi}^{\pi} f(Y \text{ to } X, \lambda)\, d\lambda. \tag{22}$$

The strict equality condition holds if $I - G_3^*(b)$ is invertible.[9] A result similar to equation (22) holds for F(X to Y).

Geweke uses the Whittle (1963) algorithm to solve the VAR system of equations (10) and (11)[10] which, if we assume that $C = \text{cov}(u_{2t}, v_{2t})$, can be used to derive estimates of the simultaneous equations transfer function system of equations (10) and (13). Specifically,

$$G_3^*(B) = G_2^*(B) - C'\Sigma_2^{-1} F_2^* \tag{23}$$

$$H_3(B) = H_2^*(B) - C'\Sigma_2^{-1} E_2^* \tag{24}$$

$$T_3 = T_2 - C'\Sigma_2^{-1} C. \tag{25}$$

The results reported in this paper follow the estimation procedure suggested by Geweke. An important problem is to determine the lag length of the VAR model. This is difficult because Geweke's program (MTSM) neither calculates the cross correlations of the estimated residuals nor is able to produce standard errors of the coefficients, because Whittle's algorithm is used. In the results reported here, the lag length of 12 months was selected by first estimating a VAR model, using OLS procedures in the WMTS-1 program for increasingly longer lags. The estimated standard errors were calculated for both models estimated[11] and the cross correlations of the residuals were calculated. Since the residuals were found to be white noise for lag length 12 the assumption was made that this lag length was sufficient to capture all the information relating the series in the VAR model. The next step was to re-estimate the VAR model, using the Geweke program, and follow the frequency decomposition procedure discussed earlier. The fact that the estimated coefficients are different is some cause for concern.[12]

---

[9]The proof of equation (22) is contained in theorem 2 in Geweke (1982a, 308). In subsequent empirical work, whenever this condition is not met, it is indicated in Tables 3 through 16.

[10]The Whittle approach involves solving the Yule-Walker equations for estimates of $E_2^*(B)$, $F_2^*(B)$, $G_2^*(B)$ and $H_2^*(B)$. Such a procedure is only asymptotically equivalent to OLS, but has the desirable property that the estimated VAR process is invertible (Geweke 1983, 35). Other advantages include computational speed, which is important in the replication step discussed later, and some reduction in computer storage requirements (from p^2n^2 to pn^2). Geweke (1983) lists as the only disadvantage of the procedure the fact that standard errors on the coefficients cannot be calculated. In testing the MTSM program, the VAR coefficients obtained by the Whittle procedure were found to be substantially different than those obtained by the OLS procedure in WMTS-1 (See Tiao et al. (1980) and Stokes (1983a)). A future task will be to discover if the results are sensitive to whether OLS or the Whittle procedure is used to estimate the VAR model.

[11]The form of the models estimated will be discussed in the next section.

[12]The loss of efficiency due to the Whittle algorithm can be seen if we compare the residual variances of the 8 models listed in Table 2 with the variances obtained with the Whittle algorithm. In Table 2 the OLS variances are given first, with the Whittle counterpart in parentheses. The variances of the two X_1 variables are very close. The variances of the X_2 variable (Y in terms of the model) are substantially larger for the Whittle approach.

Given sufficient computer resources, possibly a better approach would be to estimate equations (10) and (13) directly with a maximum likelihood procedure, rather than conditionally using the present procedure. If costs are a factor, the replication step could proceed with the Whittle approach. An alternative that uses less computer resources would be to at least estimate equations (10) and (11), using OLS before using equations (23) to (25) to obtain estimates of equations (10) and (13). If invertibility is a problem, the Whittle procedure would be a fallback option.

The study reports 25% and 75% fractiles on the proposed measures of feedback. These fractiles were estimated using a bootstrap procedure where R (in this case 40) sets of artificial data were generated having a population covariance matrix equal to the estimated sample covariance matrix of the original data. Analysis of these R sets of artificial data provides information on the distribution of the estimated measures of feedback in the frequency domain of the original data. In this study 40 replications were used, although probably a larger number would be better.[13] The replications provide some evidence about the distribution of the estimated measures of feedback. In Tables 3-16 the following conventions are observed. The raw feedback estimates are first given, followed by the percent of the variance explained in parentheses. The feedback estimates are then adjusted, following Geweke, for the small sample upward bias and the estimated variances explained are again given in parentheses. Finally, the 25% and the 75% fractiles are given.

3. ECONOMIC CONSIDERATIONS AND EMPIRICAL EVIDENCE

Rea (1983) investigated a number of alternative Phillips Curve models, two of which will be tested here. Following Sargent (1976)

$$UN_t = b_0 - b_1(p_t - p^e_t) + b_2 UN_{t-1} + e_t \tag{26}$$

where UN is the unemployment rate, p is the percent change in the implicit GNP deflator, and p^e is the expected value of p. If a Koyck transformation is applied to equation (26), to remove the unobservable expected price term, p^e, the natural rate (adaptive expectations) model, becomes

$$UN_t = b_0(1 - k) - b_1(p_t - P_{t-1}) + k + b_2 UN_{t-1} - b_2 k UN_{t-2} + e'_t. \tag{27}$$

An alternative monetarist model attributed to Stein (1978),

$$UN_t = d_0 + (1 - d_1)UN_{t-1} - d_2(m_{t-1} - P_{t-1}) + e''_t \tag{28}$$

relates unemployment to the lagged difference in the rate of growth of M2 and the rate of growth of prices.[14] A major weakness of Rea's paper is that single

---

[13]Tables 3 through 16 required 12 minutes of computer time on an IBM 4341 Model 2, using routines that were compiled with FORTRAN H extended with OPT=2.

[14]Rea (1983) tested these two models against a trade-off model and a natural rate (rational expectations) model in the period 1895-1956 and in the period 1957-1979. His conclusion was that in the latter period the monetarist model was best, in the former period the trade-off model was best while in both periods the natural rate (adaptive expectations) model outperforms the natural rate (rational expectations) model. The best two of the four models (equations 27 and 28) were selected for this study. While Rea used annual data and thus could use the implicit GNP deflator, this study uses monthly data to facilitate the study of the lags of adjustment. In keeping with current practice (see Sachs, 1980), the producer price index was used. Alternative specifications of the model will be investigated in future work.

equation OLS estimation techniques were used which assume away any feedback from price movements, or price movements and the monetary sector in the case of the monetary, to the labor market.

This study, in contrast, uses the VAR model approach, which allows for feedback, makes no assumptions on the structure of the model and, in the spirit of Sims (1980), provides a procedure whereby the relationship among the series in question can be decomposed by frequency. Except for pioneering work at the aggregate level by Geweke (1982b), to my knowledge, no other empirical study of the Phillips Curve relationship has addressed the question of what we mean by the "short" and the "long" run, except to confuse the issue by discussing the lag lengths of significant coefficients in distributed lag relationships. In the Phillips curve theoretical literature the speed of adjustment is not so much the question as whether, in fact, the labor market is fooled into reaching a level of unemployment below the natural rate for a short period of time. If the short run is assumed to involve high-frequency relationships between variables and the long run low-frequency ones, theory suggests that Phillips Curve models should show relationships predominantly at a high-frequency level (short run). If a Phillips curve relationship is found another important question is whether it is the same for different disaggregate sectors of the labor market? Table 1 gives the sources, names means and variances of the data used in this study. A VAR model of the form of equation (3) was estimated, using OLS, and the results are reported in Table 2. In the natural rate model, x_1 was assumed to be $(p_t - p_{t-1})$, while in the monetary model, x_1 was assumed to be $(m_{t-1} - p_{t-1})$. In every case the variance of the x_2 variable was lower with the monetarist model than with the natural rate model. This finding is totally consistent with Rea (1983), who found similar results for aggregate data, using OLS methods of analysis when he estimated equations (27) and (28). The residual variances for the unemployment series of the estimated model (x_2) show a marked reduction over the raw variances, indicating that the VAR model explains roughly 95% or more of the variance. Not all the explanation is in the diagonal polynomials, and significant terms are found at lags longer than those estimated by Rea (1983).[15] While the terms A_{11} and A_{22} control for the effects of lagged x_1 and x_2 in predicting x_1 and x_2, respectively, A_{12} shows feedback from x_2 and x_1, while A_{21} shows any Phillips curve relationship.[16]

Except for construction unemployment, long-duration unemployment and nondurable unemployment, there is no evidence of feedback. The A_{11} term for each model is similar across disaggregated unemployment series models, since the same x_1 variable was used in each of the respective models: however, the other terms show great diversity over the disaggregated series. The tentative conclusion is that no one Phillips curve relationship is stable across the disaggregations of the data tested, although the signs in term A_{21} are all "correct". As has been mentioned earlier, the location of the significant coefficients in terms of lag does not specifically test the Phillips curve relationship. However, since significant negative coefficients were found in the A_{21} term in 9 out of 12 VAR models estimated, a tentative conclusion is that Phillips curve right-hand-side variables, in both models, Granger (1969) cause unemployment series.

The VAR models have been reestimated, using the Whittle (1963) approach, and the variances reported in parentheses in Table 2. These models were then decomposed into the frequency domain following Geweke's procedures and the results reported in Table's 3 through 14. Tables 3 through 8 report the results of the Natural Rate (Adaptive Expectations) Model, while Tables 9 through 14 report results from the Monetary Model. Finally, a simple Phillips curve model relating P_t to the

---

[15]A crude measure of the effectiveness of the monetary model in explaining unemployment is the explained variance divided by the total variance. The resulting statistics are: LHUFR = .9683, LHUMR = .9852, LURMD = .9661, LURC = .9356, LUR15 = .9822, LURMND = .9558.

[16]The lag operator B has been left off A_{ij} in the text and in Table 2 to simplify notation.

unemployment rate for women (LURFR) and men (LURMR) was estimated and reported in Tables 15 and 16. The findings from Tables 3 through 16 are summarized in Table 17.

While both models appeared to have different patterns in the VAR time domain form, at the frequency decomposition stage, the similarities and differences of the models show up more clearly. If one generalization could be made, it would be that there is a persistent short-run relationship between the X and the Y series at a period of 6 months. All models showed feedback from the unemployment rate to price or the monetary and price variable, calling into question single equation formulations, such as Rea's (1983), which rule out feedback by assumption. Turning first to the models estimated with the natural rate model, the first impression is that there are three distinct groups. The long-run employment series (LUR15) which does not have a short-run effect, the construction unemployment series, which has short-run effects at both 6 and 12 months and substantially more feedback than the rest of the series, and the remaining series, which are all somewhat similar in the pattern of their frequency decomposition.

The models estimated with the monetary model all show long-run effects. This is in sharp contrast to the findings from the natural rate model and is not consistent with current Phillips Curve thinking, which has argued that an effect, if present, should be found only in the short run. It should be noted that the monetary model did find short-run effects for all series, except durable manufacturing.

The final two models reported in Table's 15 and 16 show a simple Phillips curve model estimated for female and male unemployment series and show a short-run relationship at 6 months, predicted by theory, and a long-run relationship at 120 and 240 months. While feedback was found at 2 and 240 months for females, no feedback was found for males.

4. CONCLUSION

This paper has used both time domain VAR models and frequency decompositions of these models to study two Phillips curve models for disaggregated data in the period 1959/3 to 1982/7. The paper argues that for the purposes of Phillips curve theory, the appropriate definition of the short run is in terms of high frequency and that the appropriate definition of the long run is in terms of low frequency. This is in contrast to the usual definition of the long run and the short run in terms of position of significant coefficients at different time periods in a distributed lag model. The present approach decomposes the distributed lag relationship into the frequency domain and hence makes use of both the position (lag length) of a significant coefficient and its magnitude. A persistent finding is a weak but significant relationship between the price variable, or the price and monetary variable (rate of growth of real balances) and the unemployment variable at period of six months. Strong feedback at low frequencies (long run) was found between the monetary variable and the unemployment series. Thirteen out of the fourteen models in the frequency domain indicated feedback, which call into question the single equation work of others who have assumed away feedback. This preliminary study, following the important new methods of Geweke, suggests that frequency decomposition of a VAR model is a useful exercise that should be routinely performed.

ACKNOWLEDGEMENT

Computer time for this study was provided by the University of Illinois at Chicago. Helpful comments were made by two referees, Barry Chiswick and other members of the UIC faculty. Diana Stokes provided editorial assistance. Professor Geweke provided the MTSM program for the spectral analysis.

Dr Milos Krofta of the Lenox Institute for Research suggested the general area of the study and provided partial funding support.

Table 1 Disaggregate Labor Market Data 1959/3 to 1982/7

Name	Description	Mean	Variance
LHUFR	Unemployment rate, women, 16 years and over	6.612	1.655
LHUMR	Unemployment rate, men, 16 years and over	5.214	2.488
LURMD	Unemployment rate, durable manufacturing	5.900	5.864
LURC	Unemployment rate, construction	11.725	12.058
LUR15	Unemployment rate, unemployed 15 weeks and over	1.414	.429
LURMND	Unemployment rate, nondurable manufacturing	6.210	3.232
(P-P(-1))	First difference percent change in producer price ind.	.143E-4	.809E-4
(GM-P)-1	Percent change in M2 - percent change in PPI	.265E-2	.631E-4

All Data were taken from the NBER/CITYBASE Data Bank. P = the percent change in the series PW, and GM is the percent change in the series FM2. P and FM2 are, respectively, the producer price index (all commodities) and money stock M2.

Table 2 Significance Pattern of Vector AR Models of the Form $A(B)X_t = e_t$ For Natural Rate and Monetary Models of Unemployment

	A_{11}	A_{12}	A_{21}	A_{22}	Res Var X_1		Res Var X_2	
Women = LHUFR								
a	-----------.	....	....	++..........	366	(379)	.0540	(.515)
b	.++.+.....++	....	...-........	++..........	403	(431)	.0524	(.506)
Men = LHUMR								
a	-----------.	....	--..........	+...........	377	(381)	.0384	(.484)
b	.++.+.....++	....	...-....-...	+...........	423	(435)	.0368	(.476)
Durable Mfg. = LURMD								
a	-----------.	....	.--.........	+.........-+	374	(380)	.208	(.987)
b	.++.+.....++	....	...-........	+.........-+	416	(431)	.199	(.962)
Construction = LURC								
a	-----------.	...........+	....	++..........	363	(374)	.819	(3.05)
b	.++.+.....++	+...........	....	++..........	414	(428)	.777	(2.98)
Long Durable = LUR15								
a	-----------.	-...........	-----......-	+..-........	358	(378)	.00783	(.056)
b	.++.+.....++	.+..........	...-.......-	+..-........	400	(427)	.00763	(.055)
Non Durable = LURMND								
a	-----------.	-......+....	--..........	+...........	363	(376)	.149	(.702)
b	++..+.....++	........-...	...-........	+...........	405	(432)	.143	(.686)

See the text for a complete discussion of the model estimated and data used. For the natural rate model $X_1 = (P-P(-1))$, where P is the percent change in the producer price index. For the monetary model $X_1 = (M-P)-1$, where M is the percent change in M2. In both models X_2 refers to the unemployment rate for women, men, etc. A coefficient is significant if absolute value of the t statistic is greater than 2. The first row under each heading a refers to the Natural Rate model, the second row refers to the monetary model. After each variance figure, the variance obtained by the Whittle algorithm is given. Variances for X_1 must be multiplied by .1 E -6. See the text for further details on the different approaches for obtaining the VAR estimates.

Table 3 Estimated Measures of Linear Feedback
X Vector: P-P(-1) Y Vector: LHUFR

	Estimate	Adjusted Estimate	25%	75%
F(Y to X)	.023 (2.3%)	.009 (.9%)	.007	.012
F(X to Y)	.019 (1.9%)	.008 (.8%)	.005	.009
F(X.Y)	.003 (.3%)	.001 (.1%)	.000	.002

	f(Y to X)				f(X to Y)			
Period	Estimate	Adjusted Estimate	25%	75%	Estimate	Adjusted Estimate	25%	75%
240	.082 (7.9%)	.044 (4.3%)	.008	.059	.007 (.7%)	.004 (.4%)	.001	.006
120	.084 (8.1%)	.056 (5.5%)	.009	.075	.007 (.7%)	.004 (.4%)	.001	.006
60	.085 (8.1%)	.065 (6.3%)	.010	.084	.007 (.7%)	.004 (.4%)	.001	.005
40	.084 (8.1%)	.070 (6.8%)	.014	.098	.007 (.7%)	.004 (.4%)	.001	.004
32	.083 (8.0%)	.073 (7.0%)	.017	.103	.007 (.7%)	.003 (.3%)	.001	.004
16	.045 (4.4%)	.034 (3.3%)	.014	.051	.005 (.5%)	.001 (.1%)	.000	.001
12	.017 (1.7%)	.006 (.6%)	.002	.008	.003 (.3%)	.000 (.0%)	.000	.000
10	.008 (.8%)	.001 (.1%)	.000	.002	.008 (.8%)	.001 (.1%)	.000	.002
8	.009 (.9%)	.001 (.1%)	.000	.002	.015 (1.5%)	.006 (.6%)	.002	.008
7	.017 (1.7%)	.005 (.5%)	.002	.006	.028 (2.8%)	.014 (1.4%)	.003	.018
6	.024 (2.4%)	.011 (1.1%)	.003	.016	.118 (11.1%)	.093 (8.9%)	.035	.130
5	.023 (2.3%)	.010 (1.0%)	.004	.014	.034 (3.3%)	.020 (2.0%)	.007	.027
4	.010 (1.0%)	.002 (.2%)	.000	.002	.033 (3.2%)	.014 (1.4%)	.004	.019
3	.001 (.1%)	.000 (.0%)	.000	.000	.016 (1.6%)	.005 (.5%)	.002	.006
2	.065 (6.3%)	.061 (5.9%)	.009	.092	.001 (.1%)	.000 (.0%)	.000	.000

Table 4 Estimated Measures of Linear Feedback
X Vector: P-P(-1) Y Vector: LHUMR

	Estimate	Adjusted Estimate	25%	75%
F(Y to X)	.018 (1.7%)	.006 (.6%)	.004	.007
F(X to Y)	.019 (1.8%)	.007 (.7%)	.004	.009
F(X.Y)	.003 (.3%)	.002 (.2%)	.000	.002

	f(Y to X)				f(X to Y)			
Period	Estimate	Adjusted Estimate	25%	75%	Estimate	Adjusted Estimate	25%	75%
240	.080 (7.7%)	.049 (4.8%)	.019	.071	.018 (1.8%)	.010 (1.0%)	.001	.015
120	.084 (8.0%)	.073 (7.1%)	.037	.091	.018 (1.8%)	.010 (1.0%)	.001	.015
60	.086 (8.2%)	.088 (8.5%)	.028	.109	.019 (1.9%)	.010 (1.0%)	.001	.015
40	.087 (8.3%)	.095 (9.1%)	.030	.116	.019 (1.9%)	.010 (1.0%)	.001	.015
32	.087 (8.3%)	.097 (9.3%)	.023	.119	.020 (2.0%)	.011 (1.1%)	.002	.016
16	.040 (3.9%)	.029 (2.8%)	.005	.035	.027 (2.7%)	.015 (1.5%)	.006	.019
12	.012 (1.1%)	.003 (.3%)	.001	.004	.026 (2.6%)	.011 (1.1%)	.003	.015
10	.005 (.5%)	.001 (.1%)	.000	.001	.010 (1.0%)	.002 (.2%)	.000	.003
8	.006 (.6%)	.001 (.1%)	.000	.002	.011 (1.1%)	.002 (.2%)	.001	.003
7	.012 (1.2%)	.003 (.3%)	.001	.004	.023 (2.2%)	.008 (.8%)	.003	.012
6	.021 (2.0%)	.006 (.6%)	.001	.011	.101 (9.6%)	.069 (6.7%)	.022	.103
5	.016 (1.6%)	.005 (.5%)	.001	.008	.020 (2.0%)	.010 (1.0%)	.003	.014
4	.004 (.4%)	.000 (.0%)	.000	.000	.006 (.6%)	.001 (.1%)	.000	.001
3	.004 (.4%)	.000 (.0%)	.000	.000	.022 (2.2%)	.008 (.8%)	.002	.010
2	.028 (2.8%)	.010 (1.0%)	.001	.012	.010 (1.0%)	.002 (.2%)	.000	.002

Table 5 Estimated Measures of Linear Feedback
X Vector: P-P(-1) Y Vector: LURMC

	Estimate	Adjusted Estimate	25%	75%
F(Y to X)	.020 (1.9%)	.007 (.7%)	.006	.009
F(X to Y)	.024 (2.4%)	.011 (1.1%)	.008	.014
F(X.Y)	.007 (.7%)	.005 (.5%)	.001	.008

	f(Y to X)				f(X to Y)			
Period	Estimate	Adjusted Estimate	25%	75%	Estimate	Adjusted Estimate	25%	75%
240	.077 (7.4%)	.046 (4.5%)	.014	.078	.018 (1.7%)	.012 (1.2%)	.002	.019
120	.077 (7.4%)	.057 (5.6%)	.020	.096	.018 (1.8%)	.012 (1.2%)	.002	.019
60	.084 (8.1%)	.071 (6.8%)	.023	.103	.019 (1.8%)	.013 (1.3%)	.003	.020
40	.097 (9.2%)	.086 (8.2%)	.030	.120	.020 (2.0%)	.015 (1.4%)	.004	.022
32	.110 (10.4%)	.100 (9.5%)	.034	.142	.022 (2.2%)	.016 (1.6%)	.005	.023
16	.065 (6.3%)	.046 (4.5%)	.015	.068	.048 (4.7%)	.043 (4.2%)	.010	.066
12	.018 (1.8%)	.008 (.8%)	.003	.011	.084 (8.1%)	.079 (7.6%)	.018	.107
10	.008 (0.8%)	.003 (.3%)	.001	.004	.042 (4.1%)	.025 (2.5%)	.008	.033
8	.011 (1.1%)	.004 (.4%)	.001	.006	.015 (1.4%)	.005 (.5%)	.001	.007
7	.025 (2.5%)	.015 (1.5%)	.006	.022	.023 (2.2%)	.009 (.9%)	.003	.012
6	.026 (2.6%)	.015 (1.5%)	.004	.021	.110 (10.5%)	.072 (7.0%)	.030	.115
5	.008 (.8%)	.002 (.2%)	.001	.002	.021 (2.1%)	.009 (.9%)	.003	.012
4	.001 (.1%)	.000 (.0%)	.000	.000	.005 (.5%)	.001 (.1%)	.000	.001
3	.000 (.0%)	.000 (.0%)	.000	.000	.027 (2.7%)	.015 (1.5%)	.005	.020
2	.008 (.8%)	.001 (.1%)	.000	.001	.007 (.7%)	.002 (.2%)	.000	.002

Table 6 Estimated Measures of Linear Feedback
X Vector: P-P(-1) Y Vector: LURC

	Estimate	Adjusted Estimate	25%	75%
F(Y to X)	.035 (3.4%)	.020 (2.0%)	.014	.026
F(X to Y)	.015 (1.5%)	.006 (.6%)	.004	.007
F(X.Y)	.007 (.7%)	.006 (.6%)	.001	.009

	f(Y to X)				f(X to Y)			
Period	Estimate	Adjusted Estimate	25%	75%	Estimate	Adjusted Estimate	25%	75%
240	.135 (12.6%)	.105 (10.0%)	.045	.133	.003 (.3%)	.001 (.1%)	.000	.001
120	.141 (13.1%)	.137 (12.8%)	.050	.185	.003 (.3%)	.001 (.1%)	.000	.001
60	.138 (12.9%)	.146 (13.6%)	.035	.189	.004 (.4%)	.001 (.1%)	.000	.001
40	.130 (12.2%)	.136 (12.8%)	.029	.180	.005 (.5%)	.001 (.1%)	.000	.002
32	.121 (11.4%)	.122 (11.5%)	.029	.176	.006 (.6%)	.002 (.2%)	.000	.003
16	.045 (4.4%)	.023 (2.3%)	.008	.034	.024 (2.4%)	.015 (1.5%)	.004	.019
12	.017 (1.7%)	.005 (.5%)	.001	.008	.052 (5.1%)	.046 (4.5%)	.016	.060
10	.012 (1.2%)	.003 (.3%)	.001	.004	.024 (2.4%)	.012 (1.2%)	.003	.014
8	.007 (.7%)	.002 (.2%)	.001	.004	.001 (.1%)	.000 (.0%)	.000	.000
7	.003 (.3%)	.000 (.0%)	.000	.001	.003 (.3%)	.000 (.0%)	.000	.000
6	.009 (.9%)	.002 (.2%)	.001	.003	.091 (8.7%)	.074 (7.1%)	.033	.106
5	.027 (2.6%)	.014 (1.4%)	.004	.018	.034 (3.3%)	.033 (3.2%)	.018	.038
4	.002 (.2%)	.000 (.0%)	.000	.000	.016 (1.5%)	.006 (.6%)	.002	.007
3	.041 (4.0%)	.021 (2.1%)	.007	.029	.005 (.5%)	.001 (.1%)	.000	.001
2	.116 (10.9%)	.155 (14.4%)	.027	.244	.000 (.0%)	.000 (.0%)	.000	.000

Table 7 Estimated Measures of Linear Feedback
X Vector: P-P(-1) Y Vector: LUR15

	Estimate	Adjusted Estimate	25%	75%
F(Y to X)	.022 (2.2%)	.009 (.9%)	.006	.011
F(X to Y)	.036 (3.5%)	.020 (2.0%)	.015	.024
F(X.Y)	.005 (.5%)	.003 (.3%)	.000	.006

	f(Y to X)				f(X to Y)			
Period	Estimate	Adjusted Estimate	25%	75%	Estimate	Adjusted Estimate	25%	75%
240	.097 (9.2%)	.058 (5.6%)	.027	.079	.082 (7.9%)	.082 (7.9%)	.031	.119
120	.106 (10.1%)	.078 (7.5%)	.031	.110	.083 (7.9%)	.082 (7.9%)	.031	.120
60	.107 (10.1%)	.084 (8.1%)	.022	.122	.084 (8.0%)	.083 (8.0%)	.032	.120
40	.101 (9.6%)	.082 (7.9%)	.021	.124	.086 (8.2%)	.085 (8.2%)	.034	.119
32	.094 (9.0%)	.076 (7.3%)	.021	.119	.088 (8.4%)	.088 (8.4%)	.034	.118
16	.026 (2.5%)	.013 (1.3%)	.004	.014	.109 (10.3%)	.106 (10.0%)	.039	.159
12	.008 (.8%)	.002 (.2%)	.001	.003	.073 (7.0%)	.048 (4.7%)	.013	.062
10	.018 (1.8%)	.009 (.9%)	.004	.013	.006 (.6%)	.001 (.1%)	.000	.001
8	.043 (4.2%)	.031 (3.0%)	.006	.039	.015 (1.4%)	.006 (.6%)	.002	.009
7	.048 (4.7%)	.033 (3.3%)	.007	.040	.029 (2.9%)	.018 (1.8%)	.004	.024
6	.033 (3.2%)	.018 (1.8%)	.007	.022	.058 (5.6%)	.040 (3.9%)	.014	.052
5	.014 (1.4%)	.004 (.4%)	.002	.005	.011 (1.1%)	.004 (.4%)	.001	.006
4	.002 (.2%)	.000 (.0%)	.000	.000	.015 (1.5%)	.004 (.4%)	.001	.006
3	.001 (.1%)	.000 (.0%)	.000	.000	.031 (3.1%)	.016 (1.6%)	.005	.021
2	.034 (3.3%)	.013 (1.3%)	.001	.018	.001 (.1%)	.000 (.0%)	.000	.000

Table 8 Estimated Measures of Linear Feedback
X Vector: P-P(-1) Y Vector: LURMND

	Estimate	Adjusted Estimate	25%	75%
F(Y to X)	.028 (2.8%)	.014 (1.4%)	.011	.017
F(X to Y)	.024 (2.4%)	.011 (1.1%)	.007	.014
F(X.Y)	.001 (.1%)	.000 (.0%)	.000	.000

	f(Y to X)				f(X to Y)			
Period	Estimate	Adjusted Estimate	25%	75%	Estimate	Adjusted Estimate	25%	75%
240	.107 (10.1%)	.057 (5.5%)	.024	.080	.010 (1.0%)	.005 (.5%)	.001	.007
120	.110 (10.4%)	.077 (7.4%)	.028	.102	.010 (1.0%)	.005 (.5%)	.001	.007
60	.111 (10.5%)	.086 (8.3%)	.027	.114	.011 (1.0%)	.005 (.5%)	.001	.007
40	.110 (10.4%)	.088 (8.4%)	.028	.128	.011 (1.1%)	.005 (.5%)	.001	.007
32	.107 (10.1%)	.086 (8.2%)	.026	.126	.011 (1.1%)	.005 (.5%)	.001	.007
16	.050 (4.9%)	.032 (3.2%)	.006	.047	.016 (1.6%)	.006 (.6%)	.002	.007
12	.017 (1.7%)	.006 (.6%)	.002	.009	.023 (2.3%)	.008 (.8%)	.002	.011
10	.007 (.7%)	.001 (.1%)	.001	.002	.020 (2.0%)	.008 (.8%)	.003	.010
8	.003 (.3%)	.000 (.0%)	.000	.000	.027 (2.7%)	.015 (1.5%)	.005	.022
7	.007 (.7%)	.001 (.1%)	.000	.002	.054 (5.2%)	.034 (3.3%)	.012	.048
6	.021 (2.1%)	.009 (.9%)	.003	.012	.216 (19.5%)	.189 (17.2%)	.082	.274
5	.031 (3.1%)	.019 (1.9%)	.007	.028	.030 (2.9%)	.021 (2.1%)	.006	.027
4	.009 (.9%)	.003 (.3%)	.001	.003	.004 (.4%)	.001 (.1%)	.000	.001
3	.011 (1.1%)	.003 (.3%)	.001	.005	.021 (2.0%)	.008 (.8%)	.002	.013
2	.062 (6.0%)	.045 (4.4%)	.005	.059	.000 (.0%)	.000 (.0%)	.000	.000

Table 9 Estimated Measures of Linear Feedback
X Vector: (GM-P)-1 Y Vector: LHUFR

	Estimate	Adjusted Estimate	25%	75%
F(Y to X)	.044 (4.3%)	.027 (2.7%)	.020	.033
F(X to Y)	.037 (3.7%)	.023 (2.2%)	.017	.028
F(X.Y)	.004 (.4%)	.002 (.2%)	.000	.003

	f(Y to X)*				f(X to Y)			
Period	Estimate	Adjusted Estimate	25%	75%	Estimate	Adjusted Estimate	25%	75%
240	.228 (20.3%)	.143 (13.3%)	.071	.205	.474 (37.7%)	.682 (49.4%)	.195	.960
120	.183 (16.7%)	.143 (13.4%)	.074	.201	.261 (23.0%)	.303 (26.1%)	.101	.436
60	.171 (15.7%)	.161 (14.9%)	.067	.215	.099 (9.5%)	.086 (8.3%)	.031	.120
40	.168 (15.4%)	.167 (15.4%)	.069	.236	.054 (5.2%)	.039 (3.8%)	.015	.053
32	.164 (15.1%)	.165 (15.2%)	.077	.252	.039 (3.8%)	.025 (2.5%)	.012	.032
16	.090 (8.6%)	.071 (6.9%)	.032	.086	.013 (1.3%)	.003 (.3%)	.001	.005
12	.041 (4.0%)	.023 (2.3%)	.010	.033	.000 (.0%)	.000 (.0%)	.000	.000
10	.032 (3.2%)	.017 (1.6%)	.006	.022	.003 (.3%)	.000 (.0%)	.000	.001
8	.043 (4.2%)	.030 (3.0%)	.011	.045	.010 (1.0%)	.003 (.3%)	.001	.005
7	.043 (4.2%)	.034 (3.4%)	.013	.050	.020 (1.9%)	.008 (.8%)	.002	.012
6	.022 (2.2%)	.012 (1.2%)	.005	.016	.062 (6.0%)	.033 (3.2%)	.010	.047
5	.006 (.6%)	.001 (.1%)	.001	.002	.048 (4.7%)	.043 (4.2%)	.015	.060
4	.010 (1.0%)	.002 (.2%)	.001	.004	.096 (9.1%)	.090 (8.6%)	.040	.128
3	.011 (1.1%)	.002 (.2%)	.000	.003	.025 (2.5%)	.012 (1.2%)	.004	.014
2	.010 (1.0%)	.003 (.3%)	.001	.004	.018 (1.8%)	.010 (1.0%)	.001	.011

*Invertibility Conditions Not Satisfied

Table 10 Estimated Measures of Linear Feedback
X Vector: (GM-P)-1 Y Vector: LHUMR

	Estimate	Adjusted Estimate	25%	75%
F(Y to X)	.037 (3.6%)	.018 (1.8%)	.013	.024
F(X to Y)	.037 (3.6%)	.023 (2.3%)	.016	.031
F(X.Y)	.010 (1.0%)	.006 (.6%)	.002	.008

	f(Y to X)*				f(X to Y)			
Period	Estimate	Adjusted Estimate	25%	75%	Estimate	Adjusted Estimate	25%	75%
240	.241 (21.4%)	.099 (9.4%)	.039	.131	.684 (49.6%)	1.065 (65.5%)	.338	1.777
120	.166 (15.3%)	.090 (8.6%)	.036	.125	.374 (31.2%)	.470 (37.5%)	.179	.697
60	.149 (13.8%)	.122 (11.5%)	.054	.158	.143 (13.3%)	.131 (12.3%)	.057	.183
40	.147 (13.7%)	.146 (13.6%)	.066	.187	.078 (7.5%)	.060 (5.8%)	.027	.082
32	.146 (13.6%)	.157 (14.5%)	.076	.211	.058 (5.6%)	.041 (4.0%)	.020	.057
16	.072 (6.9%)	.058 (5.6%)	.031	.075	.031 (3.0%)	.013 (1.3%)	.005	.017
12	.021 (2.1%)	.007 (.7%)	.003	.010	.005 (.5%)	.000 (.0%)	.000	.001
10	.017 (1.7%)	.005 (.5%)	.001	.008	.000 (.0%)	.000 (.0%)	.000	.000
8	.031 (3.0%)	.012 (1.2%)	.003	.015	.005 (.5%)	.001 (.1%)	.000	.001
7	.040 (3.9%)	.019 (1.9%)	.005	.029	.014 (1.4%)	.005 (.5%)	.003	.007
6	.023 (2.3%)	.008 (.8%)	.002	.010	.053 (5.2%)	.034 (3.3%)	.009	.053
5	.001 (.1%)	.000 (.0%)	.000	.000	.033 (3.3%)	.024 (2.4%)	.006	.036
4	.002 (.2%)	.000 (.0%)	.000	.000	.061 (5.9%)	.047 (4.6%)	.016	.064
3	.003 (.3%)	.000 (.0%)	.000	.000	.008 (.8%)	.001 (.1%)	.000	.002
2	.000 (.0%)	.000 (.0%)	.000	.000	.006 (.6%)	.001 (.1%)	.000	.001

*Invertibility Conditions Not Satisfied

Table 11 Estimated Measures of Linear Feedback
X Vector: (GM-P)-1 Y Vector: LURMD

	Estimate	Adjusted Estimate	25%	75%
F(Y to X)	.045 (4.4%)	.025 (2.5%)	.020	.030
F(X to Y)	.048 (4.7%)	.028 (2.8%)	.023	.034
F(X.Y)	.008 (.8%)	.006 (.6%)	.001	.009

	f(Y to X)*				f(X to Y)			
Period	Estimate	Adjusted Estimate	25%	75%	Estimate	Adjusted Estimate	25%	75%
240	.291 (25.3%)	.150 (14.0%)	.078	.186	1.037 (64.5%)	1.208 (70.1%)	.596	1.783
120	.170 (15.6%)	.095 (9.1%)	.045	.117	.568 (43.3%)	.596 (44.9%)	.323	.862
60	.144 (13.4%)	.107 (10.1%)	.052	.157	.222 (19.9%)	.193 (17.5%)	.108	.264
40	.151 (14.0%)	.130 (12.2%)	.057	.191	.124 (11.6%)	.097 (9.2%)	.050	.137
32	.163 (15.0%)	.151 (14.0%)	.071	.229	.093 (8.9%)	.070 (6.7%)	.034	.103
16	.107 (10.1%)	.077 (7.4%)	.023	.124	.068 (6.6%)	.043 (4.2%)	.019	.060
12	.033 (3.2%)	.014 (1.4%)	.007	.018	.030 (2.9%)	.011 (1.1%)	.003	.015
10	.024 (2.4%)	.009 (.9%)	.005	.013	.003 (.3%)	.000 (.0%)	.000	.000
8	.037 (3.6%)	.017 (1.7%)	.006	.021	.003 (.3%)	.000 (.0%)	.000	.000
7	.049 (4.8%)	.028 (2.7%)	.010	.036	.011 (1.1%)	.004 (.4%)	.001	.006
6	.013 (1.3%)	.004 (.4%)	.001	.006	.055 (5.3%)	.029 (2.8%)	.009	.035
5	.001 (.1%)	.000 (.0%)	.000	.000	.024 (2.3%)	.009 (.9%)	.002	.015
4	.005 (.5%)	.001 (.1%)	.000	.001	.045 (4.4%)	.021 (2.1%)	.008	.028
3	.013 (1.3%)	.003 (.3%)	.001	.004	.015 (1.5%)	.004 (.4%)	.001	.006
2	.002 (.2%)	.000 (.0%)	.000	.000	.005 (.5%)	.001 (.1%)	.000	.001

*Invertibility Conditions Not Satisfied

Table 12 Estimated Measures of Linear Feedback
X Vector: (GM-P)-1 Y Vector: LURC

	Estimate	Adjusted Estimate	25%	75%
F(Y to X)	.052 (5.1%)	.032 (3.2%)	.023	.040
F(X to Y)	.039 (3.8%)	.021 (2.1%)	.015	.024
F(X.Y)	.002 (.2%)	.001 (.1%)	.000	.001

	f(Y to X)*				f(X to Y)			
Period	Estimate	Adjusted Estimate	25%	75%	Estimate	Adjusted Estimate	25%	75%
240	.311 (26.7%)	.157 (14.6%)	.070	.191	.719 (51.3%)	1.048 (64.9%)	.447	1.417
120	.211 (19.1%)	.122 (11.5%)	.057	.186	.405 (33.3%)	.506 (39.7%)	.226	.618
60	.184 (16.8%)	.138 (12.9%)	.067	.196	.160 (14.7%)	.150 (13.9%)	.072	.182
40	.177 (16.2%)	.151 (14.0%)	.082	.192	.089 (8.5%)	.070 (6.8%)	.038	.089
32	.171 (15.7%)	.154 (14.3%)	.093	.190	.067 (6.5%)	.048 (4.7%)	.028	.061
16	.081 (7.8%)	.060 (5.8%)	.021	.077	.049 (4.7%)	.027 (2.6%)	.012	.035
12	.017 (1.7%)	.006 (.6%)	.002	.007	.019 (1.9%)	.006 (.6%)	.002	.007
10	.008 (.7%)	.002 (.2%)	.001	.002	.003 (.3%)	.000 (.0%)	.000	.000
8	.028 (2.8%)	.015 (1.5%)	.005	.018	.004 (.4%)	.001 (.1%)	.000	.001
7	.048 (4.7%)	.032 (3.1%)	.011	.046	.015 (1.5%)	.006 (.6%)	.002	.008
6	.045 (4.4%)	.026 (2.6%)	.006	.037	.080 (7.6%)	.046 (4.5%)	.019	.059
5	.005 (.5%)	.001 (.1%)	.000	.001	.041 (4.0%)	.022 (2.2%)	.007	.031
4	.004 (.4%)	.000 (.0%)	.000	.001	.039 (3.8%)	.019 (1.9%)	.006	.023
3	.039 (3.8%)	.027 (2.7%)	.007	.041	.001 (.1%)	.000 (.0%)	.000	.000
2	.009 (.9%)	.002 (.2%)	.000	.002	.007 (.7%)	.001 (.1%)	.000	.001

*Invertibility Conditions Not Satisfied

Table 13 Estimated Measures of Linear Feedback
X Vector: (GM-P)-1 Y Vector: LUR15

	Estimate	Adjusted Estimate	25%	75%
F(Y to X)	.050 (4.9%)	.029 (2.8%)	.022	.036
F(X to Y)	.060 (5.9%)	.042 (4.1%)	.029	.055
F(X.Y)	.019 (1.9%)	.018 (1.8%)	.008	.026

	f(Y to X)*				f(X to Y)			
Period	Estimate	Adjusted Estimate	25%	75%	Estimate	Adjusted Estimate	25%	75%
240	.294 (25.5%)	.172 (15.8%)	.091	.220	.905 (59.5%)	1.190 (69.6%)	.596	1.677
120	.208 (18.8%)	.142 (13.3%)	.068	.197	.489 (38.7%)	.568 (43.3%)	.341	.833
60	.185 (16.9%)	.153 (14.2%)	.064	.222	.215 (19.3%)	.205 (18.5%)	.111	.245
40	.179 (16.4%)	.159 (14.7%)	.071	.236	.145 (13.5%)	.128 (12.0%)	.068	.166
32	.174 (15.9%)	.157 (14.6%)	.077	.235	.126 (11.9%)	.109 (10.3%)	.056	.150
16	.087 (8.3%)	.064 (6.2%)	.019	.088	.155 (14.4%)	.138 (12.9%)	.058	.207
12	.037 (3.6%)	.019 (1.8%)	.003	.023	.078 (7.5%)	.058 (5.6%)	.021	.080
10	.054 (5.3%)	.033 (3.3%)	.011	.039	.005 (.5%)	.001 (.1%)	.000	.001
8	.107 (10.1%)	.083 (8.0%)	.038	.112	.004 (.4%)	.001 (.1%)	.000	.001
7	.112 (10.6%)	.085 (8.1%)	.043	.109	.013 (1.3%)	.005 (.5%)	.001	.005
6	.053 (5.1%)	.031 (3.0%)	.010	.038	.024 (2.4%)	.010 (.9%)	.002	.013
5	.002 (.2%)	.000 (.0%)	.000	.000	.019 (1.9%)	.008 (.8%)	.002	.010
4	.013 (1.3%)	.004 (.4%)	.001	.005	.125 (11.8%)	.102 (9.7%)	.042	.133
3	.003 (.3%)	.000 (.0%)	.000	.000	.015 (1.5%)	.003 (.3%)	.001	.005
2	.000 (.0%)	.000 (.0%)	.000	.000	.025 (2.5%)	.010 (1.0%)	.001	.012

*Invertibility Conditions Not Satisfied

Table 14 Estimated Measures of Linear Feedback
X Vector: (GM-P)-1 Y Vector: LURMND

	Estimate	Adjusted Estimate	25%	75%
F(Y to X)	.042 (4.1%)	.024 (2.4%)	.016	.030
F(X to Y)	.047 (4.6%)	.034 (3.3%)	.026	.041
F(X.Y)	.005 (.5%)	.002 (.2%)	.000	.004

	f(Y to X)*				f(X to Y)			
Period	Estimate	Adjusted Estimate	25%	75%	Estimate	Adjusted Estimate	25%	75%
240	.234 (20.9%)	.113 (10.7%)	.047	.178	.787 (54.5%)	1.171 (69.0%)	.512	1.523
120	.161 (14.9%)	.099 (9.4%)	.046	.136	.433 (35.1%)	.573 (43.6%)	.276	.828
60	.144 (13.4%)	.118 (11.1%)	.060	.150	.167 (15.4%)	.180 (16.5%)	.104	.265
40	.144 (13.4%)	.131 (12.3%)	.065	.164	.091 (8.7%)	.087 (8.4%)	.038	.125
32	.145 (13.5%)	.139 (13.0%)	.064	.170	.067 (6.5%)	.060 (5.9%)	.028	.081
16	.096 (9.2%)	.089 (8.6%)	.038	.134	.032 (3.2%)	.020 (2.0%)	.007	.031
12	.039 (3.8%)	.022 (2.2%)	.008	.038	.006 (.6%)	.001 (.1%)	.000	.001
10	.021 (2.1%)	.008 (.8%)	.002	.012	.005 (.5%)	.001 (.1%)	.000	.001
8	.018 (1.8%)	.006 (.6%)	.001	.008	.017 (1.7%)	.008 (.8%)	.002	.010
7	.017 (1.7%)	.005 (.5%)	.001	.008	.038 (3.7%)	.023 (2.3%)	.007	.030
6	.011 (1.1%)	.003 (.3%)	.001	.004	.123 (11.5%)	.112 (10.6%)	.039	.174
5	.005 (.5%)	.001 (.1%)	.000	.001	.048 (4.7%)	.037 (3.6%)	.010	.053
4	.001 (.1%)	.000 (.0%)	.000	.000	.046 (4.5%)	.024 (2.4%)	.006	.033
3	.041 (4.0%)	.022 (2.2%)	.010	.032	.016 (1.6%)	.005 (.5%)	.002	.007
2	.007 (.7%)	.001 (.1%)	.000	.001	.015 (1.5%)	.006 (.6%)	.001	.010

*Invertibility Conditions Not Satisfied

Table 15 Estimated Measures of Linear Feedback
X Vector: P Y Vector: LHUFR

	Estimate	Adjusted Estimate	25%	75%
F(Y to X)	.016 (1.6%)	.006 (.6%)	.004	.006
F(X to Y)	.030 (2.9%)	.015 (1.5%)	.010	.019
F(X.Y)	.000 (.0%)	.000 (0.0%)	.000	.000

	f(Y to X)				f(X to Y)			
Period	Estimate	Adjusted Estimate	25%	75%	Estimate	Adjusted Estimate	25%	75%
240	.122 (11.5%)	.059 (5.8%)	.012	.089	.433 (35.1%)	.488 (38.6%)	.089	.741
120	.058 (5.7%)	.022 (2.1%)	.006	.026	.197 (17.9%)	.163 (15.1%)	.035	.235
60	.035 (3.4%)	.014 (1.4%)	.004	.018	.067 (6.4%)	.039 (3.8%)	.014	.062
40	.029 (2.9%)	.014 (1.3%)	.003	.018	.035 (3.4%)	.017 (1.7%)	.007	.025
32	.026 (2.6%)	.013 (1.3%)	.003	.020	.026 (2.5%)	.011 (1.1%)	.005	.016
16	.009 (.9%)	.003 (.3%)	.001	.005	.013 (1.3%)	.003 (.3%)	.001	.003
12	.002 (.2%)	.000 (.0%)	.000	.000	.003 (.3%)	.000 (.0%)	.000	.000
10	.001 (.1%)	.000 (.0%)	.000	.000	.002 (.2%)	.000 (.0%)	.000	.000
8	.004 (.4%)	.000 (.0%)	.000	.000	.011 (1.1%)	.004 (.4%)	.001	.005
7	.009 (.9%)	.001 (.1%)	.000	.002	.032 (3.1%)	.018 (1.8%)	.005	.024
6	.014 (1.4%)	.005 (.5%)	.001	.006	.133 (12.4%)	.113 (10.7%)	.036	.165
5	.019 (1.8%)	.008 (.8%)	.002	.012	.027 (2.7%)	.017 (1.7%)	.005	.023
4	.017 (1.7%)	.006 (.6%)	.003	.008	.025 (2.5%)	.012 (1.2%)	.003	.016
3	.001 (.1%)	.000 (.0%)	.000	.000	.022 (2.2%)	.009 (.9%)	.002	.012
2	.056 (5.5%)	.051 (5.0%)	.014	.078	.000 (0.0%)	.000 (.0%)	.000	.000

Table 16 Estimated Measures of Linear Feedback
X Vector: P Y Vector: LHUMR

	Estimate	Adjusted Estimate	25%	75%
F(Y to X)	.009 (.9%)	.002 (.2%)	.001	.002
F(X to Y)	.039 (3.9%)	.021 (2.0%)	.014	.026
F(X.Y)	.000 (.0%)	.000 (.0%)	.000	.000

	f(Y to X)				f(X to Y)			
Period	Estimate	Adjusted Estimate	25%	75%	Estimate	Adjusted Estimate	25%	75%
240	.055 (5.4%)	.018 (1.7%)	.003	.026	.774 (53.9%)	.926 (60.4%)	.327	1.338
120	.036 (3.6%)	.011 (1.1%)	.002	.021	.352 (29.7%)	.342 (29.0%)	.114	.547
60	.028 (2.7%)	.011 (1.1%)	.002	.015	.130 (12.2%)	.100 (9.5%)	.031	.157
40	.025 (2.5%)	.012 (1.2%)	.002	.017	.078 (7.5%)	.055 (5.4%)	.022	.083
32	.023 (2.3%)	.011 (1.1%)	.002	.015	.063 (6.1%)	.043 (4.2%)	.018	.061
16	.007 (.7%)	.001 (.1%)	.000	.002	.056 (5.5%)	.031 (3.0%)	.010	.039
12	.001 (.1%)	.000 (.0%)	.000	.000	.032 (3.1%)	.010 (1.0%)	.002	.014
10	.000 (.0%)	.000 (.0%)	.000	.000	.003 (.3%)	.000 (.0%)	.000	.000
8	.002 (.2%)	.000 (.0%)	.000	.000	.007 (.7%)	.002 (.2%)	.000	.002
7	.005 (.5%)	.001 (.1%)	.000	.001	.027 (2.7%)	.015 (1.5%)	.005	.019
6	.011 (1.1%)	.003 (.3%)	.001	.004	.108 (10.2%)	.079 (7.6%)	.033	.097
5	.011 (1.1%)	.003 (.3%)	.001	.004	.013 (1.3%)	.004 (.4%)	.002	.007
4	.008 (.8%)	.002 (.2%)	.001	.002	.004 (.4%)	.000 (.0%)	.000	.000
3	.002 (.2%)	.000 (.0%)	.000	.000	.031 (3.0%)	.012 (1.2%)	.004	.018
2	.023 (2.3%)	.011 (1.1%)	.001	.012	.012 (1.2%)	.004 (.4%)	.000	.008

Table 17

Summary of Frequency Findings from Tables 3 through 16

Variable	Natural Rate Model		Monetary Model
LURFR	Short run 6 Feedback 2, 40, 60, 120, 240		Short run 4, 5, 6 Long run 60, 120, 240 Feedback 16, 32, 40, 60, 120, 240
LURMR	Short run 6 Feedback 32, 40, 60, 120		Short run 4, 5, 6 Long run 40, 60, 120, 240 Feedback 16, 32, 40, 60, 120, 240
LURMD	Short run 6, 12 Feedback 32, 40, 60		Weak short-run effect Long run 40, 60, 120, 240 Feedback 16, 32, 40, 60, 120, 240
LURC	Short run 6, 12 Feedback 2, 32, 40, 60, 120, 240		Short run 5, 6 Long run 40, 60, 120, 240 Feedback 16, 32, 40, 60, 120, 240
LUR15	Effect 16, 32, 40, 60, 120, 240 Feedback 32, 40, 60, 120, 240		Short run 4, 12 Long run 16, 32, 40, 60, 120, 240 Feedback 16, 32, 40, 60, 120, 240
LURMND	Short run 6 Feedback 32, 40, 60, 120		Short run 6 Long run 32, 40, 60, 120, 240 Feedback 16, 32, 40, 60, 120, 240
	Simple Phillips Curve Model		
LURFR	Short run 6 Long run 120, 240 Feedback 2, 240	LHUMR	Short run 6 Long run 120, 240 No Feedback

REFERENCES

BOX, GEORGE E. P. and GWILYM M. JENKINS (1976). Time Series Analysis: Forecasting and Control, revised edition. San Francisco: Holden Day.

ECKSTEIN, OTTO and JAMES A. GIROLA (1978). Long-term Properties of the Price-Wage Mechanism in the United States 1891-1977. Review of Economics and Statistics LX 3, 323-333.

GEWEKE, JOHN (1982a). Measurement of Linear Dependence and Feedback Between Multiple Time Series. Journal of American Statistical Association 77, 304-313.

GEWEKE, JOHN (1982b). Feedback between Monetary Policy, Labor Market Activity, and Wage Inflation, 1955-1978. In Workers, Jobs, and Inflation. Ed: Bailey, Brookings Institution, Washington, 159-198.

GEWEKE, JOHN (1982c). The Neutrality of Money in the United States, 1872-1980: An Interpretation of the Evidence. Unpublished Manuscript, University of Wisconsin, 1-38.

GEWEKE, JOHN (1983). The Superneutrality of Money in the United States: An Interpretation of the Evidence. Unpublished working paper #70-82-83, Carnegie-Mellon University, 1-42.

GRANGER, C. W. J. (1969). Investigating Causal Relationships by Econometric Methods and Cross Spectral Methods. Econometrica 37, 424-438.

GRANGER, C. W. J. and PAUL NEWBOLD (1977). Forecasting Economic Time Series. New York: Academic Press.

MELTZER, ALLAN (1977). Anticipated Inflation and Unanticipated Price Change: A Test of the Price-Specie Flow Theory and the Phillips Curve. Journal of Money, Credit, and Banking 9, 182-205.

REA, JOHN D. (1983). The Explanatory Power of Alternative Theories of Inflation and Unemployment, 1895-1979. Review of Economics and Statistics LXV 2, 183-193.

SACHS, JEFFREY (1980). The Changing Cyclical Behavior of Wages and Prices: 1890-1976. American Economic Review 70, 78-90.

SANTOMERO, ANTHONY M. and JOHN J. SEATER (1978). The Inflation-Unemployment Trade-off: A Critique of the Literature. Journal of Economic Literature XVI, 499-544.

SARGENT, THOMAS J. (1976). A Classical Macroeconometric Model for the United States. Journal of Political Economy 84, 207-237.

SIMS, CHRISTOPHER (1980). Macroeconomics and Reality. Econometrica 48, 1-48.

STEIN, JEROME (1978). Inflation, Unemployment, and Stagflation. Journal of Monetary Economics 4, 193-228.

STOKES, HOUSTON H., HUGH NEUBURGER and DONALD JONES (1975). Unemployment and Adjustment in the Labor Market. University of Chicago Department of Geography Research Paper No. 177, University of Chicago Press.

STOKES, HOUSTON AND HUGH NEUBURGER (1979). The Effect of Monetary Changes on Interest Rates: A Box-Jenkins Approach. Review of Economics and Statistics 61, 534-548.

STOKES, HOUSTON H. (1983a). The B34S Data analysis Program: A Short Writeup. College of Business Administration Working Paper Series, University of Illinois at Chicago, Report FY 77-1 revised, 1-220.

STOKES, HOUSTON H. (1983b). Adjustment of Midwest Labor Markets to External Disturbances. In Midwest Economy: Issues and Policy, Eds: R. Resek and R. Kosobud. Urbana, University of Illinois Press, 147-164.

STOKES, HOUSTON (1983c). The Relationship Between Money, Interest Rates and Prices 1867-1933: A Vector Model Approach. In Time Series Analysis: Theory and Practice 3 (Proceedings of the International Forecasting Conference held in Valencia, Spain, May 1982). Ed: O. D. Anderson, Amsterdam, North-Holland, 231-250.

TIAO, G., BOX G. E. P., GRUPE, M. R., HUDAK, G. B., BELL, W. R., and CHANG, I. (1980). The Wisconsin Multiple Time Series (WMTS-1) Program: A Preliminary Guide. University of Wisconsin, Department of Statistics.

TOBIN, JAMES (1972). Inflation and Unemployment. American Economic Review 52, 1-18.

WASSERFALLEN, WALTER (1983). The Phillips-Curve Under Rational Expectations: Some Evidence from Switzerland. In Time Series Analysis: Theory and Practice 3 (Proceedings of the International Forecasting Conference held in Valencia, Spain, May 1982) Ed: O. D. Anderson, Amsterdam, North-Holland, 131-149.

WHITTLE, P. (1963). On the Fitting of Multivariate Autoregressions, and the Approximate Canonical Factorization of a Spectral Density Matrix. Biometrika 50, 129-134.